Continued on back end papers

*Now available in a lower priced paperback edition in the Wiley Classics Library.

Statistical Intervals

Statistical Intervals

A Guide for Practitioners

GERALD J. HAHN

General Electric Company
Schenectady, New York

WILLIAM Q. MEEKER

Iowa State University
Ames, Iowa

A Wiley-Interscience Publication
JOHN WILEY & SONS, INC.
New York • Chichester • Brisbane • Toronto • Singapore

Copyright ©1991 by John Wiley & Sons, Inc.

All rights reserved. Published simultaneously in Canada.

Library of Congress Cataloging in Publication Data:
Hahn, Gerald J.
   Statistical intervals: a guide for practitioners/Gerald J. Hahn, William Q. Meeker.

      p.   cm.—(Wiley series in probability and mathematical statistics. Applied probability and statistics)
   "A Wiley-Interscience publication."
   Include bibliographical references and index.
   ISBN 0-471-88769-2 (alk. paper)
      1. Mathematical statistics.   I. Meeker, William Q.   II. Title.
III. Series.

QA276.H22 1991   91-8728
519.5—dc20   CIP

Printed in the United States of America

10 9 8 7 6 5 4 3 2 1

To Bea, Adrienne, Judy, Susan and John, and Omi
—G.J.H.
To Karen, Katherine, and My Parents
—W.Q.M.

Preface

Engineers, managers, scientists, and others often need to draw conclusions from scanty data. For example, based upon the results of a limited sample, one might need to decide whether a product is ready for release to manufacturing, to determine how reliable a space system really is, or to assess the impact of an alleged environmental hazard. Sample data provide uncertain results about the population or process of interest. Statistical intervals quantify this uncertainty by what is referred to, in public opinion polls, as "the margin of error." In this book, we show how to compute such intervals, demonstrate their practical applications, and clearly state the assumptions that one makes in their use. We go far beyond the discussion in current texts and provide a wide arsenal of tools that we have found useful in practical applications.

We show in the first chapter that an essential initial step is to assure that statistical methods are applicable. This requires the assumption that the data can be regarded as a random sample from the population or process of interest. In evaluating a new product, this might necessitate an evaluation of how and when the sample units were built, the environment in which they were tested, the way they were measured—and how these relate to the product or process of interest. If the desired assurance is not forthcoming, the methods of this book might provide merely a lower bound on the total uncertainty, reflecting only the sampling variability. Sometimes, our formal or informal evaluations lead us to conclude that the best way to proceed is to obtain added or improved data through a carefully planned investigation.

Next, we must define the specific information desired about the population or process of interest. For example, we might wish to determine the percentage of nonconforming product, the mean, or the 10th percentile, of the distribution of mechanical strength for an alloy, or the maximum noise that a customer may expect for a future order of aircraft engines.

We usually do not have unlimited data but need to extract the maximum information from a small sample. A single calculated value, such as the observed percentage of nonconforming units, can then be regarded as a "point estimate" that provides a best guess of the true percentage of nonconforming units for the sampled process or population. However, we need to quantify the uncertainty associated with such a point estimate. This can be accomplished by a statistical interval. For example, in determining whether a product design is adequate, our calculations might show that we can be "reasonably confident" that if we continue to build, use, and measure the product in the same way as in the sample, the long-run percentage of nonconforming units will be between 0.43 and 1.57%. Thus, if our goal is a product with a percentage nonconforming of 0.10% or less, the calculated interval is telling us that additional improvement is needed—since even an optimistic estimate of the nonconforming product for the sampled population or process is 0.43%. On the other hand, should we be willing to accept, at least at first, 2% nonconforming product, then initial product release might be justified (presumably, in parallel with continued product improvement), since this value exceeds our most pessimistic estimate of 1.57%. Finally, if our goal had been to have less than 1% nonconforming product, our results are inconclusive and suggest the need for additional data.

Occasionally, when the available sample is huge (or the variability is small), statistical uncertainty is relatively unimportant. This would be the case, for example, if our calculations show that the proportion nonconforming units for the sampled population or process is between 0.43 and 0.45%. More frequently, we have very limited data and obtain a relatively "huge" statistical interval, e.g., 0.43 to 37.2%. Even in these two extreme situations, the statistical interval is useful. In the first case, it tells us that, if the underlying assumptions are met, the data are sufficient for most practical needs. In the second case, it indicates that unless more precise methods for analysis can be found, the data at hand provide very little meaningful information.

In each of these examples, quantifying the uncertainty due to random sampling is likely to be substantially more informative to decision makers than obtaining a point estimate alone. Thus, statistical intervals, properly calculated from the sample data, are often of paramount interest to practitioners and their management (and are usually a great deal more meaningful than statistical significance or hypothesis tests).

Different practical problems call for different types of intervals. To assure useful conclusions, it is imperative that the statistical interval be appropriate for the problem at hand. Those who have taken one or two statistics courses are aware of confidence intervals to contain, say, the mean and the standard deviation of a sampled population or a population proportion. Some practitioners may also have been exposed to confidence and prediction intervals for regression models. These, however, are only a few of the statistical intervals required in practice. We have found that analysts are apt to use the type of interval that they are most familiar with—irrespective of whether or not it is appropriate. This can result in the right answer to the wrong question. Thus, we differentiate, at an elementary level, among the different kinds of statistical intervals and provide a detailed exposition, with numerous examples, on how to construct such intervals from sample data. In fact, this book is unique in providing a discussion in one single place not only of the "standard" intervals but also of such practically important intervals as confidence intervals to contain a population percentile, confidence intervals on the probability of meeting a specified threshold value, and prediction intervals to include the observations in a future sample.

Many of these important intervals are ignored in standard texts. This, we believe, is partly out of tradition; in part, because the underlying development (as opposed to the actual application) may require advanced theory; and, in part, because the calculations to obtain such intervals can be quite complex. We do not feel restricted by the fact that the methods are based upon advanced mathematical theory. Practitioners should be able to use a method without knowing the theory, as long as they fully understand the assumptions. (After all, one does not need to know what makes a car work to be a good driver.) Finally, we get around the problem of calculational complexity by providing comprehensive tabulations, charts, and computer routines, some of which were specially developed, and all of which are easy to use.

This book is aimed at practitioners in various fields who need to draw conclusions from sample data. The emphasis is on, and many of the examples deal with, situations that we have encountered in industry (although we sometimes disguise the problem to protect the innocent). Those involved in product development and quality assurance will, thus, find this book to be especially pertinent. However, we believe that workers in numerous other fields, from the health sciences to the social sciences, as well as teachers of courses in introductory statistics, and their students, can also benefit from this book.

We do not try to provide the underlying theory for the intervals presented. However, we give ample references to allow those who are interested to go further. We, obviously, cannot discuss statistical intervals for all possible situations. Instead, we try to cover those intervals, at least for a single population, that we have found most useful. In addition, we provide an introduction, and references to, other intervals that we do not discuss in detail.

It is assumed that the reader has had an introductory course in statistics or has the equivalent knowledge. No further statistical background is necessary. At the same time, we believe that the subject matter is sufficiently novel and important, tieing together work previously scattered throughout the statistical literature, that those with advanced training, including professional statisticians, will also find this book helpful. Since we provide a comprehensive compilation of intervals, tabulations, and charts not found in any single place elsewhere, this book will also serve as a useful reference. Finally, the book may be used to supplement courses on the theory and applications of statistical inference.

Further introductory comments concerning statistical intervals are provided in Chapter 1. As previously indicated, this chapter also includes a detailed discussion of the practical assumptions required in the use of the intervals, and, in general, lays the foundation for the rest of the book.

Chapter 2 gives a more detailed general description of different types of confidence intervals, prediction intervals, and tolerance intervals and their applications. Chapters 3 and 4 describe simple tabulations and other methods for calculating statistical intervals. These are based on the assumption of a normal distribution. Chapter 5 deals with distribution-free intervals. Chapters 6 and 7 provide methods for calculating statistical intervals for proportions and percentages, and for occurrence rates, respectively. Chapters 8, 9, and 10 deal with sample size requirements for various statistical intervals.

Statistical intervals for many other distributions and other situations, such as regression analysis and the comparison of populations, are briefly considered in Chapter 11. This chapter also gives references that provide further information, including technical details and examples of more complex intervals. Chapter 12 outlines other general methods for computing statistical intervals. These include methods that use large sample statistical theory and ones based on Bayesian concepts. Chapter 13 presents a series of case studies involving the calculation of statistical intervals; practical considerations receive special emphasis.

Appendix A gives extensive tables for calculating statistical intervals. The notation used in this book is summarized in Appendix B. Listings of some computer routines for calculating statistical intervals are provided in Appendix C.

We present graphs and tables for computing numerous statistical intervals and bounds. The graphs, especially, also provide insight into the effect of sample size on the length of an interval or bound.

Most of the procedures presented in this book can be applied easily by using figures or tables. Some require simple calculations, which can be performed with a hand calculator. When tables covering the desired range are not available (for some procedures, the tabulation of the complete range of practical values is too lengthy to provide here), factors may be available from alternative sources given in our references. However, often a better alternative is to have a computer program to calculate the necessary factors or the interval itself. We provide some such programs in Appendix C. It would, in fact, be desirable to have a computer program that calculates all the intervals presented in this book. One could develop such a program from the formulas given here. This might use available subroutine libraries [such as IMSL (1987) and NAG (1988)] or programs like those given in Appendix C, other algorithms published in the literature [see, e.g., Griffiths and Hill (1985), Kennedy and Gentle (1980), P. Nelson (1985), Posten (1982), and Thisted (1988)], or available from published libraries [e.g., NETLIB, described by Dongarra and Grosse (1985)]. A program of this type, called STINT (for STatistical INTervals), is being developed by W. Meeker; an initial version is available.

GERALD J. HAHN
WILLIAM Q. MEEKER

Acknowledgments

We would like to thank the many people who have helped. us. Our special appreciation goes to James V. Beck, Robert G. Easterling, Russel E. Hannula, J. Stuart Hunter, Wayne Nelson, Ernest M. Scheuer, and Jack L. Wood for the extensive time they spent in reading the manuscript and in providing us their insights, which, we believe, led to substantial improvements. We also wish to thank Craig A. Beam, Thomas J. Boardman, Necip Doganaksoy, Luis A. Escobar, Marie A. Gaudard, Emil H. Jebe, Jason R. Jones, Mark E. Johnson, William Makuch, Del Martin, Robert Mee, Vijayan N. Nair, Robert E. Odeh, Donald M. Olsson, Jeffrey R. Smith, William T. Tucker, Stephen B. Vardeman, and William J. Wunderlin for their careful reading of various drafts, and for providing us numerous important suggestions and comments. Robert E. Odeh was very helpful in providing the algorithms to compute Tables A10, A12, A28, and A29 and the factors in Tables A11, A13, and A14. We thank John Stuart for his excellent work in processing this manuscript, and Denise Riker and Rita Wojnar for their fine secretarial support. We thank our management at the GE Corporate Research and Development Center (Arthur C. Chen, Walter Berninger, Walter L. Robb, and their predecessors), at Iowa State University (Herbert A. David and Dean L. Isaacson), and at AT & T Bell Laboratories (Jeffrey H. Hooper) for their strong and long-standing support for this project. Finally, we would like to express our sincere appreciation to our wives, Bea Hahn and Karen Meeker, for bearing with us during the many days (and nights) we spent in putting this opus together.

GERALD J. HAHN
WILLIAM Q. MEEKER

Contents

Statistical Intervals

CHAPTER 1

Introduction, Basic Concepts, and Assumptions

1.1 STATISTICAL INFERENCE

Decisions frequently have to be made from limited sample data. For example:

- A television network uses the results of a sample of 1000 households to decide whether or not to continue a show.
- A company needs to use data from a sample of five turbines to arrive at a guaranteed efficiency for a further turbine to be delivered to a customer.
- A manufacturer will use tensile strength measurements in a laboratory test on 10 samples of each of two types of material to select one of the two materials for future production.

The sample data are often summarized by statements such as:

- 293 out of the 1000 sampled households were tuned in to the show.
- The average efficiency for the sample of five turbines was 67.4%.
- The samples using Material A had an average tensile strength 3.2 units higher than those using Material B.

The preceding "point estimates" provide a concise summary of the sample results, but they give no information about their precision. Thus, there may be big differences between such point estimates, calculated from the sample data, and what one would obtain if unlimited data were available. For example, 67.4% would seem a reasonable estimate (or prediction) of the efficiency of the next turbine. But how "good" is this estimate? From noting the variation in the observed efficiencies of the five turbines, we know that it is unlikely that the turbine to be delivered to the customer will have an efficiency of *exactly* 67.4%. We may, however, expect its efficiency to be "close to" 67.4%. But how close? Can we be reasonably confident that it will be within ±0.1% of 67.4%? Or within ±1%? Or within ±10%? We need to quantify the uncertainty associated with our estimate. An understanding of this uncertainty is an important input for decision making, such as, for example, in providing a performance warranty. Moreover, if our knowledge, as reflected by the length of the uncertainty interval, is too imprecise, we may wish to obtain more data before making a decision.

The example suggested quantifying uncertainty by constructing statistical intervals around the point estimate. This book shows how to obtain such intervals. We describe frequently required (but not necessarily well known) statistical intervals calculated from sample data, differentiate among the various types of intervals, and show their applications. Simple procedures for obtaining each of the intervals are presented and their use is illustrated. We also show how to choose the sample size so as to attain a desired degree of precision, as measured by the length of the statistical interval. Thus, this book provides a comprehensive guide and reference to the use of statistical intervals to quantify the uncertainty in the information about a sampled population or process, based upon, possibly, a small, but randomly selected sample. The concept of a "random sample" will be discussed further shortly.

1.2 DIFFERENT TYPES OF STATISTICAL INTERVALS: AN OVERVIEW

Various types of statistical intervals may be calculated from sample data. The appropriate interval depends upon the specific application. Frequently used intervals are:

- A *confidence interval* to contain an unknown characteristic of the sampled population or process. The quantity of interest might be a population property or "parameter," such as the mean or standard deviation (measuring variability) of the population or process, or a function of such parameters, such as a percentage point (or percentile) of the sampled population or process. Thus, one might calculate, for example, from measurements of tensile strengths on a random sample of test specimens from a particular process, different intervals (depending upon the question of interest) that we can claim, with a (specified) high degree of confidence, contain (1) the mean tensile strength, (2) the standard deviation of the distribution of tensile strength, (3) the 10% point of the tensile strength distribution, or (4) the proportion of "all" specimens that exceed a stated "threshold value" for the sampled process.

- A *statistical tolerance interval* to contain a specified proportion of the units from the sampled population or process. For example, based upon a past sample of tensile strength measurements, we might wish to compute an interval to contain, with a specified degree of confidence, the tensile strengths of at least 90% of the units from the sampled process. Hereafter we will generally, simply refer to such an interval as a "tolerance interval."
- A *prediction interval* to contain one or more future observations, or some function of such future observations, from a previously sampled population or process. For example, based upon a past sample of tensile strength measurements, we might wish to construct an interval to contain, with a specified degree of confidence, (1) the tensile strength of a randomly selected single future unit from the sampled process (this was of interest in the turbine efficiency example), (2) the tensile strengths for *all* of five future units, or (3) the average of the tensile strengths of the five future units.

Most users of statistical methods are familiar with confidence intervals for the population mean and for the population standard deviation (but often not for population percentage points or the probability of exceeding a specified threshold value). Some are also aware of tolerance intervals. However, despite their practical importance, most practitioners, and even many statisticians, know very little about prediction intervals except, perhaps, for their application to regression problems. Thus, a frequent mistake is to calculate a confidence interval to contain the population mean when the problem requires a tolerance interval or a prediction interval. At other times, a tolerance interval is used when a prediction interval is needed. Such confusion is understandable, because most texts on statistics generally discuss confidence intervals, occasionally make reference to tolerance intervals, but generally consider prediction intervals only in the context of regression analysis. This is unfortunate because, in actual applications, tolerance intervals, prediction intervals, and confidence intervals on population percentiles and on exceedance probabilities are needed almost as frequently as the better known confidence intervals. Moreover, the calculations for such intervals are generally no more difficult than those for confidence intervals. [See Scheuer (1990) and Vardeman (1990) for some related comments.]

1.3 THE ASSUMPTION OF SAMPLE DATA

We will be concerned in this book only with situations where uncertainty is present because the available data are from a (often small) random sample from a population or process. There are, of course, some situations where there is little or no such statistical uncertainty. This is the case when the relevant information on every unit in a finite population has been recorded (without measurement error), or when the sample size is so large that the uncertainty in our estimates due to sampling variability is negligible (as we shall see, how large is "large" depends on the specific application). Examples of situations where one is generally dealing with the entire population are:

- The given data are census information that have been obtained from all residents in a particular city (at least to the extent that the residents could be located).
- There has been 100% inspection (i.e., all units are measured) of a performance property for a critical component used in a spacecraft.
- A complete inventory of all the parts in a warehouse has been taken.
- A customer has received a one-time order of five parts and has measured each of these parts. Even though the parts are a random sample from a larger population or process, as far as the customer is concerned the five parts at hand make up the entire population of interest.

However, even in such situations, intervals to express uncertainty are sometimes needed. For example, based upon extensive data, we *know* that the weight of a product is approximately normally distributed with an average of 16.10 ounces and a standard deviation of 0.06 ounces. We wish an interval to contain the weight of a single unit randomly selected by a customer, or by a regulatory agency. The calculation of the resulting simple *probability interval* is described in books on elementary probability and statistics. Such intervals generally assume complete knowledge about the population (for example, the mean and standard deviation of a normal distribution). In this book, we are concerned with the more complicated problem where, for example, for a normal distribution the population mean and standard deviation are *not* known exactly but are estimated, subject to sampling variability, from a random sample. In particular, as we shall see, when the sample size increases, such statistical intervals converge to probability intervals for some evaluations (i.e., for tolerance intervals and prediction intervals). For other evaluations (confidence intervals), the statistical intervals converge to zero length as the sample size increases, because there is no uncertainty.

Another situation where there exists statistical uncertainty, even though the entire population has been evaluated, is that where the readings are subject to measurement error. For example, one might determine that the measurements of a particular property for 971 out of 983 parts in a production lot are within specification limits. However, due to measurment error, the actual number of parts within specifications may be more or less than 971. (Moreover, if something is known about the statistical distribution of measurement error, one can then also quantify the uncertainty associated with the estimated number of parts within specifications.)

Finally, we note that even when there is no quantifiable statistical uncertainty, there may still also be other uncertainties of the type suggested in the following discussion.

WARNING: APPLY STATISTICAL METHODS WITH GREAT CARE

1.4 THE CENTRAL ROLE OF PRACTICAL ASSUMPTIONS CONCERNING "REPRESENTATIVE DATA"

We have briefly described different statistical intervals that a practitioner might use to express the uncertainty in various estimates or predictions generated from sample data. This book presents the methodology for calculating such intervals. Before proceeding, we need, however, make clear the major practical assumptions dealing with the "representativeness" of the sample data. We do this in the following sections. Departures from these implicit assumptions are common in practice and can invalidate the entire analyses. Ignoring such assumptions can lead to a false sense of security, which, in many applications, is the weakest link in the inference chain. Thus, for example, production engineers need to question the assumption that the performance observed on prototype units produced in the lab also applies for production units, to be built much later, in the factory. Similarly, a reliability engineer should question the assumption that the results of a laboratory life test will adequately predict field failure rates. In fact, in some studies, the assumptions required for the statistical interval to apply may be so far off the mark that it would be inappropriate, and perhaps even misleading, to use the formal methods presented here.

In the best of situations, one can rely on physical understanding, or information from outside the study, to justify the practical assumptions. Such assessment is, however, principally the responsibility of the engineer—or "subject-matter expert." Often, the assessment is far from clear-cut and an empirical evaluation may be impossible. In any case, one should keep in mind that the intervals described in this book reflect only the statistical uncertainty, and, thus, provide a *lower bound* on the true uncertainty. The generally nonquantifiable deviations of the practical assumptions from reality provide an added *unknown* element of uncertainty. If there were formal methods to reflect this further uncertainty (occasionally there are, but often there are not), the resulting interval, expressing the *total* uncertainly, would clearly be longer than the statistical interval alone. This observation does, however, lead to a rationale for calculating a statistical interval for situations where the basic assumptions are questionable. In such cases, if it turns out that the calculated statistical interval is long, we then know that our estimates have much uncertainty—even *if* the assumptions were all correct. On the other hand, a narrow statistical interval would imply a small degree of uncertainty *only if* the required assumptions hold.

Because of their importance, we feel it appropriate to review, in some detail, the assumptions and limitations underlying the use and interpretation of statistical intervals before proceeding with the technical details of how to calculate such intervals.

1.5   ENUMERATIVE VERSUS ANALYTIC STUDIES

Deming (1953, 1975) emphasizes the important differences between "enumerative" and "analytic" studies [a concept that he briefly introduced earlier, e.g., Deming (1950)]. Despite its central role in making inferences from the sample data, many traditional textbooks in statistics have, by and large, been slow in giving this distinction the attention that it deserves. [Two exceptions, using different terminology, are Chapters 1 to 3 of Box, Hunter, and Hunter (1978) and pages 15 and 16 of Snedecor and Cochran (1967). Also, this distinction is explicitly discussed in detail in the recent book by Gitlow, Gitlow, Oppenheim, and Oppenheim (1989).]

To point out the differences between these two types of studies, and some related considerations, we return to the examples of Section 1.1. As indicated, the statements there summarize the sample data. In general, however, investigators are concerned with making inferences or predictions *beyond* the sample data. Thus, in these examples, the real interest was, not in the sample data *per se*, but in:

1. The proportion of households in the *entire country* that were tuned to the show.
2. The efficiency of the, *as yet not manufactured*, turbine to be sent to the customer.
3. A comparison of the average tensile strengths of the *production units to be built* in the factory during the *forthcoming year* using Material A and Material B.

In the first example, our interest centers on a finite identifiable collection of units, or population, from which the sample was drawn. This population, consisting of all the households in the country, exists at the time of sampling. Deming uses the term "enumerative study" to describe situations like this where there exists a specific population concerning which inferences are to be made, based upon a random sample. More specifically, Deming (1975) defines an enumerative study as one in which "action will be taken on the material in the frame studied," where he uses the conventional definition of a frame as "an aggregate of identifiable units of some kind, any or all of which may be selected and investigated. The frame may be lists of people, areas, establishments, materials, or other identifiable units that would yield useful results if the whole content were investigated." Thus, the frame results in a finite list, or other identification, of distinct (nonoverlapping) and exhaustive sampling units. The frame defines the population to be sampled in an enumerative study.

Some further examples of enumerative studies are:

- Public opinion polls; in this case, the population of interest might be the entire adult U.S. population, or some defined segment thereof, such as all registered voters.
- Sample audits to assess the correctness of last month's bills; in this case, the population of interest consists of all of last month's bills.
- Product acceptance sampling; in this case, the population of interest consists of all units in the production lot being sampled.

In an enumerative study, one wishes to draw conclusions about an existing well-defined *population* that is being sampled directly. In such a study, the correctness of statistical inferences requires a random sample from the target population. Such a sample, is, at least in theory, generally attainable in such a study; see Section 1.6.1.

In contrast, the second two examples of Section 1.1 (dealing, respectively, with the efficiency of a future turbine and the comparison of the two manufacturing processes next year) illustrate what Deming calls "analytic studies." We no longer have an existing, well-defined, finite population. Instead, we want to draw inferences, or make predictions, about a future *process*, which is not available for sampling. We can, however, obtain data from the existing (most likely somewhat different) process.

Specifically, Deming (1975) defines an analytic study as one "in which action will be taken on the process or cause-system...the aim being to improve practice in the future.... Interest centres in future product, not in the materials studied." He cites as examples "tests of varieties of wheat, comparison of machines, comparisons of ways to advertise a product or service, comparison of drugs, action on an industrial process (change in speed, change in temperature, change in ingredients)." Similarly, we may wish to use data from an existing process to predict the characteristics of future output from the same or a similar process. Thus, in a prototype study of a new part, the process (or conceptual population) of interest consists of parts of that type that may be manufactured in the future.

These examples are representative of many encountered in practice, especially in engineering, medical, and other scientific investigations. It is inherently more complex to draw inferences from analytic studies than from enumerative studies; analytic studies require the important (and often unverifiable) added assumption that the process about which one wishes to make inferences is statistically identical to that from which the sample was selected.

What one wishes to do with the results of the study is often a major differentiator between an enumerative and an analytic study. Thus, if one's interest is limited to describing an existing population, one is dealing with an enumerative study. On the other hand, if one is concerned with a process that is still to be improved, or otherwise to be acted upon, perhaps as a result of the study, then we are clearly dealing with an analytic study.

1.5.1 Differentiating between Enumerative and Analytic Studies

Deming (1975) presents the following "simple criterion to distinguish between enumerative and analytic studies. A 100 per cent sample of the frame answers the question posed for an enumerative study, subject of course to the limitations of the method of investigation. In contrast, a 100 per cent sample . . . is still inconclusive in an analytic problem." This is because for an analytic study our real interest is in a process that is not available for sampling. Thus, a 100% sample of the process that *is* available misses the mark.

Deming's rule can be useful in situations when the differentiation between an analytic and an enumerative study does not seem clear-cut. For example, a public opinion "exit poll" to estimate the proportion of voters who have voted (or, at least, would assert that they have voted) for a particular candidate, based upon a random sample of individuals leaving the polling booth, is an example of an enumerative study. In this case, a 100% sample provides perfect information (assuming 100% correct responses). However, estimating, before the election, the proportion of voters who will *actually* go to the polls and vote for the candidate is an analytic study, because it deals with a future process. Thus, between the time of the survey and election day, some voters may change their minds, perhaps as a result of some important external event. Also adverse weather conditions on election day (not contemplated on the sunny day on which the survey was conducted) might deter many from going to the polls—and the "stay-at-homes" are likely to differ demographically, and, therefore, differ in their voting preferences, from those that do vote. Thus, even if we had taken a 100% sample of every eligible voter prior to the election, we still would not be able to predict the outcome of the election with certainty, since we do not know who will actually vote and who will change their minds in the intervening period. (Special considerations in sampling people are discussed in Section 1.9.) Taking another example, it is sometimes necessary to sample from inventory to make inferences about a product population or process. If interest focuses on the inventory, the study is enumerative. If, however, interest focuses on the future output from the process, the study is analytic. Finally, drawing conclusions about the performance of a turbine to be manufactured in the future, based upon data on five turbines built in the past, involves, as we have indicated, an analytic study. If, however, the five measured turbines *and* the future turbine to be shipped were all independently and randomly selected from inventory (unlikely to be the case in practice), one would be dealing with an enumerative study.

1.5.2 Statistical Inference for Analytic Studies

We differ from the views of some [e.g., page 558 of Gitlow, Gitlow, Oppenheim, and Oppenheim (1989)] who imply that statistical inference methods, such as statistical intervals, have no place whatsoever in analytic studies. Indeed, such methods have been used successfully in statistical studies in science and industry for decades and most of these studies have been analytic. Instead, we feel that the decision of whether or not to use statistical intervals need be made on a case by case basis (see subsequent discussion). In addition, when such methods are used for analytic studies, they require assumptions different from those required for an enumerative study.

We will now consider, in further detail, the basic assumptions underlying inferences from enumerative studies and then comment on the additional assumptions made in analytic studies.

1.6 BASIC ASSUMPTIONS FOR ENUMERATIVE STUDIES

1.6.1 Definition of Target and Sampled Population

In every enumerative study there is some "target population" about which it is desired to draw inferences. An important first step—though one that is sometimes omitted by analysts—is that of explicitly and precisely defining this target population. For example, the target population may be all the automobile engines of a specified model manufactured on a particular day, or in a specified model year, or over any other defined time interval. In addition, one need also make clear the specific characteristic(s) to be evaluated. This may be a measurement or other reading on an engine, or the time to failure of a part on a life test, where a "failure" is precisely defined. Finally, in many applications, and especially those involving manufactured products, one must clearly state the operating environment in which the defined characteristic is to be evaluated. For a life test, this might be "normal operating conditions," where exactly what constitutes such conditions also needs to be clearly stated.

As indicated in the Deming quote, the next step is that of establishing a frame from which the sample is to be taken, i.e., developing a specific listing, or other enumeration of the population from which the sample can subsequently be selected. Examples of such frames may be the serial numbers of all the automobile engines built over the specified time period, the complete listing in a telephone directory for a community, the schedule of incoming commercial flights into an airport on a given day, or a tabulation of all invoices billed during a calendar year. Often, the frame is *not* identical to the target population. For example, a telephone directory generally lists households, rather than individuals, and omits those who do not have a telephone, people with unlisted phones, new arrivals in the community, etc.—and also includes businesses, which are not always clearly identified as such. If one were trying to conduct an evaluation of the proportion of listed telephones in working order at a given time, a complete listing of telephones (available to the phone company) will probably coincide with the target population. However, for most other studies, there may be an important difference between the telephone directory listing and the target population of real interest.

The listing provided by the frame will henceforth be referred to as the "sampled population." Clearly, the inferences from a study, as expressed by statistical intervals, will be on this sampled population and—when the two differ—not on the target population. Thus, our third step—after defining the target population and the sampled population—is that of evaluating the differences between the two, and the possible consequences of this difference on the results of the study. Moreover, it needs to be stated emphatically that these differences introduce uncertainties above and beyond those generally quantified by the statistical intervals provided in this book. (If well understood, these differences can sometimes be dealt with.)

1.6.2 The Assumption of a Random Sample

The data are assumed to be a random sample from the sampled population. Because we will only deal with random samples in this book, we will sometimes use the term "sample" to denote a random sample. In enumerative studies, we will be concerned principally with the most common type of random sampling, namely *simple random sampling* (although we will briefly describe other types of random samples in Section 1.6.3).

Simple random sampling gives every possible sample of n units from the population the same probability of being selected. A simple random sample can, at least in theory, be obtained from a population of size N by numbering each unit in the population from 1 to N, placing N balls bearing the N numbers into a bin, thoroughly mixing the balls, and then randomly drawing n balls from the bin. The units to be sampled are those with numbers corresponding to the n selected balls. In practice, tables of random numbers, such as those given in tabulations of such numbers [e.g., Rand (1955)] and generated by computer algorithms [e.g., IMSL (1987) and Kennedy and Gentle (1980)] and by statistical computing software (e.g., MINITAB), provide easier ways of obtaining a random sample. [The "complete randomness" of some computer algorithms has been challenged—see, e.g., Kennedy and Gentle (1980)—and varies from one random number generation scheme to the next. We believe, however, that most available tables and modern computer routines are usually good enough to serve the purpose of selecting a simple random sample.]

The assumption of random sampling is critical. This is because the statistical intervals reflect only the randomness introduced by the random sampling process and do not take into consideration biases that might be introduced by a nonrandom sample. It is especially important to recognize this, since in many real world studies one does not have a random sample (see Section 1.9 for further discussion).

1.6.3 More Complicated Random Sampling Schemes

There are also other random sampling schemes, notably stratified random sampling, cluster random sampling, and systematic random sampling. These are used frequently in such applications as survey sampling of human populations and inventory estimation. For such samples, rather than every possible sample of *n* units from the sampled population having the *same* probability of being selected, each such sample has a *known* probability of being selected. Statistical intervals can also be constructed for such more complicated sampling schemes; these intervals are generally more complex than the ones described in this book. The interested reader is referred to Cochran (1977), Scheaffer, Mendenhall, and Ott (1979), Sukhatme, Sukhatme, Sukhatme, and Asok (1984), and Williams (1978).

The more complicated *random* sampling schemes, described briefly below, need to be differentiated from various *nonrandom* sampling schemes, which we describe in Section 1.8 under the general heading of "convenience sampling."

1.6.3.1 Stratified Random Sampling

In some sampling applications, the population is naturally divided into nonoverlapping groups or *strata*. For example, a population might be divided according to gender, job rank, age group, geographic region, manufacturer, etc. It may be important or useful to take account of these strata in the sampling plan because:

1. Some important questions may focus on individual strata; e.g.,information is needed by geographical region, as well as for the entire country.

2. Cost, methods of sampling, access to sample units, or resources may differ among strata, e.g., salary data is generally more readily available for individuals in the public sector than for those working in the private sector.

3. Stratified random sampling provides more precise estimates of population parameters than one would obtain from a simple random sample of the same size—if the response variable has less variability within a

stratum than across strata. The increase in precision results in shorter (but more complex) statistical intervals concerning the entire population.

In stratified random sampling, one takes a simple random sample from each stratum in the population. The methods presented in this book can be used with data from a single stratum to compute statistical intervals for that stratum. However, when stratified random sample data are combined across strata (e.g., to obtain a confidence interval for the population mean or total) special methods are needed. These are discussed in the previously indicated standard texts on survey sampling. These books also discuss how to allocate units across strata and how to choose the sample size within each stratum.

1.6.3.2 Cluster Random Sampling

In some studies it is less expensive to obtain samples by using "clusters" of elements that are conveniently located together in some manner, rather than taking a simple random sample. For example, when items are packed in boxes, it is often much easier to take a random sample of boxes of items, and either evaluate all items in each selected box, or, possibly, take a second random sample within each of the selected boxes rather than to randomly sample individual items, irrespective of the box in which they are contained. Also, it is often more natural to sample some or all members of a randomly selected family rather than a randomly selected individual. Finally, it is often easier to find a frame for, and sample groups of, individuals clustered in a random sample of locations, rather than taking a simple random sample of the same number of individuals spread over, say, an entire city. In other cases, a listing of clusters, but not of elements (i.e., individuals or items) may be available; clusters may, for this reason, be the natural sampling unit. In each of theses cases, we have to define the clusters, obtain a frame that lists all clusters, and then take a random sample of clusters. Typically, a response is then obtained for each element in the cluster, although, sometimes, subsampling within clusters is conducted.

The value of the information for each additional sample unit within a cluster can be appreciably less than that for a simple random sample, especially if the items in a cluster tend to be similar and many units are chosen from each cluster. However, if the elements in the clusters are a "well mixed" representation of the population, the loss in precision may be slight. Often, the *lower per element* cost of cluster random sampling will more than make up for the loss of statistical efficiency. Thus, given a specified total cost to conduct a study, the net result of cluster random sampling can be an improvement in the overall statistical efficiency of the study, as evidenced by shorter statistical intervals, compared to those for a simple random sample involving a smaller total sample size. (Sometimes, the investigator has a say in choosing the cluster size. In such cases, the loss of efficiency can be mitigated by taking a larger sample of smaller clusters; of course, when clusters contain only one element, one is back to simple random sampling.) Chapter 8 of Cochran (1977) discusses cluster sample design, including choice of cluster size, and gives detailed methods for analyzing data from cluster random samples.

1.6.3.3 Systematic Random Sampling

It is often much easier to choose the sample in a systematic manner than to take a simple random sample. For example, it would be difficult to take a simple random sample of all the customers who come into a store on a particular day, because there is no readily available frame. It would, however, be relatively simple to sample every 10th (or other preselected number) person entering the store. Similarly, it would be much easier to have a clerk examine every 10th item in a file cabinet, instead of choosing a simple random sample of items from all of the files in the cabinet. In both examples, the information/cost ratio with systematic sampling might be much higher than with simple random sampling. In some situations, in fact, a systematic sample may be the only feasible alternative.

Systematic samples are random samples as long as one uses a random starting point. Special methods and formulas, however, are needed to compute statistical intervals. Also, particular caution is needed in choosing the systematic pattern that is to be used in sampling. Serious losses of efficiency or biases are possible if there are periodicities in the population and if these periodicities are in phase with the systematic sampling scheme. For example, if a state motor vehicle bureau measures traffic volume each Wednesday (where Wednesday was a randomly selected weekday), the survey results would, undoubtedly, provide a biased estimate of average weekday traffic volume. On the other hand, if such sampling took place every sixth day, the resulting estimate would likely be a lot more reasonable. Also, strictly speaking, for the sample to be random, the initial day would have had to be selected randomly—although, from a practical viewpoint, this probably would not be of much consequence in *this* example. Books on survey sampling, such as those previously mentioned, discuss various aspects of systematic random sampling.

1.7 ADDITIONAL ASPECTS OF ANALYTIC STUDIES

1.7.1 Analytic Studies

In an enumerative study, one generally wishes to draw inferences by sampling from a well-defined existing population, the members of which can usually be enumerated, at least conceptually—even though, as we have seen, difficulties can arise in obtaining a random sample from the target population. In contrast, as previously indicated, in an analytic study one wishes to draw conclusions about a process which often does not even exist at the time of the study. As a result, the process that is sampled may differ, in various ways, from the one about which it is desired to draw inferences. As we have indicated, sampling prototype units, made in the lab or on a pilot line, to draw conclusions about subsequent full-scale production is one common and very obvious example of an analytic study.

1.7.2 The Concept of "Statistical Control"

A less evident example of an analytic study arises if, in dealing with a mature production process, one wishes to draw inferences about future production, based upon sample data from current or recent production. Then, if the process is in so-called "statistical control," *and remains so*, the current data may be used to draw valid inferences about the future performance of the process. The concept of statistical control means, in its simplest form, that the process is stable or unchanging. It implies that the statistical distribution of the characteristic of interest for the current process is identical to that for the process in the future. It also suggests that units selected consecutively from production are no more likely to be alike than units selected, say, a day, a week, a month, or, even a year, apart. All of this, in turn, means that the only sources of variability are "common cause" within the system, and that variation due to "assignable" or "special" causes, such as differences between raw material lots, operators, ambient conditions, etc., have been removed. The concept of statistical control is an ideal state that, in practice, may exist only infrequently, though it may often provide a useful working approximation. In such cases, the statistical intervals provided in this book might yield reasonable inferences about the process of real interest. On the other hand, when the process is not in, or near statistical control, the applicability of the statistical intervals given here for characterizing the process, etc., may be undermined by trends, shifts, cycles, and other variations (unless these are accounted for in a more comprehensive model).

1.7.3 Other Analytic Studies

Although analytic studies frequently require projecting from the present to a future time period, this is not the only way an analytic study arises. For example, production constraints, concerns for economy, and a variety of other considerations may lead one to conduct a laboratory scale or pilot line assessment, rather than perform direct on-line evaluations, even though production is up and running. In such cases, it is, sometimes, possible to perform "verification studies" to compare the results of the sampled pilot process with the actual production process.

1.7.4 Random Samples in Analytic Studies

In applying statistical inference methods in an analytical study, one assumes that observations are independently and identically distributed. This assumption is, sometimes, referred to as having a "random sample" and can be achieved most readily for a process "in statistical control." For analytical studies of processes that change over time, the timing of the selection of sample units becomes critical, as indicated in Section 1.7.6.

1.7.5 How to Proceed

The following operational steps appear appropriate for many analytic studies:

- Have the engineer or subject matter expert define the process of interest.
- Determine the possible sources of data that will be useful for making the desired inferences about the process of interest, i.e., define the process to be sampled or evaluated.
- Have the engineer or subject matter expert clearly state the assumptions that are required for the results of the study on the sampled process to be translatable to the process of interest.
- Collect well-targeted data and, to the extent possible, check the model and the other assumptions.
- Jointly decide, in light of the assumptions and the data, and an understanding of the underlying cause mechanism, whether there is still value in calculating a statistical interval, or whether this might lead to a false sense of security, and should, therefore, be avoided.
- If it is decided to obtain a statistical interval, ensure that the underlying assumptions are fully recognized and make clear that the length of this interval represents only a lower bound on the total uncertainty. That is, it deals only with the uncertainty associated with the random sampling—and does not include uncertainties due to the differences between the sampled process and the process of real interest.

Some further comments on these steps are provided in the next subsection.

1.7.6 Planning and Conducting an Analytic Study

In conducting an analytic study, because the process of real interest may not be available for sampling, one often has the opportunity and, indeed the responsibility, of defining the specific process that is to be sampled. As Deming (1975) and others [e.g., Gitlow, Gitlow, Oppenheim, and Oppenheim (1989), Nolan (1988) and Provost (1989)] emphasize, in conducting analytic studies, one should generally aim to consider as broad an environment as possible. For example, one should include the wide spectrum of raw materials, operating conditions, etc. that might be encountered in the future. If this is done, one generally has to make fewer assumptions in subsequently using the results of the sampled process to draw inferences about the process of interest. Moreover, in sampling a process over time, it is usually advisable to sample over relatively long periods, because observations taken over a short period of time are less likely to be representative (with regard to both average and long-run variability) than those obtained over a longer time period (unless the process is in strict statistical control). For example, in a study of the properties of a new alloy, specimens produced closely together in time may be more alike than those produced over longer time intervals due to variations in ambient conditions, raw material, operators, the condition of the machines, the measuring equipment, etc.

In some analytic studies, one might deliberately wish to make evaluations under extreme conditions. In fact, Deming (1975) feels that in the early stages of an investigation, "it is nearly always the best advice to start with strata near the extremes of the spectrum of possible disparity in response, as judged by the expert in the subject matter, even if these strata are rare" (in their occurrence in practice). He cites an example that involves the comparison of "speed of reaching equilibrium" for different types of thermometers. He advocates, in this example, to perform an initial study on two groups of people—those with normal temperature and those with (high) fever. Moreover, whenever possible, data on concommitant variables such as operator, raw material, etc., should be obtained (in time order), along with the response(s) of primary interest, for possible later (graphical) analysis.

1.8   CONVENIENCE AND JUDGMENT SAMPLES

In practice, it is sometimes difficult, or even impossible, even in an enumerative study, to obtain a random sample. Often, it is much more *convenient* to sample without strict randomization. Say, for example, that it is practical to sample a manufactured product to characterize its properties only after it has been packaged in boxes. If the product is ball bearings, it might be easy to thoroughly mix the contents of a box and sample randomly. On the other hand, suppose the product is made up of fragile ceramic plates, stacked in large boxes. In this case, it is much easier to sample from the top of the box than to obtain a random sample from among all of the units in the box. Similarly, if the product is produced in rolls of material, it is often simple to cut a sample from either the beginning or the end of the roll, but often impractical to sample from any place else. This is not a systematic random sample, because there is not a random starting point. In other cases, one might wish to sample from a process during production, or select product immediately after production. In this case, it is often more practical to sample material periodically, say every 2 hours during an 8-hour shift, than to select material at four different randomly selected times during each shift. (In this, and other applications, a further justification for such periodic sampling, in addition to convenience, is the need to consistently monitor the process for changes by the use of control charts, etc.)

Selection of product from the top of the box, from either end of a roll, or at prespecified periodic time intervals for a production process (without a random starting point) are examples of what is sometimes referred to as "convenience sampling." Such samples are generally *not* strictly random; for example, some units (e.g., those not at either end of the roll) have no chance of being selected.

Because one is not sampling randomly, statistical intervals, strictly speaking, are not applicable for convenience sampling. In practice, however, one uses experience and understanding of the subject matter to decide on the applicability of applying statistical inferences to the results of convenience sampling. Frequently, one might conclude that the convenience sample will provide data that, for all practical purposes, are as "random" as those that one would obtain by a simple random sample. This might, for example, be the case if, even though samples were selected only from the top of the box, the items were thoroughly mixed before placement into the box. Also, sampling from an end of a roll might yield information equivalent to that from random sampling *if* production is continuous, the process is in statistical control, and there is no "end effect." Similar assumptions apply in drawing conclusions about a process from scheduled periodic samples from production. Our point is that, treating a convenience sample as if it were a random sample *may sometimes* be reasonable from a practical point of view. However, the fact that this assumption is being made needs to be recognized, and the validity of using statistical intervals, as if a random sample had been selected, needs to be critically assessed based upon the specific circumstances.

Similar considerations apply in "judgment" or "pseudo-random" sampling. This occurs when personal judgment is used to choose "representative" sample units. For example, when a foreman, by eyeball, takes what appears to be a "random" selection of production units, without going through the necessary formalities that we have described for selecting a random sample. In many cases, this might yield results that are essentially equivalent to those of a random sample. Sometimes, however, this procedure will, either deliberately or nondeliberately, result in a higher probability of selecting, for example, conforming (or nonconforming) units. In fact, studies have shown that what might be called "judgment" can actually lead, even unintentionally, to seriously biased samples and, therefore, invalid or misleading results. Thus, strictly speaking, the use of such judgment, in place of random selection of sample units, invalidates the probabilistic basis for statistical inference methods and could render statistical intervals to be meaningless, in practice.

Judgment is, of course, important in planning studies, but it needs to be applied carefully in the light of available knowledge and practical considerations. Moreover, where possible, judgment should *not* be used as a substitute for random sampling or other randomization needed to make probabilistic inferential statements. Thus, returning to Deming's example of comparing the "speed of reaching equilibrium" for different type thermometers, it might well be advantageous to make comparisons for strata of people with normal temperature and with high fever. However, within these two strata, we should select patients at random. Also, thermometers should be randomly selected (possibly within strata) to the greatest degree possible for each type to be evaluated. This will provide the opportunity for valid statistical inferences, even though these inferences may be in a severely limited domain. Deming (1975, 1976) provides additional discussion on the use of judgment in analytic studies.

1.9 SAMPLING PEOPLE

Additional considerations frequently arise in sampling human populations, such as in a public opinion poll or a television rating survey. In this case, in contrast to sampling a manufactured product, the subject selected for the study generally chooses whether or not to respond, and whether or not to provide a truthful response. As a result, response rates on mail surveys, without special inducements, have been found to be extremely low, and the results correspondingly unreliable. (Moreover, although special inducements might increase the sample size, they might also lead to a less "representative" sample.) "On site" surveys, such as at shopping malls, might result in higher response rates, but such haphazard sampling often results in a nonrandom selection from the population of interest, by tending to exclude the very elderly, poor people, wealthy people, and others who rarely visit shopping malls. (The data from such studies are sometimes analyzed by taking the disproportionality of the sampling into consideration, e.g., by performing the analysis as if a stratified study had been conducted.)

Telephone surveys might obtain a "more random" selection of households, but require respondents to be home and might result in a biased selection of certain family members. In TV rating surveys, one is interested in determining the viewing habits of, say, the entire viewing audience. However, a particular survey generally provides information for only those individuals or households who can be induced, perhaps, in part, by financial rewards, to participate in the survey.

Nonresponse, and related problems, defeat the goal of strict random sampling, and, thus, again compromise the use of statistical intervals that are calculated under the assumption of such a sample. These problems are likely to introduce biases if, as is often the case, the respondents and the nonrespondents have different views. For example, willingness to participate in a TV rating survey is likely to be correlated with the respondent's programming preferences. These difficulties are well known to experts who conduct such surveys, and various procedures have been developed to mitigate them or to compensate for them. These include:

- A follow-up sample from among the initial nonrespondents, and a comparison of the results with those from the initial respondents.
- Comparison between respondents and nonrespondents with regard to "demographics," or other variables, that are likely to be related to the response variable (and possible adjustments based thereon).

1.10   INFINITE POPULATION ASSUMPTIONS

The methods for calculating statistical intervals discussed in this book (except for those in Section 11.3.17) are based on the assumption that the sampled process or population is infinite, or, at least, very large relative to the sample size. (This is because, as previously indicated, it is assumed that the population or process is stable, i.e., unchanging. With a finite population, however, the sampling itself changes the population available for further sampling, by depletion and, therefore, samples are no longer truly independent. For example, if the population consisted of 1000 parts, selection of the first sampled unit reduces the population available for the second sample to 999 units.)

Sometimes, in enumerative studies, the assumption of an infinite population or process is not met. In addition, statistical methods have also been developed for making inferences about finite populations. These, however, are somewhat more complicated.

In practice, however, if the sample size is a small proportion of the population (10% or less is a commonly used figure), standard statistical methods will give results that are approximately correct. Thus, in many situations, it is not unreasonable to assume an infinite population in calculating a statistical interval. Moreover, books on survey sampling [e.g., Cochran (1977)] show how to use a "finite population correction" to adjust, at least approximately for the finite population size. This generally results in a shorter statistical interval than one would obtain without such a correction. Thus, ignoring the fact that one is sampling from a finite population usually results in conservative intervals (i.e., intervals that are longer than needed) because it tends to inflate the actual uncertainty. In an analytic study, the population is conceptually infinite; thus, no finite population correction is needed.

The preceding discussion leads to a closely related, and frequently misunderstood, point concerning sampling from a finite (large or small) population. In particular, precision of the results from a statistical study depends principally on the *absolute*—and, not the *relative*—sample size. For example, consider an analyst who wants to make a statement about mean strength of units produced from two lots from a stable production process. Lot 1 consists of 10,000 units and Lot 2 consists of 100 units. Then, a simple random sample of 100 units from Lot 1 (a 1% sample) will provide *much* more information about Lot 1 than a random sample of size 10 from Lot 2 (a 10% sample) provides about Lot 2. For further discussion, see Hahn (1979). We will describe methods of choosing the sample size in Chapters 8, 9, and 10.

1.11 PRACTICAL ASSUMPTIONS: OVERVIEW

In Figure 1.1 we summarize the major points of our discussion in Sections 1.3 to 1.10 and suggest a possible approach for evaluating the assumptions underlying the calculation of the statistical intervals described in this book. It is, of course, not possible to consider all possible situations in such a diagram. However, we feel that Figure 1.1 provides a useful guide to practitioners on how to proceed for many situations.

The following comments are keyed to the numbers in parentheses in Figure 1.1:

(1) Is the purpose of the study to act on or draw conclusions about an existing population (enumerative study), or is it to act on or draw conclusions about a conceptual population or process (analytic study)?

(2) Statistical intervals apply to the process under evaluation when it was sampled—but not necessarily to the future state of the process or some future "similar" process. When the process changes, e.g., over time, the length of the interval calculated from the sampled process provides only a lower bound on the total uncertainty.

(3) Statistical intervals apply to the population defined by the frame from which the sample is taken. When the frame does not correspond to the target population, inferences could be biased, and the calculated interval may not accurately reflect the uncertainty.

(4) The statistical intervals in this book assume a simple random sample from the existing population or identically distributed independent samples from the process of interest.

(5) More complex statistical intervals than those described in this book apply for random samples other than simple random samples; see Cochran (1977), Scheaffer, Mendenhall and Ott (1979), Sukhatme, Sukhatme, Sukhatme, and Asok (1984), and Williams (1978).

(6) Statistical intervals do not apply for nonrandom samples. If they are calculated, their length generally provides only a lower bound on the total uncertainty.

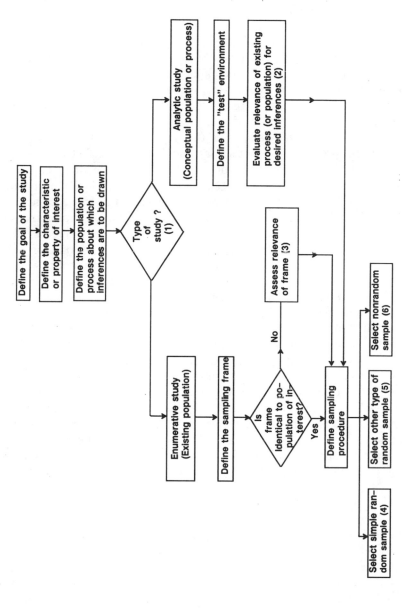

Figure 1.1 Possible approach to evaluating assumptions underlying the calculation of statistical intervals.

1.12   PRACTICAL ASSUMPTIONS: FURTHER EXAMPLE

As a further example, we cite, taking some liberties, a study conducted by the World Health Organization (WHO) to evaluate the effectiveness of self-examination by women as a means of early detection of breast cancer. The study was conducted on a sample of factory workers in Leningrad and Moscow. We assume that this group was selected for such practical reasons as the ready listing of potential participants and the willingness of factory management and workers to cooperate. We assume, for the purpose of discussion, that a major characteristic of interest is the time that self-examination saves in the detection of breast cancer.

Assume, initially, that the goal was the very narrow one of drawing conclusions (about breast cancer detection times) for female factory workers in Leningrad and Moscow at the time of study. The frame for this (enumerative) study is the (presumably complete, current, and correct) listing of female factory workers in Leningrad and Moscow. In this case, the frame coincides with the target population, and it may be possible to obtain a simple random sample from this frame. We assume further that the women selected by the random sample participate in the study and provide correct information and that the sample size is small (i.e., less than 10%), relative to the size of the population. Then the statistical intervals, provided in this book, apply directly for this (very narrow) target population. (It is, of course, possible in an enumerative study to define the target population so narrowly that it becomes equivalent to the "sample." In that case, one has complete information about the population and, as previously indicated, the confidence intervals presented in this book degenerate to zero length, and the tolerance and prediction intervals become probability intervals.)

Extending our horizons slightly, if we defined the target population to be all women in Moscow and Leningrad at the time of the study, the frame (of female factory workers) is more restrictive than the target population. As we have indicated, the statistical uncertainty, as expressed by the appropriate statistical interval, applies only to the sampled population (i.e., the female factory workers), and its relevance to the actual target population needs to be assessed.

In actuality, the World Health Organization is likely to be interested in a much wider group of women and a much broader period of time. In fact, the basic purpose of the study could well be that of drawing inferences about the effects of encouraging self-examination for women throughout the world, not only during the period of study, but, say, for the next 25 years. In this case, not only is the frame highly restrictive, but we are, in fact, dealing with an analytic study. In addition to the projection into the future, we need be concerned with such matters as the equivalence of learning skills and discipline, alternative ways of detecting breast cancer, the possibility of different manifestations of breast cancer, etc. In practice, the unquantifiable uncertainty involved in translating the results from the sampled population or process (female factory workers in Moscow and Leningrad today) to the (conceptual) population of major interest (e.g., all women throughout the world in the next 25 years) may well be much greater than the quantifiable statistical uncertainty.

If it were decided that the population of interest were all the women in the world *today*, one could debate whether this is an analytic study (requiring inferences about a population different from the one being sampled) or an enumerative study (for which the sampled population is a small subset of the target population). Such discussion would, however, be somewhat academic because, irrespective of one's preferences in terminology, the bottom line is still the need to assess the relevance of the results of the sample to the much larger population or process of actual interest.

We hasten to add that our comments are, in no way, a criticism of the WHO study, the major purpose of which appears to be that of assessing whether, under a particular set of circumstances and over a particular period of time, self-examination can be beneficial. It is, moreover, highly important to have a careful statistical plan for such a study (see discussion in the next section). We only bring out this study as an extreme example of a case in which statistical intervals, such as those discussed in this book, describe only part of the total uncertainty, and may, in fact, be of limited, if any, relevance.

Fortunately, not all studies are as global in nature and inference as this one. However, it seems safe to say that, in applications, the simple case of an enumerative study in which one is randomly sampling from the target population, is the exception, rather than the rule. Instead it is more common to encounter situations that have at least one of the following properties:

- One wishes to draw inferences concerning a future process (and, thus, is dealing with an analytic, rather than an enumerative, study).
- The sampled population differs from the target population.
- The sampling is not (strictly) random.

As indicated, in each of these cases, one need be concerned with the implications in generalizing one's conclusions beyond what is warranted from statistical theory alone—or, as we have repeatedly stated, the calculated statistical interval provides an optimistic lower bound on the total uncertainty, reflecting only the sampling variability. Thus, in studies like the WHO breast cancer detection evaluation, the prudent analyst needs to decide whether to calculate statistical intervals at all and stress their limitations as lower bounds on the total uncertainty, or to refrain from calculating such intervals under the belief that they may do more harm than good. (In any case, these intervals may be secondary to the use of statistical graphics to describe the data.)

1.13 PLANNING THE STUDY

A logical conclusion from the preceding discussion is that it is of prime importance to properly plan the study, if at all possible. Such planning is, of course, suggested by Figure 1.1 and by our discussion in Section 1.7.6 dealing with analytic studies. Planning the study helps assure that:

- The target population or process of interest is defined initially.
- The frame or sampling scheme is established to match the target population or process as closely as practical.
- A sampling procedure as close to random as feasible is used.

Unfortunately, studies are not always conducted in this way. In some cases, in fact, the analyst is handed the results of a study and asked to analyze the data. This requires *retrospectively* defining the target population or process of interest and the population or process that was actually sampled, determining how well these match, and deciding whether, and, to what degree, it is reasonable to assume that the sample was randomly selected. This is often a highly frustrating, or even impossible, job, and the necessary information is not always available. In fact, one may sometimes, conclude that in light of the deficiencies of the investigation, or the lack of knowledge about the sampling method, it might be misleading to employ statistical intervals.

The moral is clear. If one wishes to perform statistical analyses of the data from a study, including calculation of the intervals described here, it is most important to plan the investigation statistically in the first place. One element of planning the study is determining an appropriate sample size, see Chapters 8, 9, and 10. This technical consideration is, however, usually secondary to the more fundamental issues described in this chapter. Further details on planning studies are provided in texts on survey sampling (dealing mainly, but not exclusively, with enumerative studies) and books on experimental design (dealing mainly with analytic studies of processes)—see the categorized bibliography by Hahn and Meeker (1984).

1.14 THE ROLE OF STATISTICAL DISTRIBUTIONS

Many of the statistical intervals described in this book assume a distributional model, such as a normal distribution, for the measured variable (possibly after some transformation of the data, see Section 4.11). Frequently, the assumed model only approximately represents the population, although this approximation is often adequate for the problem at hand. With a sufficiently large sample (say, 40 or more observations), it should be possible to detect important departures from the assumed model and, if the departure is large, to decide whether there is a need to reject or refine the model. For more pronounced departures, fewer observations are needed for such detection. (We will discuss this topic further in Section 4.11.) When there is not enough data to detect important departures from the assumed model, the model's correctness must be justified from an understanding of the physical situation and/or past experience. (These, of course, need enter the assessment, irrespective of the sample size.) Some intervals, notably confidence intervals to include the population mean, are relatively insensitive to the assumed distribution; other intervals, unfortunately, strongly depend on this assumption. We will indicate the importance of distributional assumptions in our discussion of specific intervals. Also, Hahn (1971) discusses "How Abnormal is Normality?" and numerous books on statistics, including Hahn and Shapiro (1967), describe graphical methods and statistical tests to evaluate the assumption of normality. We provide a brief introduction to, and example of, this subject in Section 4.11.

Statistical intervals that do not require any distributional assumptions are described in Chapter 5. Such "distribution-free" or "nonparametric" intervals have the obvious advantage that they do not require one to assume a specific distribution. They are, therefore, especially appropriate for those situations where the results are sensitive to assumptions about the distributional model. Their disadvantage is that they tend to be longer, i.e., less precise, than the corresponding interval under an assumed distributional model. They also still require the other important assumptions discussed in the preceding sections.

1.15   THE INTERPRETATION OF A STATISTICAL INTERVAL

Because statistical intervals are based upon limited sample data that are subject to random sampling variation, they will sometimes not contain the quantity of interest that they were calculated to contain, even when all the necessary assumptions hold. Instead, they can be claimed to be correct only a specified percentage (e.g., 90%, 95%, or 99%) of the time they are calculated, that is, they are correct with a specified "degree of confidence." The percentage of such statistical intervals that contain what they claim to contain is known as the confidence level associated with the interval. The selection of a confidence level is discussed in Section 2.6.

The confidence level, at least from a traditional point of view, describes the process of constructing a statistical interval, and not the computed interval itself. The confidence level is the probability that, in any given study, the random sample will result in a correct interval. For example, a particular confidence interval to contain the mean of a population *cannot* correctly be described as being an interval that contains the true population mean with a specified probability. This is because the mean is an unknown fixed characteristic of the population which, in a given situation, is either contained within the interval or is not. All that we can say is that in calculating many different confidence intervals to contain population means from different (independent) random samples, the calculated confidence interval will actually contain the true population mean with a specified probability (the confidence level), and, due to the vagaries of chance, will not contain the population mean the other times. We shall provide further elaboration in Sections 2.2.5, 2.3.6, and 2.4.3.

1.16   COMMENTS CONCERNING SUBSEQUENT DISCUSSION

The assumptions that we have emphasized in this chapter apply throughout this book and warrant restatement each time that we present an interval or an example. However, we have decided to relieve the reader from such repetition. Thus, frequently, we limit ourselves to saying that the resulting interval applies "to the sampled population (or process)." However, the reader needs to keep in mind, in all applications, the underlying assumptions and admonitions stated in this chapter.

CHAPTER 2

Overview of Different Types
of Statistical Intervals

Different kinds of confidence intervals, prediction intervals, and tolerance intervals and their applications are introduced in this chapter. A discussion of the mechanics for constructing such intervals is postponed to subsequent chapters. We begin with some general comments about the choice of a statistical interval.

2.1 CHOICE OF A STATISTICAL INTERVAL

The appropriate statistical interval for a particular application depends on the application. Thus, the analyst must determine which interval(s) to use, based on the needs of the problem. The following comments give brief guidelines for this selection. Table 2.1 categorizes the statistical intervals discussed in this book and provides some examples. The intervals are classified according to (1) the general purpose of the interval and (2) the characteristic of interest.

2.1.1 Purpose of the Interval

In selecting an interval, one must decide whether the main interest is in *describing the population or process* from which the sample has been selected or in *predicting the results of a future sample* from the same population. Intervals that describe the sampled population (or enclose population parameters) include confidence intervals for the population mean and for the population standard deviation and tolerance intervals to contain a specified proportion of a population. In contrast, prediction intervals for a future mean and for a future standard deviation and prediction intervals to include all of m future observations deal with predicting (or containing) the results of a future sample from a previously sampled population.

27

Table 2.1   Examples of Some Statistical Intervals

Characteristic of Interest	General Purpose of the Statistical Interval	
	Description	Prediction
Location	Confidence interval for a population mean or median or a specified distribution percentile.	Prediction interval for a future sample mean, future sample median, or a particular ordered observation from a future sample.
Spread	Confidence interval for a population standard deviation.	Prediction interval for the standard deviation of a future sample.
Enclosure interval	Tolerance interval to contain (or cover) at least a specified proportion of a population.	Prediction interval to contain all or most of the observations from a future sample.
Probability of an event	Confidence interval for the probability of being less than (or greater than) some specified value.	Prediction interval to contain the proportion of observations in a future sample that exceed a specified limit.

2.1.2   Characteristic of Interest

There are various characteristics of interest that one may wish to enclose with a high degree of confidence. The intervals considered in this book may be roughly classified as dealing with:

- The *location* of a distribution or sample as measured, for example, by its mean or a specified distribution percentile. Specific examples are a confidence interval for the population mean and a prediction interval for the mean of a future sample.
- The *spread* of a distribution or sample as measured, for example, by its standard deviation. Specific examples are a confidence interval for the population standard deviation and a prediction interval to contain the standard deviation of a future sample.
- An *enclosure interval*. Specific examples are a tolerance interval to contain a specified proportion of the population and a prediction interval to contain all, or most, of the observations from a future sample.
- The *probability of an event*. A specific example is a confidence interval for the probability that a measurement will exceed a specified threshold value.

We shall now describe some specific intervals in greater detail, considering in turn, confidence intervals, prediction intervals, and tolerance intervals, and complete this overview chapter with some important general comments.

2.2 CONFIDENCE INTERVALS

2.2.1 Confidence Interval for a Population Parameter

Confidence intervals quantify our knowledge, or lack thereof, about a parameter or some other characteristic of a population, based upon a random sample. For example, asserting that a 95% confidence interval to contain the mean life of a particular brand of light bulb is 800 to 900 hours is considerably more informative than simply stating that the mean life is approximately 850 hours.

A frequently used type of confidence interval is one to contain the population mean. Sometimes, however, one desires a confidence interval to include some other parameter, such as the standard deviation of a normal distribution. For example, our knowledge, based upon past data, about the precision of an instrument that measures air pollution might be expressed by a confidence interval for the standard deviation of the instrument's measurement error. Confidence intervals for parameters of other statistical distributions, such as for the "failure rate" of an exponential distribution, or for the shape parameter of a Weibull distribution, are also sometimes desired.

2.2.2 Confidence Interval for a Distribution Percentile

Frequently, our major interest centers on one or more percentiles (also known as percentage points or quantiles) of the distribution of the sampled population, rather than on the distribution parameters directly. For example, in evaluating the tensile strength of an alloy, it might be desired to estimate, using data from a random sample of test specimens, the loading that would cause 1% of such specimen to fail. Thus, one would wish to construct a confidence interval for the first percentile of the tensile strength distribution.

2.2.3 Confidence Interval on the Probability of Meeting Specifications

Many practical problems involve stated performance or specification limits that a product is required to meet. The sample data are then used to assess the probability of meeting the specification. Some examples are:

- A machined part needs to be between 73.151 and 73.157 centimeters in diameter in order to fit correctly with some other part.
- The noise level of an engine must be below 80 decibels to satisfy a government regulation.
- The mean life of a light bulb must be at least 800 hours to meet an advertising claim.
- The net contents of a bottled soft drink must be at least 32 fluid ounces to conform with the product labeling.

These situations are characterized by the fact that a limit is stated, rather than having to be determined, and the population proportion (or probability of) meeting this stated limit is to be evaluated. If the distribution parameters of the sampled population are known, elementary methods, described in introductory probability and statistics texts, can be used to determine the exact probability of meeting the specification limit. A common example arises in dealing with a normal distribution with a *known* mean and standard deviation. When, however, the available data are limited to a random sample from the population, as is frequently the case, the probability of meeting the specification limit cannot be found exactly. One can, however, construct a confidence interval for the unknown population proportion. For example, from the available data, one can construct a confidence interval for the proportion of light bulbs from a specified population that will survive 800 hours of operation without failure or, equivalently, the probability that any randomly selected bulb will operate for at least 800 hours. This probability is the reliability at 800 hours. Confidence intervals for reliability are widely used in product life analyses (see Nelson, 1982).

2.2.4   One-Sided Confidence Bounds

Much of the discussion so far has implied a two-sided confidence interval; i.e., an interval with *both* a (finite) lower limit and a (finite) upper limit, usually with an equal degree of uncertainty associated with the parameter or other quantity of interest being located outside each of the two interval endpoints. A two-sided interval is generally appropriate, for example, when one is dealing with specifications on dimensions to allow one part to fit with another. In many other problems, the major interest, however, is restricted to the lower limit alone or to the upper limit alone. This is frequently the case for problems dealing with product quality or reliability. In such situations, one is generally concerned with questions concerning "how bad might things be?" and not "how good might they be?" This calls for the construction of a one-sided lower confidence bound or a one-sided upper confidence bound, depending upon the specific application, rather than a two-sided confidence interval. For example, the results of a reliability demonstration test for a system might be summarized by stating, "The estimated reliability for a mission time of 1000 hours of operation for the system is 0.987, and a lower 95% confidence bound on the reliability is 0.981." This "pessimistic" bound would be of particular interest when a minimum reliability must be met.

A one-sided lower confidence bound may be regarded as a confidence interval, where the upper limit of the interval is the largest possible value of interest and all of the uncertainty is concentrated on the lower limit of the interval. A one-sided upper confidence bound is similarly interpreted. For this reason, one-sided confidence bounds are sometimes referred to as "one-sided confidence intervals." We generally use the terms "two-sided confidence interval" and "one-sided confidence bound."

2.2.5   Interpretation of Two-Sided Confidence Intervals and One-Sided Confidence Bounds

Due to random sampling variation, a sample does not provide perfect information about the sampled population. However, using the results of a random sample, a confidence interval for some parameter or other fixed, unknown characteristic of the sampled population, provides limits which one can claim, with a specified degree of confidence, contain the true value of that parameter or characteristic.

A $100(1 - \alpha)\%$ confidence interval for an unknown quantity θ may be characterized as follows: "If one repeatedly calculates such intervals from many independent random samples, $100(1 - \alpha)\%$ of the intervals would, in the long run, correctly bracket the true value θ [or, equivalently, one would, in the long run, be correct $100(1 - \alpha)\%$ of the time in claiming that the true value of θ is contained within the confidence interval]." More commonly, but less precisely, a two-sided confidence interval is described by a statement such as, "we are 95% confident that the interval $\underset{\sim}{\theta}$ to $\tilde{\theta}$ contains the unknown true parameter value θ." In fact, the observed interval either contains θ or does not. Thus the 95% refers to *the procedure* for constructing a statistical interval, and not to the observed interval itself.

2.3   PREDICTION INTERVALS

2.3.1   Prediction Interval to Contain a Single Future Observation

A prediction interval for a single future observation is an interval that will, with a specified degree of confidence, contain the next (or some other prespecified) randomly selected observation from a population. Such an interval would interest the purchaser of a single unit of a particular product and is generally more relevant to such an individual than, say, a confidence interval to contain average performance. For example, the purchaser of a new automobile might wish to obtain, from the data on a previous sample of five similar automobiles, an interval that contains, with a high degree of confidence, the gasoline mileage that the new automobile will obtain under specified driving conditions. This interval is calculated from the sample data under the important assumption that the previously sampled automobiles and the future one(s) can be regarded as random samples from the same population; this assumes identical production processes and similar driving conditions. In many applications, the population may be conceptual, as per our discussion of analytic studies in Chapter 1.

2.3.2   Prediction Interval to Contain All of m Future Observations

A prediction interval to contain (the values of) all of m future observations generalizes the concept of a prediction interval to contain a single future observation. For example, a traveler who must plan a specific number of trips is generally not very interested in the amount of fuel that will be needed on the average in the population of all trips. Instead, this person would want to determine the amount of fuel that will be needed to complete each of one, or three, or five future trips.

Prediction intervals to contain all of m future observations are often of interest to manufacturers of large equipment who produce only a small number of units of a particular type product. For example, a manufacturer of gas turbines might wish to establish an interval that, with a high degree of confidence, will contain the performance values for all three units in a future shipment of such turbines, based upon the observed performance of similar past units. In this example, the past units and those in the future shipment would conceptually be thought of as random samples from the population of all turbines that the manufacturer might build (see discussion in Chapter 1).

Prediction intervals are especially pertinent to a user of one or a small number of units of a product. This person is generally more concerned with the performance of a specific sample of one or more units, rather than with that of the entire process from which the sample was selected. For example, based upon the data from a life test of 10 systems, one might wish to construct an interval that would have a high probability of including the lives of all of three additional systems that are to be purchased. Prediction intervals to contain all of m future observations are often referred to as *simultaneous* prediction intervals, because we are concerned with simultaneously containing *all* of the m observations within the calculated interval (with the associated level of confidence).

2.3.3   Prediction Interval to Contain k out of m Future Observations

A generalization of a prediction interval to contain all m future observations is one to contain k out of m of the future observations. We will refer to this type interval again in the next section.

2.3.4   Prediction Interval to Contain the Mean or Standard Deviation of a Future Sample

Sometimes one desires an interval to contain the *mean* of a future sample of m observations, rather than one to contain *all* of the future sample values. Such an interval would be pertinent, for example, if acceptance or rejection of a particular design were to be based upon the average of a future sample from a previously sampled process.

Consider the following example: A manufacturer of a high voltage insulating material must provide a potential customer "performance limits" to contain average breakdown strength of the material, estimated from a destructive test on a sample of 10 devices. Here "average" is understood to be the sample mean of the readings on the devices. The tighter these limits, the better are the chances that the manufacturer will be awarded a forthcoming contract. However, the manufacturer has to provide the customer 5 randomly selected devices for a test. If the average for these 5 devices does not fall within the performance limits stated by the manufacturer, the product is automatically disqualified. The manufacturer has available a random sample of 15 devices from production. Ten of these devices will be randomly selected and tested by the manufacturer to establish the desired limits. The remaining 5 devices will be shipped to and tested by the customer. Based on the data from the sample of 10 devices, the manufacturer will establish limits for the average of the five future readings so as to be able to assert with 95% confidence that the average of the 5 devices to be tested by the manufacturer will lie in the interval. A 95% prediction interval to contain the future mean provides the desired limits.

Assume now, that in the preceding example the major concern is uniformity of performance, as measured by the standard deviation, rather than average performance. In this case, one might wish to compute a prediction interval, based upon measurements on a random sample of 10 devices, to contain the standard deviation of 5 further randomly selected devices.

2.3.5 One-Sided Prediction Bounds

Some applications call for a one-sided prediction bound, instead of a two-sided prediction interval, to contain future sample results. An example is provided by a manufacturer who needs to warranty the efficiency of all units (or of their average) for a future shipment of three motors, based upon the results of a sample of five previously tested motors from the same process. This problem calls for a one-sided lower prediction bound, rather than a two-sided prediction interval.

2.3.6 Interpretation of Prediction Intervals and Bounds

If all the parameters of a probability distribution are known, one can use elementary methods to compute a probability interval to contain the values of a future sample. For example, for a normal distribution with mean μ and standard deviation σ, the probability is 0.95 that a single future observation will be contained in the interval $\mu \pm 1.96\sigma$. In the more usual situation where one has only sample data, one can construct a $100(1 - \alpha)\%$ prediction interval to contain the future observation with a specified degree of confidence. Such an interval may be characterized as follows: "If from many independent pairs of random samples, a $100(1 - \alpha)\%$ prediction interval is computed from the data of the first sample to contain the value(s) of the second sample, $100(1 - \alpha)\%$ of the intervals would, in the long run, correctly bracket the future value(s)" [or, equivalently, one would, in the long run, be correct $100(1 - \alpha)\%$ of time in claiming that the future value(s) will be contained within the prediction interval]. The requirement of independence holds both with regard to the different pairs of samples and the observations within each sample.

As with confidence intervals, the $100(1 - \alpha)\%$ refers to *the procedure* used to construct the prediction interval and not any particular interval that is computed. The actual probability that a prediction interval will contain the value that it is supposed to contain is unknown because the probability depends on the unknown parameters.

2.4 STATISTICAL TOLERANCE INTERVALS

2.4.1 Tolerance Interval to Contain a Proportion of a Population

As indicated, prediction intervals are needed by a manufacturer (or user) who wishes to predict the performance of one, or a small number, of future units. Consider now the case where one wishes to draw conclusions about the performance of a relatively large number of future units, based upon the data from a random sample from the population of interest. Conceptually, one can also construct prediction intervals for such situations, e.g., a prediction interval to contain all 100, 1000, or any number m, future units. Such intervals would, however, often be very long. Also, the exact number of future units of interest is sometimes not known or may be conceptually infinite. Moreover, rather than requiring that the calculated interval contain all of a specified number of units, it is generally sufficient to construct an interval to contain a *large proportion* of such units.

As indicated in Section 2.3.3, there are procedures for calculating prediction intervals to contain at least k out of m future observations, where $k \le m$. More frequently, however, applications call for the construction of intervals to contain a specified proportion, p, of the entire sampled population. (Such a population may again be conceptual.) This leads to the concept of a tolerance interval. Specifically, a tolerance interval is an interval that one can claim to contain at least a specified proportion, p, of the population with a specified degree of confidence, $100(1 - \alpha)\%$. Such an interval would be of particular interest in setting limits on the process capability for a product manufactured in large quantities. This is in contrast to a prediction interval which, as noted, is of greatest interest in making predictions about a small number of units of a particular type. Assume, for example, that measurements of the diameter of a machined part have been obtained on a random sample of 25 units from a production process. A tolerance interval calculated from such data provides limits that one can claim, with a specified degree of confidence (e.g., 90%), contains the (measured) diameters of at least a specified proportion (e.g., 95%) of units from the sampled population. The two percentages in the preceding statement should not create any confusion when one recognizes that one (95% in the example) refers to the percentage of the population to be contained, and the other (90%) deals with the degree of confidence associated with the claim [that the interval encloses (at least) 95% of the population].

Henceforth, we employ the terminology "a $100(1 - \alpha)\%$ tolerance interval to contain $100p\%$ of the population" to describe a tolerance interval that one can claim with $100(1 - \alpha)\%$ confidence encloses $100p\%$ of the population.

2.4.2 One-Sided Tolerance Bounds

Practical applications often require the construction of one-sided tolerance bounds. For example, in response to a request by a regulatory agency, a manufacturer has to make a statement concerning the maximum noise that, under specified operating conditions, is met by a high proportion of units, such as 95%, of a particular model of a jet engine. The statement is to be based upon measurements from a random sample of 10 engines and is to be made with 90% confidence. In this case, the manufacturer desires a one-sided upper 90% tolerance bound that will be met (i.e., not exceeded) by at least 95% of the population of jet engines, based upon the previous test results. A one-sided upper tolerance bound is appropriate here because the regulatory agency is concerned principally with how noisy, and not how quiet, the engines might be.

A one-sided tolerance bound is the inverse of a one-sided confidence bound for the probability of meeting a one-sided specification limit (see Section 2.2.3). In the previous case, the specification limit was provided and the analyst desired to use the available data to compute a confidence bound on the proportion of product that meets the limit. In the case of a one-sided upper (lower) tolerance bound, the data are to be used to construct a bound to exceed (or be exceeded by) a specified population proportion. Also, a one-sided tolerance bound is equivalent to a one-sided confidence bound on a distribution percentile (see Section 2.2.2). For example, a one-sided upper 95% tolerance bound to be met by at least 99% of the population values is the same as an upper 95% confidence bound on the 99th percentile of the population distribution.

2.4.3 Interpretation of Tolerance Intervals and Bounds

If the parameters of a distribution are known, one can use elementary techniques to compute a probability interval to contain a specified proportion of the population values. For example, for a normal distribution with mean μ and standard deviation σ, 95% of the population values are contained within the interval $\mu \pm 1.96\sigma$. In most practical situations, however, the distribution parameters are unknown and the available information is limited to a sample. The lack of perfect information about the population is taken into consideration by the confidence statement associated with the tolerance interval. Thus, a tolerance interval will bracket at least a certain proportion of the population with a specified degree of confidence.

   The following characterizes a tolerance interval that one can claim contains a proportion p of the population with $100(1 - \alpha)\%$ confidence: "If one calculated such intervals from many independent groups of random samples, $100(1 - \alpha)\%$ of the intervals would, in the long run, correctly include at least $100p\%$ of the population values [or, equivalently, one would, in the long run, be correct $100(1 - \alpha)\%$ of the time in claiming that the true proportion of the population contained in the interval is at least p]."

   As with confidence and prediction intervals, the $100(1 - \alpha)\%$ refers to *the procedure* of constructing the tolerance interval and not any particular interval that is computed. The actual proportion of the population contained within the tolerance interval is unknown because this proportion depends on the unknown parameters.

2.5   WHICH STATISTICAL INTERVAL DO I USE?

In the previous sections, we differentiated among the various types of statistical intervals. It should be clear from this discussion that the appropriate interval depends upon the problem at hand. That is, the specific questions that need to be answered will determine whether a confidence interval, a tolerance interval, or a prediction interval is needed and, for that matter, which subtype of the selected interval is appropriate (e.g., a confidence interval to include the population mean, the standard deviation, a specified distribution percentile, etc.). We can differentiate among these intervals on the basis of the questions that they answer and we can provide methods for obtaining the intervals. The analyst, however, has to decide which type of interval is needed in a specific application.

   Moreover, as detailed as the discussion in this book may seem, we can provide information on only a relatively small number of specific statistical intervals. A particular problem may call for a statistical interval which, in some way or other, is different from any of those described. We hope, however, that our discussion will be sufficient to help the analyst identify when this is the case, and thus, if necessary, call upon a professional statistician for guidance in developing the needed interval.

2.6   CHOOSING A CONFIDENCE LEVEL

All statistical intervals have an associated confidence level. Very loosely speaking, the confidence level is the degree of assurance one desires that the calculated statistical interval contains the value of interest. (See Sections 2.2.5, 2.3.6, and 2.4.3 for a more precise definition.) For a particular problem, the analyst must determine the confidence level, based upon what seems to be an acceptable degree of assurance for the specific application. Thus, one has to trade off the risk of not including the correct parameter, percentile, future observation, etc., in the interval, against the fact that when the degree of assurance is raised, the statistical interval becomes longer.

2.6.1 Further Elaboration

Statistical consultants are frequently requested by their clients to construct 95% confidence intervals to contain, say, the means of different populations from which random samples have been taken. In 95% of such cases, the calculated interval will include the true population mean—if the various assumptions stated in Chapter 1 hold. However, due to chance, the client is "misled" in 5% of the cases; that is, 5% of the time the computed interval will not contain the population mean. Clients who are afraid of being among the unlucky 5%, and desire added protection, can request a higher level of confidence, such as 99%. This reduces to one in a hundred the chances of obtaining an interval that does not contain the population mean. The client, however, pays a price for the higher level of confidence. For a fixed sample size, increasing the confidence level results in an increase in the length of the calculated interval. For example, the extreme length of a 99.9% confidence interval may be quite sobering. Besides this, in most cases, the only restriction in selecting a confidence level is that it be less than 100%; to obtain 100% confidence, the entire population must be observed. (This, of course, is not possible if our concern is with the output of a future process.)

2.6.2 Problem Considerations

The importance of any decision that is to be made from a statistical interval needs to be taken into consideration in selecting a confidence level. For example, at the outset of a research project, one might be willing to take a reasonable chance of drawing incorrect conclusions about the mean of the distribution of a population characteristic and use a relatively low confidence level, such as 90%, or even 80%, because the initial conclusions will presumably be corroborated by subsequent analyses. On the other hand, if one is about to report the final results of a project, and perhaps recommend building an expensive new plant, or releasing a new drug for general use, one might wish a higher degree of confidence in the correctness of one's claim.

2.6.3 Sample Size Considerations

As we shall see in subsequent chapters, shorter statistical intervals are expected with larger samples for a fixed level of confidence. This is especially the case for confidence intervals to contain a population parameter, because the lengths of such intervals generally shrink to zero (around the parameter) as the sample size increases. It is also true, but less dramatically so, for prediction intervals and for tolerance intervals, as these intervals reduce only to those which one would obtain if the model parameters were known with certainty. In Chapters 8, 9, and 10 we give simple methods for evaluating the effect that sample size has on the (expected) length of a statistical interval and for finding the sample size that is needed to provide an interval that has a specified (expected) length. This information can be used to evaluate the trade-offs between sample size and choice of confidence level.

Sometimes, one might wish to use relatively high confidence levels (e.g., 99%) with large samples, and lower confidence levels (e.g., 90 or 80%) with small samples. This implies that the more data one has, the surer one would like to be of one's conclusions. Also, such practice accords recognition to the fact that, for small sample sizes, the calculated statistical interval is often so long that it has little practical meaning and, therefore, one may wish to obtain a somewhat shorter interval at the cost of reducing one's confidence in its correctness.

2.6.4 An Added Practical Consideration

As indicated in Chapter 1, statistical intervals take into consideration only the vagaries of random statistical fluctuations due to sampling variation. The sample is assumed to be randomly selected from the population or process of interest. Moreover, unless a distribution-free procedure is used, a statistical model, such as a normal distribution, is assumed. As we have indicated, these assumptions hold only approximately, if at all, in many real world problems. Errors due to deviations from these assumptions are not included in the confidence level. A high confidence level, thus, may provide a false sense of security. A user of a statistical interval with 99.9% associated confidence might forget that the interval could fail to include the value of interest because of such other factors. Thus, the "true" confidence is typically less than 99.9%. A similar point could be made for an interval with 90% associated confidence, but users usually regard 90% intervals with less reverence than 99.9% intervals.

2.6.5 Further Remarks

The 95% confidence level appears to be used more frequently in practice than any other level, perhaps out of custom as much as anything else; 90 and 99% confidence levels seem to be next in popularity. However, confidence levels such as 50, 75, 80, 99.9, and 99.99% are also used. The lower confidence levels are, perhaps, used more frequently with prediction and tolerance intervals, than with confidence intervals. This is so because, as we shall see, prediction and tolerance intervals often tend to be longer than confidence intervals, especially for small sample sizes.

Thus, there is no single, straightforward answer to the question, "What confidence level should I select?" Much depends on the specific situation and the desired trade-off between risk and interval length. Of course, one can dodge the issue by presenting intervals for a number of different confidence levels, and this in fact, is often a reasonable approach.

2.7 STATISTICAL INTERVALS VERSUS SIGNIFICANCE TESTS

Statistical significance (or hypothesis) tests are frequently used to assess the correctness of a hypothesis about a population parameter or characteristic based upon the sample data. For example, a consumer protection agency might wish to test a manufacturer's claim (or hypothesis) that the average life of a brand of light bulbs is at least 800 hours. Statistical significance tests are designed such that the small probability of incorrectly rejecting the hypothesis when it is true is a specified probability, known as the "significance level," often denoted by α.

There is a close relationship between confidence intervals and significance tests. In fact, a confidence interval can often be used to test a hypothesis. For the light bulb example, suppose the one-sided $100(1 - \alpha)\%$ upper confidence bound for the true mean life, calculated from the data, exceeds the claimed mean life of 800 hours. Then one would conclude that the claim has not been contradicted by the data at the $100\alpha\%$ level of significance. On the other hand, if the one-sided $100(1 - \alpha)\%$ upper confidence bound is less than 800 hours, there is evidence at the $100\alpha\%$ significance level to reject the claim. Thus, a one-sided $100(1 - \alpha)\%$ confidence bound generally gives the accept or reject information provided by a one-sided hypothesis test at a $100\alpha\%$ level of significance. There is a similar relationship between a two-sided $100(1 - \alpha)\%$ confidence interval and a two-sided hypothesis test at a $100\alpha\%$ significance level.

Confidence intervals generally provide much more information than a simple hypothesis test. This is because confidence intervals provide quantitative bounds that express the uncertainty, instead of a mere accept or reject statement. Also, whether or not statistical significance is achieved is highly dependent on sample size. This, of course, is also true for the length of a confidence interval. However, for such an interval, the consequences of a small sample are evident from noting the length of the interval, while this is not the case for a significance test. Thus, confidence intervals are usually more meaningful than statistical hypothesis tests. In fact, one can argue that in some practical situations, there is really no reason for the statistical hypothesis to hold exactly. For example, two different processes generally would not be expected to have identical means. Thus, whether or not the hypothesis of equal means for the two processes is rejected depends upon the sample sizes, as well as the magnitude of the true difference between the process means.

The length of a confidence interval, as we have indicated, depends upon the sample size, and the effect of sample size can be seen directly from the length of the interval. Thus, a long confidence interval, which, incidentally involves the acceptance of some stated null hypothesis, might reflect the fact that only a relatively small sample size is available. Thus, rather than concluding that the null hypothesis is correct, one should conclude that the results are inconclusive and that additional data should be obtained, if at all possible. Similarly, a short confidence interval, which, incidentally results in the rejection of the hypothesis, might not necessarily imply that an effect or a difference of practical importance exists. Instead, it might just be a consequence of the fact that we were fortunate enough to have been able to secure a great deal of precision by having a large sample. See Hahn (1974) for further discussion of these topics.

In any case, we maintain that, in most practical situations, statistical intervals, in general, and confidence intervals, in particular, avoid some of the pitfalls inherent in many statistical hypothesis tests. Moreover, they are generally a great deal more informative and easier to explain to those with no formal training in statistics. [Although we believe that most applied statisticians now agree with this viewpoint, it is not universally accepted, especially among those involved in medical clinical trials. For a different viewpoint, see Hall (1989) and Gardner and Altman (1989).]

CHAPTER 3

Constructing Statistical Intervals Assuming a Normal Distribution Using Simple Tabulations

3.1 INTRODUCTION

3.1.1 Overview

This chapter presents and compares factors for calculating the following kinds of intervals, based upon a sample of size n from a normal population with unknown mean μ and unknown standard deviation σ:

- One-sided and two-sided confidence intervals to contain the population mean μ and the population standard deviation σ.
- One-sided and two-sided tolerance intervals to contain at least a proportion p of the population values for $p = 0.90, 0.95$, and 0.99.
- One-sided and two-sided prediction intervals to contain the values of all of $m = 1, 2, 5, 10, 20$, and $m = n$ randomly selected future observations, and the sample mean and standard deviation of $m = n$ future observations.

For each kind of interval, the tables provide factors for both 95 and 99% confidence levels.

3.1.2 The Normal Distribution

The normal distribution is the best known and most frequently used statistical model. Its theoretical justification is often based on the "Central Limit Theorem"; see Section 4.10 and Hahn and Shapiro (1967). In practice, the normal distribution may or may not provide an adequate description of one's data. The sensitivity of an interval to the assumption of normality depends on the specific type of interval, the population or process being studied, and the

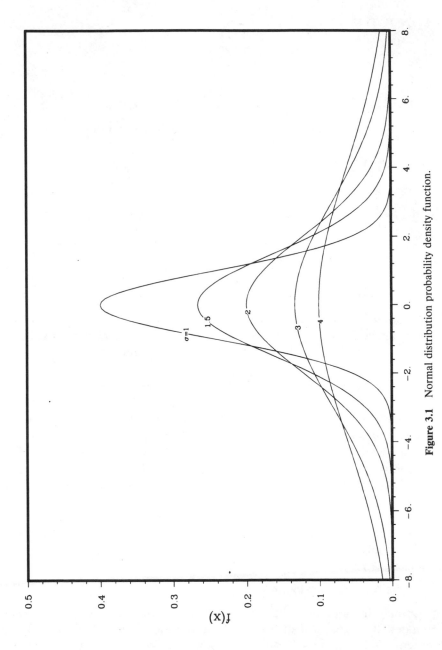

Figure 3.1 Normal distribution probability density function.

sample size (see Section 4.11 for further discussion of this issue). The normal distribution density function is

$$f(x;\mu,\sigma) = \frac{1}{\sqrt{2\pi}\,\sigma} \exp\left[-\frac{1}{2}\left(\frac{x-\mu}{\sigma}\right)^2\right],$$

$$-\infty < x < -\infty, \quad -\infty < \mu < \infty, \quad \sigma > 0$$

where μ is the mean and σ is the standard deviation of the distribution. This function is graphed in Figure 3.1 for mean $\mu = 0$ (which is also the distribution median and mode) and several values of σ. The corresponding cumulative distribution function for the population proportion below x' is

$$F(x';\mu,\sigma) = \Pr(x \leq x') = \int_{-\infty}^{x'} f(t;\mu,\sigma)\, dt.$$

This can also be computed from

$$\Pr(x \leq x') = \Phi\left(\frac{x'-\mu}{\sigma}\right)$$

where $\Phi(z) = F(z;0,1)$ is the cumulative distribution function of the standard normal distribution (i.e., the normal distribution with mean $\mu = 0$ and standard deviation $\sigma = 1$), which is provided by Table A.5. When the normal distribution is used as a model for a population, μ and σ are usually unknown and must be estimated from sample data. Section 4.1.2 introduces other related notation.

3.1.3   Using the Simple Factors

The tables of factors described in this chapter were computed from the more familiar t, χ^2, and F distribution tables, using standard formulas, which we review in Chapter 4. The simpler factors given here serve several purposes:

- Statistical intervals can be computed with slightly less computation than required with the more general formulas given in Chapter 4.
- Interval lengths can be compared directly, providing insight into the differences among the various types of statistical intervals.
- One can easily assess the effect of sample size and confidence level on the length of a particular statistical interval.

In the rest of this chapter we will illustrate the construction, describe the interpretation, and compare the lengths of some of the most commonly used statistical intervals for a normal distribution. Chapter 4 discusses, in more detail and generality, the computations for these and some other statistical intervals.

To construct intervals from these factors, it is assumed that a random sample of size n with values x_i, $i = 1, \ldots, n$ has been taken from the population or process of interest, and that the sample mean $\bar{x}$ and the sample standard deviation s have been computed from these observations using the expressions

$$\bar{x} = \frac{\sum_{i=1}^{n} x_i}{n} \quad \text{and} \quad s = \left(\frac{\sum_{i=1}^{n} (x_i - \bar{x})^2}{n - 1} \right)^{1/2}.$$

3.2 NUMERICAL EXAMPLE

3.2.1 Problem and Data

A manufacturer wanted to characterize empirically the voltage outputs for a new electronic circuit pack design. Five prototype units were built and the following measurements were obtained: 50.3, 48.3, 49.6, 50.4, and 51.9 volts. The sample size for this study was small, as it often is in practice, because of the high cost of manufacturing such prototype units. In this case, the experimenters felt that a sample of five units would be sufficient to answer some initial questions. Other questions might require more units; see Chapters 8, 9, and 10 for a discussion of the choice of sample size.

From previous experience with the construction and evaluation of similar circuit packs and from engineering knowledge about the components being used in the circuit, the experimenters felt it would be reasonable to assume that the voltage measurements follow a normal distribution (with an unknown mean and standard deviation).

Figure 3.2 is a dot plot of the five voltage measurements.

From these data, the sample mean and standard deviation were computed to be

$$\bar{x} = \frac{(50.3 + 48.3 + \cdots + 51.9)}{5} = 50.10 \quad \text{and}$$

$$s = \left(\frac{(50.3 - 50.10)^2 + \cdots + (51.9 - 50.10)^2}{5 - 1} \right)^{1/2} = 1.31.$$

```
         .              .            ..                      .
  -+---------+---------+---------+---------+---------+-----
  48.30     49.00     49.70     50.40     51.10     51.80   Volts
```

Figure 3.2 Dot plot of circuit pack output voltages.

3.2.2 Validity of Inferences

The experimenters were most interested in making inferences about the population or process of circuit packs that would be manufactured in the future. Thus, in the terminology of Chapter 1, this is an analytic study. The statistical intervals assume that the five sample prototype circuit packs *and* the future circuit packs of interest come from the same stable production process and that the unit-to-unit variability in the voltage readings can be adequately described by a normal distribution with a constant mean μ and standard deviation σ. If the process changes (e.g., because of a component substitution) after the sample units are made, inferences about the new state of the process or predictions about future units might not be valid. (This is also a reason for not insisting on a larger size sample at this time.) The validity of this stability assumption need be assessed by the experimenter, based on knowledge of the process.

The assumption of normality can often be assessed from the sample data. This, however, generally requires a sample of at least 20 to 30 observations to have any reasonable degree of sensitivity. Thus, for the present example, there is not enough data to properly check the assumption of normality, and, again its justification needs to be based on the experimenter's knowledge of the process and previous experience in similar situations. However, one can, and should, evaluate the sensitivity of the conclusions to this assumption and repeat the analysis under alternative assumptions. For example, transforming the response variable, or comparing inferences made with and without distributional assumptions, and with different distributional assumptions can be informative, as illustrated at the end of Chapter 4.

3.3 TWO-SIDED STATISTICAL INTERVALS

3.3.1 Simple Tabulations for Two-Sided Statistical Intervals

Tables A.1a and A.1b provide factors $c_{(1-\alpha;n)}$, where n is the number of observations in the given sample, and $100(1-\alpha)\%$ is the confidence level associated with the calculated interval, so that the two-sided interval

$$\bar{x} \pm c_{(1-\alpha;n)}s$$

is (in turn):

- A two-sided confidence interval for the population mean μ.
- A two-sided tolerance interval to contain at least a specified proportion p of the population for $p = 0.90, 0.95$, and 0.99.
- A two-sided simultaneous prediction interval to contain all of m future observations from the previously sampled normal population for $m = 1, 2, 5, 10, 20$, and $m = n$.
- A two-sided prediction interval to contain the mean of $m = n$ future observations.

Tables A.2a and A.2b provide factors $c_{L(1-\alpha;n)}$ and $c_{U(1-\alpha;n)}$, where n is the number of observations in the given sample and $100(1-\alpha)\%$ is the associated confidence level so that the interval

$$[c_{L(1-\alpha;n)}s, c_{U(1-\alpha;n)}s]$$

is (in turn):

- A two-sided confidence interval for the population standard deviation σ.
- A two-sided prediction interval to contain the standard deviation of $m = n$ future observations.

The tabulated values are for $n = 4(1)12, 15(5)30, 40, 60$, and ∞. Tables A.1a and A.2a are for confidence levels of $100(1-\alpha)\% = 95\%$ and Tables A.1b and A.2b are for $100(1-\alpha)\% = 99\%$. Methods for obtaining factors for other values of n and for other confidence levels are given in Chapter 4.

We will use $[\underset{\sim}{\theta}, \tilde{\theta}]$ to denote a two-sided statistical interval that is constructed for θ, some quantity of interest. Thus, $\underset{\sim}{\theta}$ is the lower endpoint of the interval and $\tilde{\theta}$ is the upper endpoint of the interval. If only a lower or an upper statistical bound is desired, we compute either $\underset{\sim}{\theta}$ or $\tilde{\theta}$, which we call a one-sided lower or upper bound for θ. For example, $[\underset{\sim}{s}_m, \tilde{s}_m]$ denotes a two-sided prediction interval to contain the sample standard deviation for a future sample of size m and $\tilde{\mu}$ denotes an upper confidence bound for the population mean μ.

3.3.2 Examples

Using the sample values $\bar{x} = 50.10$ and $s = 1.31$, based on $n = 5$ observed voltage measurements, we can use the factors given in Table A.1a to construct the following statistical intervals:

- A two-sided 95% confidence interval for the mean μ of the population of sampled circuit packs is

$$\left[\underset{\sim}{\mu}, \tilde{\mu}\right] = 50.10 \pm 1.24(1.31) = [48.5, 51.7].$$

Thus, we are 95% confident that the interval 48.5 to 51.7 volts contains the unknown mean μ of the population of circuit pack voltage readings. It is important to remember, as explained in Section 2.2.5, that "95% confident" describes the success rate of the procedure (i.e., the percentage of time that a claim of this type is correct).

- A two-sided 95% tolerance interval to contain at least 99% of the sampled population of circuit packs is

$$\left[\underset{\sim}{T}_{0.99}, \tilde{T}_{0.99}\right] = 50.10 \pm 6.60(1.31) = [41.5, 58.7].$$

Thus, we are 95% confident that the interval 41.5 to 58.7 volts contains at least 99% of the population of circuit pack voltage readings.

- A two-sided (simultaneous) 95% prediction interval to contain the voltage readings of all of 10 additional circuit packs randomly sampled from the same population is

$$\left[\underline{y}_{10}, \tilde{y}_{10}\right] = 50.10 \pm 5.23(1.31) = [43.2, 57.0].$$

Thus, we are 95% confident that the voltage readings of all 10 additional manufactured circuit packs will be contained within the interval 43.2 to 57.0 volts.

- A two-sided 95% prediction interval to contain the mean of the voltage readings of five additional circuit packs randomly sampled from the same population is

$$\left[\underline{\bar{y}}_5, \tilde{\bar{y}}_5\right] = 50.10 \pm 1.76(1.31) = [47.8, 52.4].$$

Thus, we are 95% confident that the mean of the voltage readings of five additional circuit packs will be in the interval 47.8 to 52.4 volts.

Using the factors in Table A.2a, we can construct the following 95% statistical intervals:

- A two-sided 95% confidence interval for the standard deviation σ of the population of sampled circuit packs is

$$\left[\underline{\sigma}, \tilde{\sigma}\right] = [0.60(1.31), 2.87(1.31)] = [0.8, 3.8].$$

Thus, we are 95% confident that the interval 0.8 to 3.8 volts contains the unknown standard deviation σ of the population of circuit pack voltage readings.

- A two-sided 95% prediction interval to contain the standard deviation of the voltage readings of five additional circuit packs randomly sampled from the same population is

$$\left[\underline{s}_5, \tilde{s}_5\right] = [0.32(1.31), 3.10(1.31)] = [0.4, 4.1].$$

Thus, we are 95% confident that the standard deviation of the voltage readings of five additional circuit packs will be in the interval 0.4 to 4.1 volts.

3.3.3 Comparison of Two-Sided Statistical Intervals

Figure 3.3 provides a comparison of some of the preceding factors for computing two-sided 95% statistical intervals. It illustrates the large differences in interval length between the various interval types. Thus, Figure 3.3

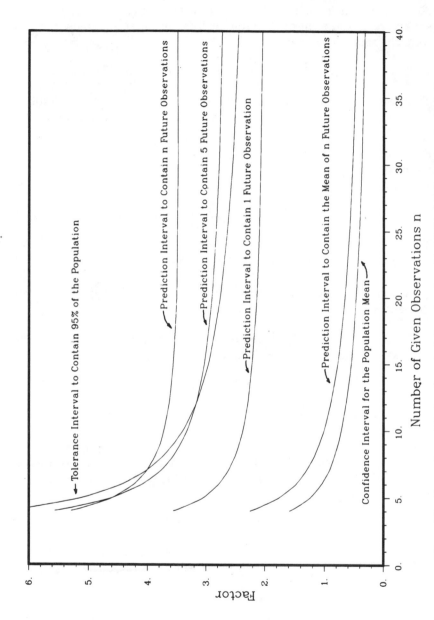

Figure 3.3 Comparison of factors for calculating some two-sided 95% statistical intervals. A similar figure first appeared in Hahn (1970b). Adapted with permission of the American Society for Quality Control.

Tolerance Interval to Contain 95% of the Population

Prediction Interval to Contain n Future Observations

Prediction Interval to Contain 5 Future Observations

Prediction Interval to Contain 1 Future Observation

Prediction Interval to Contain the Mean of n Future Observations

Confidence Interval for the Population Mean

Number of Given Observations n

Factor

and Tables A.1a and A.1b show that, for a given sample size and confidence level, a confidence interval for the population mean is always smaller than a prediction interval to contain all m future observations or a tolerance interval to contain 90, 95, or 99% of the population. On the other hand, whether or not a particular prediction interval is smaller than a particular tolerance interval depends on the number of future observations to be contained in the prediction interval and the proportion of the population to be contained in the tolerance interval. Moreover, the relative sizes of different prediction intervals depends on the number of future observations to be contained in the prediction interval, and the relative sizes of different tolerance intervals depends on the proportion of the population to be contained in the tolerance interval, for a given sample size and confidence level. Finally, confidence intervals to contain the population mean and the population standard deviation are smaller, respectively, than the corresponding prediction intervals to contain the mean and standard deviation of a future sample of any size.

3.4 ONE-SIDED STATISTICAL BOUNDS

3.4.1 Simple Tabulations for One-Sided Statistical Bounds

Tables A.3a and A.3b provide factors for calculating one-sided statistical bounds similar to those in Tables A.1a and A.1b for calculating two-sided statistical intervals. Thus, Table A.3a [for $100(1 - \alpha)\% = 95\%$] and Table A.3b [for $100(1 - \alpha)\% = 99\%$] provide factors $c'_{(1-\alpha; n)}$, where n is the number of observations in the given sample, and $100(1 - \alpha)\%$ is the associated confidence level so that the one-sided lower bound

$$\bar{x} - c'_{(1-\alpha; n)}s$$

or the one-sided upper bound

$$\bar{x} + c'_{(1-\alpha; n)}s$$

is (in turn):

- A lower (or upper) confidence bound for the population mean μ.
- A lower (or upper) tolerance bound to be exceeded by (exceed) at least a specified proportion p of the population for $p = 0.90, 0.95$, and 0.99.
- A one-sided lower (or upper) simultaneous prediction bound to be exceeded by (exceed) all of m future observations from the previously sampled normal population for $m = 1, 2, 5, 10, 20$, and $m = n$.
- A one-sided lower (or upper) prediction bound for the mean of $m = n$ future observations.

Tables A.4a and A.4b provide factors for calculating one-sided statistical bounds similar to those in Table A.2a and Table A.2b for calculating two-sided intervals. Thus, Table A.4a [for $100(1 - \alpha)\% = 95\%$] and Table A.4b [for $100(1 - \alpha)\% = 99\%$] provide factors $c'_{L(1-\alpha;n)}$ and $c'_{U(1-\alpha;n)}$, where n is the number of observations in the given sample and $100(1 - \alpha)\%$ is the associated confidence level, so that the one-sided lower bound

$$c'_{L(1-\alpha;n)}s$$

or the one-sided upper bound

$$c'_{U(1-\alpha;n)}s$$

is (in turn):

- A lower (or upper) confidence bound for the population standard deviation σ.
- A lower (or upper) prediction bound for the standard deviation of $m = n$ future observations.

The tabulations are again for $n = 4(1)12$, $15(5)30$, 40, 60, and ∞. Methods for obtaining factors other than those tabulated are given in Chapter 4.

3.4.2 Examples

Using the sample values $\bar{x} = 50.10$ and $s = 1.31$, based on $n = 5$ observations, we can use the factors given in Table A.3a to construct the following one-sided statistical bounds:

- A one-sided lower 95% confidence bound for the mean μ of the population of sampled circuit packs is

$$\underset{\sim}{\mu} = 50.10 - 0.95(1.31) = 48.9 \,.$$

  Thus, we are 95% confident that the unknown mean μ of the population of circuit pack readings exceeds the lower confidence bound of 48.9 volts.

- A one-sided upper 95% tolerance bound to exceed at least 99% of the sampled population of circuit packs is

$$\tilde{T}_{0.99} = 50.10 + 5.74(1.31) = 57.6 \,.$$

  Thus, we are 95% confident that at least 99% of the population of circuit packs have voltage readings less than 57.6 volts.

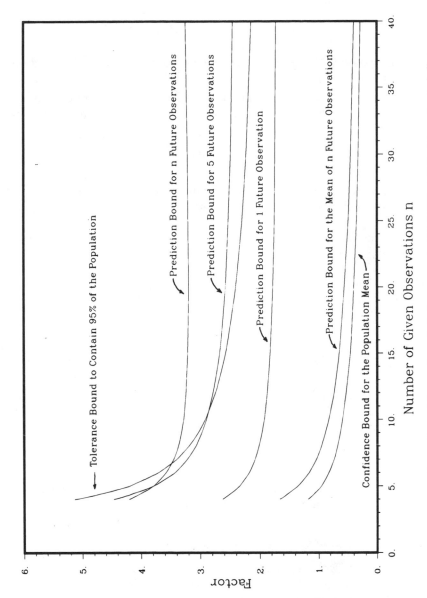

Figure 3.4 Comparison of factors for calculating some one-sided 95% statistical bounds. A similar figure first appeared in Hahn (1970b). Adapted with permission of the American Society for Quality Control.

- A one-sided lower 95% prediction bound to be exceeded by the voltage readings of all of 10 additional circuit packs randomly sampled from the same population is

$$\underline{y}_{10} = 50.10 - 4.42(1.31) = 44.3.$$

Thus, we are 95% confident that the voltage readings of all 10 additional circuit packs will exceed 44.3 volts.

- A one-sided lower 95% prediction bound to be exceeded by the mean of the voltage readings of 5 additional circuit packs from the same population is

$$\underline{\bar{y}}_5 = 50.10 - 1.35(1.31) = 48.3.$$

Thus, we are 95% confident that the mean of the voltage readings of five additional circuit packs will exceed 48.3 volts.

Using the factors in Table A.4a, we can construct the following one-sided 95% statistical bounds:

- A one-sided upper 95% confidence bound for the standard deviation σ of the population of sampled circuit packs is

$$\tilde{\sigma} = 2.37(1.31) = 3.1.$$

Thus, we are 95% confident that the unknown standard deviation σ of the population of circuit packs is less than 3.1 volts.

- A one-sided upper 95% prediction bound to exceed the standard deviation of the voltage readings of five additional circuit packs randomly sampled from the same population is

$$\tilde{s}_5 = 2.53(1.31) = 3.3.$$

Thus, we are 95% confident that the standard deviation of the voltage readings of five additional circuit packs will be less than 3.3 volts.

3.4.3 Comparison of One-Sided Bounds

Figure 3.4 provides a comparison of some of the preceding factors for calculating one-sided 95% bounds. Inspection of this figure and the tabulations leads to conclusions about the relative magnitudes of the factors for calculating different types of one-sided bounds that are similar to those for the two-sided case discussed previously.

CHAPTER 4

Methods for Calculating Statistical Intervals for a Normal Distribution

4.1 INTRODUCTION

4.1.1 Overview

This chapter extends the results of Chapter 3 and gives general methods for calculating various statistical intervals for samples from a population or process that can be approximated with a normal distribution. These are of interest for cases not covered in the tabulations in Chapter 3, for developing computer programs that will construct statistical intervals, and for gaining general understanding. In addition, we describe confidence intervals for normal distribution percentiles and for the population proportion less (greater) than a specified value. In each case, we describe a two-sided interval, a one-sided lower bound, and a one-sided upper bound. We also include references to additional tabulations and formulas for approximating the required factors. To illustrate the intervals, we continue to use the numerical example from Chapter 3. We also discuss the importance of the assumption of normality, methods for assessing this assumption, and ways of dealing with nonnormality, especially by transforming the data. We also provide references for additional information on the theory for the intervals.

4.1.2 Notation

The following notation will be used in this chapter:

- $\Phi(z')$ = the standard normal cumulative distribution function, giving $\Pr(z < z')$, where z is a normally distributed random variable with mean $\mu = 0$ and standard deviation $\sigma = 1$. This cumulative distribution function is given in Table A.5. Computer programs for obtaining $\Phi(z')$ are also available [e.g., ANORDF in IMSL (1987)].

53

- $z_{(\gamma)}$ = the 100γth percentile of the standard normal distribution (i.e., the value below which a normally distributed variate with mean $\mu = 0$ and standard deviation $\sigma = 1$ falls with probability γ). For example, $z_{(0.05)} = -1.645$ and $z_{(0.95)} = 1.645$. Standard normal distribution percentiles are given in Table A.6. Computer programs for obtaining $z_{(\gamma)}$ are also available [e.g., ANORIN in IMSL (1987)].

- $t_{(\gamma; r)}$ = the 100γth percentile of Student's t distribution with r degrees of freedom. For example, $t_{(0.05; 5)} = -2.015$ and $t_{(0.95; 5)} = 2.015$. Student's t distribution percentiles are given in Table A.7. Computer programs for obtaining $t_{(\gamma; r)}$ are also available [e.g., TIN in IMSL (1987)].

- $\chi^2_{(\gamma; r)}$ = the 100γth percentile of the chi-square (χ^2) distribution with r degrees of freedom. For example, $\chi^2_{(0.05; 5)} = 1.145$ and $\chi^2_{(0.95; 5)} = 11.07$. Chi-square distribution percentiles are given in Table A.8. Computer programs for obtaining $\chi^2_{(\gamma; r)}$ are also available [e.g., CHIIN in IMSL (1987)].

- $F_{(\gamma; r_1, r_2)}$ = the 100γth percentile of the F distribution with r_1 numerator and r_2 denominator degrees of freedom. For example, $F_{(0.95; 5, 2)} = 19.30$. F distribution percentiles are given in Tables A.9a to A.9f. Computer programs for obtaining $F_{(\gamma; r_1, r_2)}$ are also available [e.g., FIN in IMSL (1987)].

We will, as in Chapter 3, assume that the data are a random sample of size n from a normal distribution with sample mean $\bar{x}$ and sample standard deviation s.

4.2 CONFIDENCE INTERVAL FOR THE MEAN OF A NORMAL DISTRIBUTION

4.2.1 Method

A two-sided $100(1 - \alpha)\%$ confidence interval for the mean μ of a normal distribution is

$$\left[\underline{\mu}, \tilde{\mu}\right] = \bar{x} \pm t_{(1 - \alpha/2; n - 1)} \frac{s}{\sqrt{n}}. \tag{4.1}$$

One-sided $100(1 - \alpha)\%$ confidence bounds are obtained by replacing $\alpha/2$ by α (and $\pm$ by either $+$ or $-$) in the above expressions. The factors $c_{(1-\alpha; n)}$ and $c'_{(1-\alpha; n)}$ for confidence intervals and bounds for μ given in Tables A.1a, A.1b, A.3a, and A.3b were computed from these expressions: $c_{(1-\alpha; n)} = t_{(1-\alpha/2; n-1)}/\sqrt{n}$ and $c'_{(1-\alpha; n)} = t_{(1-\alpha; n-1)}/\sqrt{n}$. When $n - 1$ is large (say greater than 60), $t_{(1-\alpha; n-1)} \approx z_{(1-\alpha)}$. Thus, normal distribution percentiles provide a generally adequate approximation for t distribution percentiles when n is large and $1 - \alpha/2$ is not too large. For example, $t_{(0.975; 60)} = 2.000$ and $z_{(0.975)} = 1.960$.

Most elementary texts on statistical methods discuss confidence intervals for the mean of a normal distribution. The underlying theory involves Student's t distribution and is given in numerous books on mathematical statistics.

4.2.2   Example

For the example, $n = 5$, $\bar{x} = 50.10$ volts, $s = 1.31$ volts, and from Table A.7, $t_{(0.975; 4)} = 2.776$, and $t_{(0.95; 4)} = 2.132$. Then $c_{(0.95; 5)} = t_{(0.975; 4)}/\sqrt{5} = 1.241$ and $c'_{(0.95; 5)} = t_{(0.95; 4)}/\sqrt{5} = 0.953$. These are the values given in Tables A.1a and A.3a. A two-sided 95% confidence interval for μ is

$$\left[\underset{\sim}{\mu}, \tilde{\mu}\right] = 50.10 \pm (1.241)1.31 = [48.47, 51.73].$$

A lower 95% confidence bound for μ is

$$\underset{\sim}{\mu} = 50.10 - (0.953)1.31 = 48.85.$$

4.3   CONFIDENCE INTERVAL FOR THE STANDARD DEVIATION OF A NORMAL DISTRIBUTION

4.3.1   Method

A two-sided $100(1 - \alpha)\%$ confidence interval for the standard deviation σ of a normal distribution is

$$\left[\underset{\sim}{\sigma}, \tilde{\sigma}\right] = \left[s\left\{\frac{n - 1}{\chi^2_{(1-\alpha/2; n-1)}}\right\}^{1/2}, s\left\{\frac{n - 1}{\chi^2_{(\alpha/2; n-1)}}\right\}^{1/2}\right].$$

One-sided $100(1 - \alpha)\%$ confidence bounds are obtained by replacing $\alpha/2$ by α in the above expressions. The factors $c_{L(1-\alpha; n)}$, $c_{U(1-\alpha; n)}$, $c'_{L(1-\alpha; n)}$, and $c'_{U(1-\alpha; n)}$ for obtaining confidence intervals and bounds for σ given in Tables A.2a, A.2b, A.4a, and A.4b were computed from these expressions. For example, $c_{L(1-\alpha; n)} = [(n - 1)/\chi^2_{(1-\alpha/2; n-1)}]^{1/2}$.

Many introductory texts on statistical methods discuss confidence intervals for the standard deviation (or variance, i.e., the square of the standard deviation) of a normal distribution. The underlying theory involves the chi-square distribution and is given in numerous books on mathematical statistics.

4.3.2 Example

For the example, $n = 5$, $s = 1.31$ volts, and from Table A.8, $\chi^2_{(0.975; 4)} = 11.14$, $\chi^2_{(0.025; 4)} = 0.484$, and $\chi^2_{(0.05; 4)} = 0.711$. Then $c_{L(0.95; 5)} = \{4/\chi^2_{(0.975; 4)}\}^{1/2} = 0.60$, $c_{U(0.95; 5)} = \{4/\chi^2_{(0.025; 4)}\}^{1/2} = 2.87$, and $c'_{U(0.95; 5)} = \{4/\chi^2_{(0.95; 4)}\}^{1/2} = 2.37$. These are the values given in Tables A.2a and A.4a. A two-sided 95% confidence interval for σ is

$$\left[\underline{\sigma}, \tilde{\sigma}\right] = [1.31(0.60), 1.31(2.87)] = [0.79, 3.76].$$

An upper 95% confidence bound for σ is $\tilde{\sigma} = 1.31(2.37) = 3.10$.

4.4 CONFIDENCE INTERVAL FOR A PERCENTILE OF A NORMAL DISTRIBUTION

4.4.1 Method

A two-sided $100(1 - \alpha)\%$ confidence interval for Y_p, the $100p$th percentile of a normal distribution, is

$$\left[\underline{Y}_p, \tilde{Y}_p\right] = \left[\bar{x} - g'_{(1-\alpha/2; 1-p, n)}s, \bar{x} - g'_{(\alpha/2; 1-p, n)}s\right]$$

for $0.00 < p < 0.50$ and

$$\left[\underline{Y}_p, \tilde{Y}_p\right] = \left[\bar{x} + g'_{(\alpha/2; p, n)}s, \bar{x} + g'_{(1-\alpha/2; p, n)}s\right]$$

for $0.50 \leq p < 1.0$, where the factors $g'_{(\gamma; p, n)}$ are given in Table 1 of Odeh and Owen (1980) for all combinations of

$p = 0.75, 0.90, 0.95, 0.975, 0.99$, and 0.999,
$n = 2(1)100(2)180(5)300(10)400(25)650(50)1000, 1500$, and 2000, and
$\gamma = 0.995, 0.99, 0.975, 0.95, 0.9, 0.75, 0.50, 0.25, 0.10, 0.05, 0.025, 0.01$, and 0.005.

Our Tables A.12a to A.12d provide $g'_{(\gamma; p, n)}$ for a smaller set of entries. One-sided lower and upper $100(1 - \alpha)\%$ confidence bounds for Y_p are obtained by substituting α for $\alpha/2$ in the preceding lower and upper limits, respectively.

The $g'_{(\cdot)}$ factors are also used to obtain one-sided tolerance bounds for a normal distribution. This relationship and the tabulations are discussed further in Section 4.6. Confidence intervals for percentiles are not considered in most statistical texts. The underlying theory involves the noncentral t distribution and is given by Odeh and Owen (1980).

4.4.2   Example

For the example, $n = 5$, $\bar{x} = 50.10$ volts, and $s = 1.31$ volts. The manufacturer wanted a two-sided 95% confidence interval for $Y_{0.10}$, the 10th percentile of the distribution of the voltage outputs for the population of circuit packs. Thus, $p = 0.10$, and from Table A.12b, $g'_{(0.025; 0.90, 5)} = 0.389$ and from Table A.12d, $g'_{(0.975; 0.90, 5)} = 4.166$ and $g'_{(0.95; 0.90, 5)} = 3.407$. A two-sided 95% confidence interval for $Y_{0.10}$ is

$$\left[\underline{Y}_{0.10}, \tilde{Y}_{0.10}\right] = [50.10 - (4.166)1.31, 50.10 - (0.389)1.31] = [44.64, 49.59].$$

A lower 95% confidence bound for $Y_{0.10}$ is

$$\underline{Y}_{0.10} = 50.10 - (3.407)1.31 = 45.64.$$

4.5   CONFIDENCE INTERVAL FOR THE PROPORTION LESS (GREATER) THAN A SPECIFIED VALUE

The probability of an observation from a normal distribution with known mean μ and known standard deviation σ being less than a specified value y' can be computed as $p_L = \Pr(y \le y') = \Phi[(y' - \mu)/\sigma]$, where $\Phi(z')$ is given in Table A.5. Similarly, the normal distribution probability of such an observation being greater than a specified value y' is $p_G = \Pr(y > y') = 1 - \Phi[(y' - \mu)/\sigma]$. Equivalently, p_L and p_G are the proportions of the population less than y' and more than y', respectively. When μ and σ are not known, we obtain point estimates $\hat{p}_L$ and $\hat{p}_G$ by substituting $\bar{x}$ for μ and s for σ, respectively, in these formulas.

4.5.1   Method

A two-sided $100(1 - \alpha)\%$ confidence interval for $p_L = \Pr(y \le y') = \Phi[(y' - \mu)/\sigma]$, the population proportion p_L less than a specified value, y', is

$$\left[\underline{p}_L, \tilde{p}_L\right] = \left[h_{(1-\alpha/2; -K, n)}, 1 - h_{(1-\alpha/2; K, n)}\right]$$

where $K = (\bar{x} - y')/s$ and the factors $h_{(\gamma; K, n)}$ are given in Table 7 of Odeh and Owen (1980) for all combinations of

$K = -3.0(0.20)3.0$,
$n = 2(1)18(2)30, 40(20)120, 240, 600, 1000, 1200$, and
$\gamma = 0.50, 0.75, 0.90, 0.95, 0.975, 0.99, 0.995$.

Because of space limitations, we do not give tables of these factors here.

Similarly, a two-sided $100(1 - \alpha)\%$ confidence interval for $p_G = \Pr(y > y') = 1 - p_L$, the population proportion greater than the value y', is

$$\left[\underset{\sim}{p}_G, \tilde{p}_G \right] = \left[1 - \tilde{p}_L, 1 - \underset{\sim}{p}_L \right] = \left[h_{(1-\alpha/2; K, n)}, 1 - h_{(1-\alpha/2; -K, n)} \right].$$

One-sided lower and upper $100(1 - \alpha)\%$ confidence bounds for p_L or p_G are obtained by replacing $\alpha/2$ by α in the appropriate part of the preceding expressions. These types of intervals are not considered in most statistical texts. The underlying theory involves the noncentral t distribution and is given by Odeh and Owen (1980).

4.5.2 Example

For the example, $n = 5$, $\bar{x} = 50.10$ volts, and $s = 1.31$ volts. The manufacturer wanted to obtain a two-sided 95% confidence interval to contain p_G, the proportion of units in the population with voltages greater than $y' = 48.0$ volts (or equivalently, the probability that the voltage of a single randomly selected unit will exceed $y' = 48.0$ volts). A point estimate for this proportion, using Table A.5, is $\hat{p}_G = \Pr(y > 48) = 1 - \Phi[(48 - \bar{x})/s] = 1 - \Phi(-1.60) = 0.9452$. Then, using $K = (50.10 - 48.0)/1.31 = 1.60$ and taking $h_{(0.975; 1.60, 5)} = 0.57455$, $h_{(0.975; -1.60, 5)} = 0.00160$, and $h_{(0.95; 1.60, 5)} = 0.64929$ from Table 7 of Odeh and Owen (1980), a two-sided 95% confidence interval for p_G is

$$\left[\underset{\sim}{p}_G, \tilde{p}_G \right] = [0.57455, 1 - 0.00160] = [0.57, 0.9984],$$

and a one-sided lower 95% confidence bound for p_G is $\underset{\sim}{p}_G = 0.65$. A one-sided upper 95% confidence bound for p_L, the proportion of units less than $y' = 48$ volts, is $\tilde{p}_L = 1 - \underset{\sim}{p}_G = 1 - 0.65 = 0.35$.

4.6 STATISTICAL TOLERANCE INTERVALS TO CONTAIN A PROPORTION OF A POPULATION

4.6.1 Two-Sided Tolerance Interval (to Control Center of Distribution)

A two-sided $100(1 - \alpha)\%$ tolerance interval to contain at least a proportion, p, of a population described by a normal distribution, is

$$\left[\underset{\sim}{T}_p, \tilde{T}_p \right] = \bar{x} \pm g_{(1-\alpha; p, n)} s$$

where $g_{(\gamma; p, n)}$ is given in Table 3 of Odeh and Owen (1980) for all combinations of

$p = 0.75, 0.90, 0.95, 0.975, 0.99,$ and $0.995,$

$n = 2(1)100(2)180(5)300(10)400(25)650(50)1000, 1500, 2000, 3000, 5000,$
    $10000,$ and $\infty,$ and

$\gamma = 0.50, 0.75, 0.90, 0.95, 0.975, 0.99,$ and $0.995.$

Less extensive tabulations of these factors are given in Tables A.1a, A.1b, and A.10. Two-sided tolerance intervals are considered in some statistical texts. Odeh and Owen (1980) also describe the theory behind the computation of these intervals.

4.6.2 Two-Sided Tolerance Interval (to Control Both Tails of Distribution)

The two-sided tolerance interval in Section 4.6.1 and Chapter 3 contains, with specified confidence, at least a certain proportion, p, of the population between its two limits, irrespective of how the proportion of the population below the lower limit and above the upper limit is apportioned. Sometimes, it is more appropriate to use an interval that will separately control the population proportions in *each* of the two tails of the distribution. This type of tolerance interval would, for example, be useful if the cost of being above the upper limit were different from that of being below the lower limit. The two specified population proportions need not (and often will not) be equal. A $100(1 - \alpha)\%$ tolerance interval to control both tails of the distribution will be denoted by $[\underset{\sim}{T}_{p_L}, \tilde{T}_{p_U}]$. Then

$$\Pr\left(y < \underset{\sim}{T}_{p_L}\right) < p_L \quad \text{and} \quad \Pr\left(y > \tilde{T}_{p_U}\right) < p_U$$

with $100(1 - \alpha)\%$ confidence, where p_L is the desired maximum proportion of the population in the lower tail of the distribution, and p_U is the desired maximum proportion of the population in the upper tail of the distribution.

The tolerance interval is

$$\left[\underset{\sim}{T}_{p_L}, \tilde{T}_{p_U}\right] = \left[\bar{x} - g''_{(1-\alpha/2; p_L, n)}s, \bar{x} + g''_{(1-\alpha/2; p_U, n)}s\right]$$

where $g''_{(\gamma; p, n)}$ is given in Table 4 of Odeh and Owen (1980) for all combinations of

$\gamma = 0.995, 0.99, 0.975, 0.95, 0.90, 0.75, 0.50,$

$p = 0.125, 0.10, 0.05, 0.025, 0.01, 0.005, 0.0005,$ and

$n = 2(1)100(2)180(5)300(10)400(25)650(50)1000, 1500, 2000, 3000, 5000,$
    $10000,$ and $\infty.$

Tables A.11a and A.11b give $g''_{(\gamma; p, n)}$ for a smaller set of entries. Odeh and Owen (1980) also describe the theory for these intervals.

4.6.3 One-Sided Tolerance Bounds

A one-sided lower $100(1 - \alpha)\%$ tolerance bound to be exceeded by at least $100p\%$ of a normal population is

$$\underset{\sim}{T}_p = \bar{x} - g'_{(1 - \alpha; p, n)}s$$

and a one-sided upper $100(1 - \alpha)\%$ tolerance bound to exceed at least $100p\%$ of the population is

$$\tilde{T}_p = \bar{x} + g'_{(1 - \alpha; p, n)}s,$$

where $g'_{(\gamma; p, n)}$ is given in Table 1 of Odeh and Owen (1980) for the same values of p, n, and γ as for the two-sided factors. Tables A.12a to A.12d, A.3a, and A.3b contain less extensive tabulations of these factors. One-sided tolerance bounds are considered in a few statistical texts. Odeh and Owen (1980) review the theory, based upon the noncentral t distribution, and describe the numerical methods that were used in the computations.

4.6.4 Example

For the example, $n = 5, \bar{x} = 50.10$ volts, and $s = 1.31$ volts. Assume now that the manufacturer wanted a two-sided 95% tolerance interval to contain 90% of the population values and an upper 95% tolerance bound to exceed at least 90% of the population values.

    We use $g_{(0.95; 0.90, 5)} = 4.291$ (from Table A.10b), $g''_{(0.95; 0.05, 5)} = 4.847$ (from Table A.11b) and $g'_{(0.95; 0.90, 5)} = 3.407$ (from Table A.12d). (Also, $c_{(0.95; 5)} = 4.29$ from Table A.1a, and $c'_{(0.95; 5)} = 3.41$ from Table A.3a.) A standard two-sided 95% tolerance interval to contain at least 90% of the sampled population is

$$\left[\underset{\sim}{T}_{0.90}, \tilde{T}_{0.90}\right] = 50.10 \pm (4.291)1.31 = [44.5, 55.7].$$

A two-sided 95% tolerance interval to have less than 5% of the population in each distribution tail is

$$\left[\underset{\sim}{T}_{0.05}, \tilde{T}_{0.05}\right] = 50.10 \pm (4.847)1.31 = [43.8, 56.4].$$

This interval is longer than the preceding one that does not control each tail of the distribution. This is because simultaneously controlling both tails is a more stringent requirement than just controlling the total proportion of the distribution outside limits.

A one-sided upper 95% tolerance bound to exceed at least 90% of the sampled population is

$$\tilde{T}_{0.90} = 50.10 + (3.407)1.31 = 54.6.$$

4.6.5 Relationship between One-Sided Tolerance Bounds and Confidence Bounds on Percentiles

The factors $g'_{(\gamma;\,p,\,n)}$ for computing one-sided tolerance bounds for a normal population are the same as those used in Section 4.4 for computing a one-sided confidence bound for a normal distribution percentile. This is because a one-sided lower (upper) $100(1 - \alpha)\%$ tolerance bound to be exceeded by (to exceed) at least $100p\%$ of a normal population is equivalent to a one-sided lower (upper) $100(1 - \alpha)\%$ confidence bound for the $100p$th percentile of the normal distribution.

4.7 PREDICTION INTERVAL TO CONTAIN A SINGLE FUTURE OBSERVATION OR THE MEAN OF m FUTURE OBSERVATIONS

4.7.1 Method

A two-sided $100(1 - \alpha)\%$ prediction interval to contain the mean of m future, independently and randomly selected observations, based upon the results of a previous independent random sample of size n from the same population (or process) described by a normal distribution, is

$$\left[\underline{\bar{y}}_m, \bar{\bar{y}}_m\right] = \bar{x} \pm t_{(1 - \alpha/2;\,n - 1)}\left(\frac{1}{m} + \frac{1}{n}\right)^{1/2} s. \tag{4.2}$$

An important special case arises when $m = 1$; this results in a prediction interval to contain a single future observation. One-sided lower and upper $100(1 - \alpha)\%$ prediction bounds are obtained by replacing $\alpha/2$ by α (and $\pm$ by either $+$ or $-$)in the appropriate part of the preceding expression. The factors for two-sided prediction intervals given in Tables A.1a and A.1b, and the factors for one-sided prediction bounds in Tables A.3a and A.3b for the cases when $m = 1$ and $m = n$ were computed from these formulas.

Prediction intervals, in general, are not considered in most statistical texts, except in the context of regression analysis. [One exception is Guttman, Wilks, and Hunter (1982).] The theory underlying these methods is given by Hahn (1969).

4.7.2 Example

Assume now that the manufacturer needs to predict the mean voltage of a future random sample of $m = 3$ circuit packs. From the previous sample, $n = 5$, $\bar{x} = 50.10$ volts, and $s = 1.31$ volts. Also, from Table A.7, $t_{(0.975; 4)} = 2.776$ and $t_{(0.95; 4)} = 2.132$. Thus, a two-sided 95% prediction interval to contain the mean voltage of the sample of three future units is

$$\left[\bar{y}_3, \bar{\bar{y}}_3 \right] = 50.10 \pm 2.776\left(\tfrac{1}{3} + \tfrac{1}{5} \right)^{1/2}(1.31) = [47.44, 52.76].$$

An upper 95% prediction bound to exceed the mean of the three future units is

$$\bar{\bar{y}}_3 = 50.10 + 2.132\left(\tfrac{1}{3} + \tfrac{1}{5} \right)^{1/2}(1.31) = 52.14.$$

4.8 PREDICTION INTERVAL TO CONTAIN ALL OF m FUTURE OBSERVATIONS

4.8.1 Two-Sided Prediction Interval

A two-sided $100(1 - \alpha)\%$ simultaneous prediction interval to contain the values of *all* of m future randomly selected observations from a previously sampled population (or process) that can be described by a normal distribution is

$$\left[y_m, \tilde{y}_m \right] = \bar{x} \pm r_{(1 - \alpha; m, n)}s$$

where $r_{(\gamma; m, n)}$ is given in Table A.13. The factors for obtaining two-sided prediction intervals to contain all m future observations given in Tables A.1a and A.1b were taken from the preceding tabulations.

A conservative approximation for $r_{(1 - \alpha; m, n)}$ is

$$r_{(1 - \alpha; m, n)} \cong \left(1 + \frac{1}{n} \right)^{1/2} t_{(1 - \alpha/(2m); n - 1)}.$$

This approximation may be used for nontabulated values and for constructing a simple computer program to perform the calculations. It is based on a Bonferroni inequality [see Section 11.1.1 or Miller (1981)] and was suggested by Chew (1968), who also describes a second approximation. Both approximations were evaluated by Hahn (1969). The one given above was found to be satisfactory for most practical purposes, except for combinations of small n, large m, and small γ. Also, the above expression is exact for the special case of $m = 1$, i.e., a prediction interval to contain a single future observation (see Section 4.7.1).

The theory for prediction intervals to contain all of m future observations is given by Hahn (1969, 1970a). General concepts of prediction intervals and additional references are given by Hahn and Nelson (1973).

4.8.2   One-Sided Prediction Bounds

One-sided lower and upper $100(1 - \alpha)\%$ (simultaneous) prediction bounds to be exceeded by and to exceed *all* of m future observations from a previously sampled normal population are, respectively,

$$\underline{y}_m = \bar{x} - r'_{(1-\alpha;\,m,\,n)} s$$

and

$$\tilde{y}_m = \bar{x} + r'_{(1-\alpha;\,m,\,n)} s$$

where $r'_{(\gamma;\,m,\,n)}$ is given in Table A.14. The factors for one-sided prediction bounds to contain all of m future observations given in Tables A.3a and A.3b were taken from these tabulations. Hahn (1970a), provides the theory for these intervals.

A conservative approximation for $r'_{(1-\alpha;\,m,\,n)}$ is

$$r'_{(1-\alpha;\,m,\,n)} \cong \left(1 + \frac{1}{n}\right)^{1/2} t_{(1-\alpha/m;\,n-1)}.$$

This expression was evaluated by Hahn (1970a) with results similar to those for the approximate expression for the two-sided prediction interval. The expression is again exact for $m = 1$.

4.8.3   Example

For the example, $n = 5$, $\bar{x} = 50.10$ volts, and $s = 1.31$ volts. The manufacturer wants a 95% prediction interval to contain the voltages of all $m = 10$ future circuit packs randomly selected from the same process and an upper 95% prediction bound to exceed the voltages of all $m = 10$ future units. From Table A.13, $r_{(0.95;\,10,\,5)} = 5.229$ and from Table A.14, $r'_{(0.95;\,10,\,5)} = 4.418$. A two-sided 95% prediction interval to contain all 10 future voltages is

$$\left[\underline{y}_{10}, \tilde{y}_{10}\right] = 50.10 \pm (5.229)1.31 = [43.25, 56.95].$$

A one-sided upper 95% prediction bound to exceed all 10 future voltages is

$$\tilde{y}_{10} = 50.10 + (4.418)1.31 = 55.89.$$

4.9 PREDICTION INTERVAL TO CONTAIN THE STANDARD DEVIATION OF m FUTURE OBSERVATIONS

4.9.1 Method

A two-sided $100(1 - \alpha)\%$ prediction interval to contain s_m, the standard deviation of m future observations based upon the results of a previous independent, random sample of size n from the same normally distributed population (or process), is

$$\left[\underline{s}_m, \tilde{s}_m \right] = \left[s \left\{ \frac{1}{F_{(1-\alpha/2;\, n-1,\, m-1)}} \right\}^{1/2}, s \left\{ F_{(1-\alpha/2;\, m-1,\, n-1)} \right\}^{1/2} \right].$$

One-sided lower and upper $100(1 - \alpha)\%$ prediction bounds are obtained by replacing $\alpha/2$ by α in the appropriate part of the preceding expression. The factors for two-sided prediction intervals and one-sided prediction bounds to contain the standard deviation of a future sample given in Tables A.2a, A.2b, A.4a, and A.4b were computed from these expressions. The theory underlying these methods is given by Hahn (1972a).

4.9.2 Example

For the example, the manufacturer now wishes a 95% prediction interval to contain the standard deviation of the voltages for a future random sample of $m = 3$ circuit packs from the same process. From the previous sample, $n = 5$, and $s = 1.31$ volts. Also, from Table A.9e, $F_{(0.975;\, 4,\, 2)} = 39.25$, $F_{(0.975;\, 2,\, 4)} = 10.65$, and from Table A.9d $F_{(0.95;\, 2,\, 4)} = 6.944$. Then, a two-sided 95% prediction interval to contain s_3, the standard deviation of the voltage of three future units is

$$\left[\underline{s}_3, \tilde{s}_3 \right] = \left[1.31 \left(\frac{1}{39.25} \right)^{1/2}, 1.31 \left\{ 10.65 \right\}^{1/2} \right] = [0.21, 4.27].$$

An upper 95% prediction bound to exceed the sample standard deviation of the three future units is

$$\tilde{s}_3 = 1.31 (6.944)^{1/2} = 3.45.$$

4.10 THE ASSUMPTION OF A NORMAL DISTRIBUTION

The tabulations and formulas for constructing the statistical intervals in this and the preceding chapters are based upon mathematical theory that assumes that the given sample is selected randomly from a normal distribution.

Moreover, in the case of a prediction interval, it is also assumed that the future sample is selected randomly from the same normal distribution as, and independently of, the first sample. Thus, the intervals are strictly valid only under these assumptions.

A confidence interval for the population (or process) mean, however, is quite insensitive to deviations from normality in the sampled population, unless the sample is very small and the deviation from normality is pronounced. (In statistical terminology, a confidence interval to contain the population mean is said to be "robust" to deviations from normality.) This is the result of the well-known Central Limit Theorem, which states that the distribution of a sample mean is approximately normal if the sample size is not too small and if the underlying distribution is not too skewed [see, for example, Hahn and Shapiro, (1967)]. Thus, a confidence interval for the population mean, assuming a normal distribution, may be used to construct a confidence interval in many practical situations, even if the assumption of normality is not strictly met. In such cases, instead of being an exact $100(1 - \alpha)\%$ confidence interval for the population mean, the resulting interval is an approximate $100(1 - \alpha)\%$ confidence interval. Everything else being equal, the approximation becomes worse as α decreases (the desired level of confidence increases). Detailed discussion and specific evaluations of the effect of nonnormality are given by Scheffé (1959).

For similar reasons, prediction intervals to contain the mean of a future sample are also relatively insensitive to deviations from normality, unless either the given sample or the future sample is very small or the deviation from normality is pronounced. This limitation, however, includes the important case of a prediction interval for a single future observation, i.e., $m = 1$. Thus, the prediction interval given here to contain a single future observation may be appreciably off when sampling from a nonnormal population (or process).

Unfortunately, one cannot use the Central Limit Theorem to justify the use of normal distribution based methods for most other types of statistical intervals. For example, the procedure in Section 4.3 for obtaining a confidence interval for the population (or process) standard deviation is highly sensitive to the assumption of normality, even for large sample sizes. Therefore, serious errors could result in using the methods given here when the sampled population (or process) is not normal. The same holds for a prediction interval to contain the standard deviation of a future sample.

Tolerance intervals to contain a specified proportion p of a distribution and prediction intervals to contain all of m future observations tend to involve inferences or predictions about the tails of a distribution. Deviations from normality are generally most pronounced in the distribution tails. For this reason, such intervals could be seriously in error when the underlying distribution is not normal, especially for high confidence levels, and for tolerance intervals when the proportion of the population to be contained within the interval is close to 1.0. Similar conditions apply to confidence intervals for an extreme percentile and confidence intervals to contain the probability of exceeding an extreme specified value. For example, an estimate of the 5th percentile under the assumption of a normal distribution, and, even more so, an estimate of the 1st percentile, based on only 5 observations, is strongly dependent on the assumption of normality in the distribution tail. See Hahn (1971) and Canavos and Kautauelis (1984) for further discussion.

4.11   ASSESSING DISTRIBUTION NORMALITY AND DEALING WITH NONNORMALITY

4.11.1   Probability and Q–Q Plots

Before using intervals that depend heavily on the normality assumption, one should assess the adequacy of the normal distribution as a model. Normal distribution probability plots of the data are a simple and effective tool for doing this, especially if there are 20 or more observations. A probability plot is constructed by plotting the ith ordered (from smallest to largest) observation against the proportion $p_i = (i - 0.5)/n$ on specially prepared *normal probability paper*. Normal distribution probabilities $\Pr(y < y')$ plot as a straight line against y' on such paper. Similar probability papers are also available for other commonly used distributions, like the Weibull. Hahn and Shapiro (1967) and Nelson (1982) give theory, methods, and examples for probability plots.

Chambers, Cleveland, Kleiner, and Tukey (1983), and others, discuss Q–Q (short for quantile–quantile) plots, which serve the same purpose. To obtain a Q–Q plot, one plots, on linear scales, the ith-ordered observation against $z_{(p_i)}$, the $100p_i$th normal distribution percentile, where $p_i = (i - 0.5)/n$. The only difference between a Q–Q plot and a probability plot is that the probability plot has a probability scale for the p_i while the Q–Q plot has a linear scale for the corresponding normal distribution percentile values $z_{(p_i)}$. The probability scales on probability plots are easier to interpret. However, because they use only linear scales, Q–Q plots are easy to produce with standard graphics and statistical computer packages, or with ordinary graph paper.

4.11.2   Interpreting Probability and Q–Q Plots

If the data points plotted on a normal probability, or a Q–Q plot, deviate appreciably from a straight line, the adequacy of the normal distribution as a model for the population is in doubt. Judging departures from a straight line requires one to allow for the variability in the sample data. For many models, including the normal distribution, one expects the ordered observations in the tails of the distribution to vary more than those in the center of the distribution. To help judge this variability, Hahn and Shapiro (1967) give normal probability plots for simulated data with different sample sizes and with data from both normal and nonnormal distributions. (The analyst may wish to do similar simulations, or, possibly, incorporate these simulations into a computer program for routine analysis.)

4.11.3   Example

Table 4.1 gives sample data from a life test on ball bearings, first reported by Lieblein and Zelen (1956). Figures 4.1a and 4.1b give stem-and-leaf and box plots, respectively, of these data. These and other useful graphical displays are described, for example, in Velleman and Hoaglin (1981). These displays were produced using the MINITAB® statistical analysis software package. The $n = 23$ life times are in millions of revolutions to failure. Figure 4.2a is a normal probability plot of these data. The plotted points tend to scatter around a curve, rather than a straight line. This suggests that the normal distribution does *not* provide an adequate model for the population from which the data were obtained.

Table 4.1   Ball Bearing Failure Data
(millions of revolutions)

17.9	28.9	33.0	41.5	42.1
45.6	48.5	51.8	51.9	54.1
55.5	67.8	68.6	68.6	68.9
84.1	93.1	98.6	105.1	105.8
127.9	128.0	173.5		

```
Leaf Unit = 10

   1    0 1
   3    0 23
  11    0 44445555
 (4)    0 6666
   8    0 899
   5    1 00
   3    1 22
   1    1
   1    1 7
```

Figure 4.1a   Stem-and-leaf display of ball bearing cycles to failure (in millions of cycles).

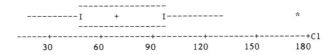

Figure 4.1b   Box plot of ball bearing cycles to failure (in millions of cycles).

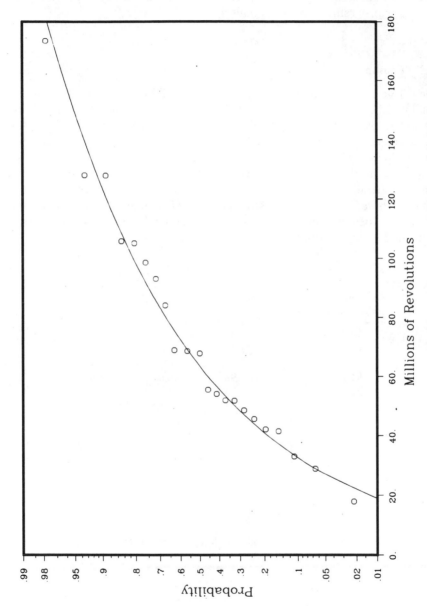

Figure 4.2a Normal distribution probability plot for ball bearing failure data: original observations.

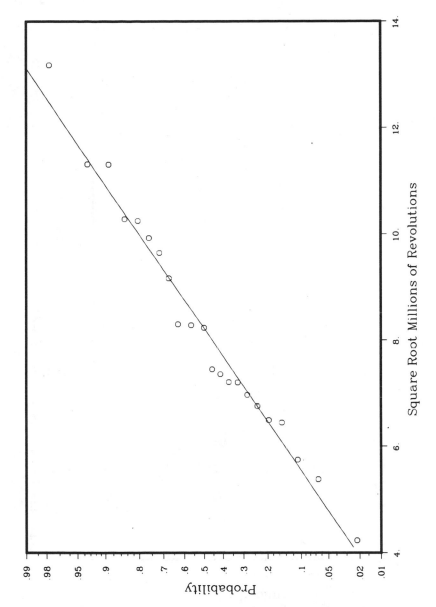

Figure 4.2b Normal distribution probability plot for ball bearing failure data: square root of observations.

69

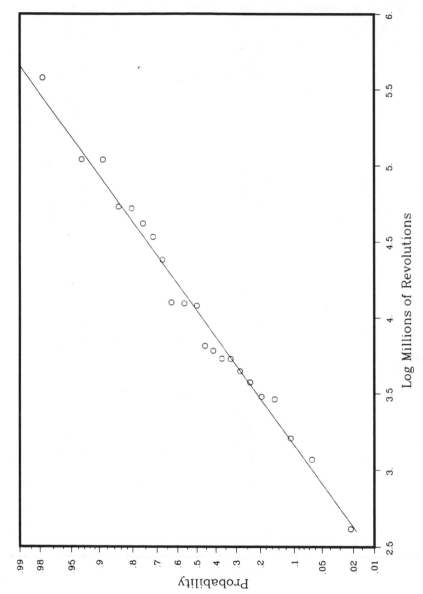

Figure 4.2c Normal distribution probability plot for ball bearing failure data: log of observations.

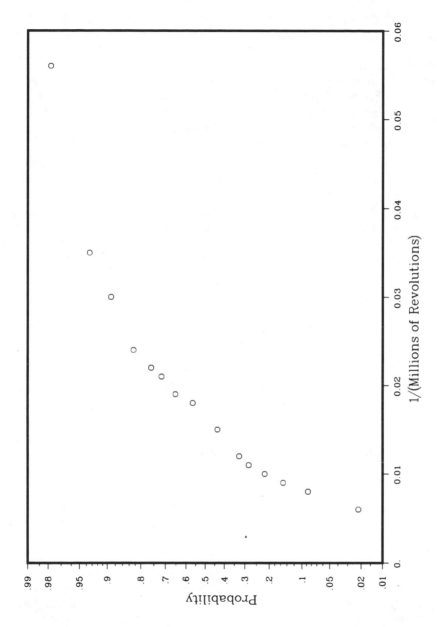

Figure 4.2d Normal distribution probability plot for ball bearing failure data: inverse of observations.

4.11.4 Dealing with Nonnormal Data

If one constructs a probability plot and determines that the normal distribution is not an appropriate model, there are several ways to proceed:

- A smooth curve drawn through the points on a probability plot will provide a simple and easy-to-understand *distribution-free* graphical estimate of the cumulative distribution function. For example, from Figure 4.2a, we estimate that approximately 30% of the ball bearings will fail by the end of 50 million·revolutions. In Chapters 5 and 6, we present special methods to construct "distribution-free" statistical intervals.

- It may be possible to find some other distribution (e.g., the lognormal, gamma, or Weibull) that provides an adequate representation for the data. (Such an alternative model may also be justified from physical considerations.) In Chapter 11 we provide references and further information about using statistical distributions other than the normal distribution to model data and to compute confidence intervals. For example, Lieblein and Zelen (1956) used the Weibull distribution as a model for the ball bearing failure data.

- It may be possible to transform the data in a manner that allows the normal distribution to provide an adequate model for the population and a basis for making inferences (using the transformed data). This is equivalent to fitting an alternative distribution. For example, analyzing the logarithms of a set of data as if they were normally distributed is equivalent to fitting a (2-parameter) lognormal distribution to the original data (see Section 11.3.3). In the next section we will illustrate the use of transformed data to compute statistical intervals.

4.12 INFERENCES FROM TRANSFORMED DATA

4.12.1 Box-Cox Transformations

If the original sample does not appear to have come from a normal distribution, it may still be possible to find a transformation that will allow the data to be adequately represented by a normal distribution. For example, Figures 4.2b, 4.2c, and 4.2d give normal probability plots, respectively, for the square roots, logarithms (base e), and the reciprocals of the ball bearing failure data. From these plots we might conclude that the normal distribution appears to provide a good fit to both the square roots and the logarithms of the data. These transformations are members of the Box–Cox (1964) family of transformations which is defined as follows:

$$x^{(\gamma)} = \begin{cases} x^{\gamma} & \text{if } \gamma \neq 0 \\ \log(x) & \text{if } \gamma = 0, \end{cases}$$

where γ is a parameter that defines the transformation. This parameter is generally not known. Thus, in practice, one tries to find a value (or range of values) for γ that leads to approximate normality. In particular, one may try different values of γ (e.g., $\gamma = 1$, 0.5, 0.333, 0, and -1, corresponding to no transformation, square root, cube root, log, and reciprocal transformations, respectively) to try to find a value (or range of values) that gives a probability plot that is nearly linear. In some cases physical considerations or experience may suggest such a value. Also analytical methods have been proposed to help determine the "best" value for γ; [see, for example, Draper and Smith (1981, page 225)].

4.12.2 Computing Statistical Intervals from Transformed Data

The sample mean and standard deviation of the logarithms (base e) of the ball bearing data are, respectively, $\bar{x}^{(0)} = 4.150$ and $s^{(0)} = 0.533$ log millions of revolutions. These can be used to compute statistical intervals on the log scale that, in turn, can be translated into intervals on the original data scale.

For example, the method in Section 4.4, gives the following lower 95% confidence bound for the 10th percentile of the distribution on the log scale:

$$\underline{Y}^{(0)}_{0.10} = \bar{x}^{(0)} - g'_{(0.95; 0.90, 23)} s^{(0)} = 4.150 - (1.869)0.533 = 3.15.$$

Taking *antilogs* gives the desired lower 95% confidence bound for the 10th percentile on the original data scale, i.e.,

$$\underline{Y}_{0.10} = \exp\left(\underline{Y}^{(0)}_{0.10}\right) = \exp(3.15) = 23.3$$

millions of revolutions.

Similar calculations yield confidence intervals for other percentiles, tolerance intervals, and prediction intervals on the original data scale. To use the method in Section 4.5 to compute a confidence interval on a normal distribution tail probability, one can simply compute $K = [\bar{x}^{(0)} - \log(y')]/s^{(0)}$ (because $\Pr(y \leq y') = \Pr[\log(y) \leq \log(y')]$) and proceed as before.

We note, however, that a confidence interval for the mean of the distribution of the transformed data does *not* translate into a confidence interval for the mean on the original scale. It does, however, translate into a confidence interval for the median (i.e., 50th percentile) of the original distribution. Similarly, a confidence interval for the standard deviation of the distribution of the transformed data does *not* translate into a confidence interval for the standard deviation on the original scale.

4.12.3 Comparison of Inferences Using Different Transformations

Table 4.2 gives one-sided 95% confidence bounds, computed for several different percentiles of the ball bearing life distribution, using the original untransformed data, three alternative, reasonable data transformations, and

Table 4.2   Comparison of One-Sided 95% Confidence Bounds for Various Percentiles from Different Transformations for Ball Bearing Life Data

Transformation	γ	Sample Statistics		One-sided 95% Confidence Bounds			
		$\bar{x}^{(\gamma)}$	$s^{(\gamma)}$	$\underset{\sim}{Y}_{0.01}$	$\underset{\sim}{Y}_{0.10}$	$\underset{\sim}{Y}_{0.50}$	$\tilde{Y}_{0.50}$
No transformation	1	72.21	37.50	-48.0	2.11	58.8	85.6
Square root	$\frac{1}{2}$	8.234	2.146	1.82	17.8	55.8	81.0
Cube root	$\frac{1}{3}$	4.048	0.707	5.64	20.3	54.9	79.5
Log	0	4.150	0.533	11.5	23.3	52.5	76.7
Distribution-free	—	—	—	—	—	51.8	84.1

Note: γ is the Box–Cox transformation parameter.

a distribution-free method. (Recall that the normal probability plots indicated that the original untransformed observations did not appear to follow a normal distribution. The square root and log transformations did, however, provide reasonably good normal distribution fits. The cube root transformation is intermediate to these and would also provide a good fit.) The *distribution-free* intervals are described in Chapter 5; these make no assumptions about the form of the underlying distribution. (Because there are only 23 observations, it is not possible to find distribution-free lower 95% confidence bounds for the smaller percentiles. These require extrapolation outside the range of the data, and this is possible only by assuming a distribution form. See Chapter 5 for further discussion.)

All of the transformations give similar lower confidence bounds and similar upper confidence bounds for the 50th percentile. The lower confidence bounds for the 1st percentile, however, differ substantially. Notice also, the physically impossible negative lower confidence bound for the 1st percentile, obtained from fitting a normal distribution to the original data.

This comparison provides two general messages about fitting parametric models to data. First, it suggests that it is useful to analyze the data under alternative reasonable models. We could also have plotted the lower 95% confidence bounds for various percentiles versus γ for the range of reasonable values of γ (based on probability plots of the data). Second, it shows that alternative models that fit the data well, as well as distribution-free methods, are likely to give similar results for inferences that are *within* the range of the data (i.e., for inferences that only involve interpolation among data points). When the desired inferences require extrapolation outside the range of the data, inferences are, in general, much more sensitive to the assumed model. Although a poorly fitting model may provide credible inferences within the range of the data, it will often give unreasonable answers (e.g., the negative value for $\underset{\sim}{Y}_{0.01}$ in Table 4.2) for extrapolations outside the data range.

CHAPTER 5

Distribution-Free Statistical Intervals

5.1 INTRODUCTION

This chapter shows how to calculate "distribution-free" two-sided statistical intervals and one-sided bounds. Such intervals and bounds do not require one to assume a particular distribution (such as the normal distribution). However, the important assumption that sampling is random from the population (or process) of interest and, more generally, the assumptions discussed in Chapter 1, still pertain.

One might ask "When should I use distribution-free statistical methods?" The answer, we assert, is "Whenever possible." If one can do a study with minimal assumptions, then the resulting conclusions are based on a more solid foundation. Moreover, it is often appropriate to have the data analysis begin with a distribution-free approach (as, for example, provided by the probability plots discussed in Section 4.11). Then, if needed, one can proceed to a more structured analysis involving a particular parametric distribution, such as the normal distribution (perhaps following a transformation chosen so that the distribution better represents the data).

It is interesting and useful to compare conclusions drawn from analyses that use and do not use an assumed distribution (as well as ones that use different assumed distributions). If an assumed distribution fits the data well, the point estimates obtained from the analyses (e.g., percentile estimates within the range of the data) often do not differ much from those obtained from a distribution-free approach. If an assumed distribution does not fit the data well, such point estimates could differ substantially. (In this situation it may, however, be inappropriate to calculate a distribution-dependent interval.) As we shall see, a distribution-free interval (if one exists) will generally be longer than a corresponding interval based on a particular distribution. This is part of the price that one pays for not making the distributional assumption, and sometimes this may limit the value of the distribution-free approach.

In particular, let $x_1, x_2, \ldots, x_n$ represent n independent observations from any continuous distribution and let $x_{(1)}, x_{(2)}, \ldots, x_{(n)}$ denote the same observations, ordered from smallest to largest. These ordered observations are commonly called the "order statistics" of the sample. The distribution-free confidence intervals, tolerance intervals, and prediction intervals discussed here use selected order statistics as interval endpoints. A two-sided distribution-free interval requires the use of two order statistics, such as the smallest and largest observations. The one-sided distribution-free bounds discussed here use a single order statistic as the endpoint of the bound.

Because the endpoints of distribution-free statistical intervals are restricted to the observed order statistics, it is generally not possible to obtain an interval with precisely the desired associated confidence level. Therefore, a frequent practice is to take a somewhat higher (i.e., conservative) confidence level than that originally specified. A more serious problem is that for small samples it is not always possible to obtain a distribution-free interval with as high a confidence level as is desired, even if one uses extreme observations (i.e., the smallest and/or largest order statistics) as the interval endpoints. Then one must settle for an interval with the highest achievable confidence level, even though this might be less than the desired level. It will, moreover, sometimes be impossible to obtain a meaningful distribution-free interval altogether. For example, with a sample of size 50, it is impossible to calculate a lower confidence bound for the 1st percentile of the distribution at any reasonable confidence level without making an assumption about the form of the underlying distribution. On the other hand, if the sample size can be made sufficiently large, it is usually possible to approximately achieve the desired confidence level.

Another shortcoming of distribution-free intervals is that they can be much longer than distribution-dependent ones. Thus, relatively large samples are often needed to attain an acceptable interval length or even to provide an interval at the desired confidence level, even if the extreme order statistic(s) are used.

Thus, distribution-free intervals have some severe limitations. We must reiterate, however, that such intervals still warrant serious consideration. This is because the alternative of using methods that require the assumption of a particular distribution (such as the normal distribution) can lead to seriously incorrect intervals if the assumption is incorrect.

The theory underlying distribution-free statistical intervals, other details, and further references can be found, for example, in David (1981), Fligner and Wolfe (1976), Gibbons (1971, 1975), Guttman (1970), and Randles and Wolfe (1979).

5.1.1 Overview

This chapter describes and illustrates the calculation of the following distribution-free two-sided intervals:

- Confidence intervals for a distribution percentile, such as the median.
- Tolerance intervals to contain (at least) a specified percentage of a population.

- Prediction intervals to contain at least k of m future observations.
- Prediction intervals to contain a specified ordered observation in a future sample.

We also discuss corresponding one-sided bounds. These distribution-free intervals and bounds involve distribution percentiles and ordered observations. They do not involve distribution "moment" parameters, such as the mean and standard deviation, or their sample counterparts. For this reason, these and similar intervals based on order statistics are also called "nonparametric." However, we will continue to use the term "distribution-free."

To obtain a two-sided distribution-free statistical interval from a random sample from a specified population one generally proceeds as follows:

1. Specify the desired confidence level for the interval.
2. Determine (from tabulations or calculations) the order statistics that provide the statistical interval with at least the desired confidence level for the sample size. The actual confidence level will usually be higher than the specified confidence level. If no such order statistics exist, use the extreme order statistics to obtain the interval endpoints (because these will come closest to providing the desired confidence level), and determine the associated confidence level. (Sometimes this level is so low or the interval is so long that the interval will have little practical value.)
3. Use the selected order statistics as the endpoints of the distribution-free interval.

One-sided distribution-free bounds are obtained in a similar manner, except that only one order statistic is used as the desired lower or upper bound.

The endpoint(s) of a distribution-free statistical bound or interval should be determined before noting the values of the observations. Thus, for example, one might decide, initially, based upon the sample size and the desired confidence level, to use the extreme order statistic(s) (i.e., the smallest and/or largest observations) as endpoint(s). These observations are then the interval endpoints, irrespective of what their observed values turn out to be. This might result in an interval that is so long that it is of limited practical usefulness (and might suggest the need for a larger sample).

For samples from a population with a continuous distribution (often the case with physical measurements), the actual confidence level for a statistical interval is exact (although, as indicated, generally higher than the originally specified confidence level). If the distribution is discrete, the resulting interval is conservative. That is, the true confidence level of the distribution-free interval is somewhat higher than the computed confidence level (which, in turn, is again generally higher than the originally specified confidence level).

5.1.2 Notation and Tabulations

This section outlines basic notation and references tabulations and computer algorithms for some probability distributions used in calculating the distribution-free intervals and bounds provided in this chapter. The notation pertains to probability distributions with dichotomies, i.e., populations that contain units that belong to one class or another. To follow standard quality control terminology, we generically refer to these units as "conforming" and "non-conforming."

5.1.2.1 The Binomial Distribution
The function

$$B(x'; n, p) = \Pr(x \le x') = \sum_{i=0}^{x'} \binom{n}{i} p^i (1 - p)^{n-i}$$

gives the (cumulative binomial) probability of obtaining x' or fewer nonconforming units in a random sample of n units from a specified process or population, where the probability of a nonconforming unit equals p for each unit. Here

$$\binom{n}{i} = \frac{n!}{i!(n - i)!}$$

is the "binomial coefficient." The binomial model is appropriate if the probability p of a nonconforming unit is constant, for each sampled unit. For a sampled process, this requires stationarity, i.e., p does not vary over time, etc. For a population, p will be constant if sampling is random and "with replacement" or approximately constant if n is small relative to the size of the population (e.g., if n is less than 10% of the population). There are computer algorithms that will compute these probabilities [e.g., BINDF in IMSL (1987)].

Limited tabulations and approximations for the cumulative binomial distribution function are given in many standard statistics texts. Table A.22 provides cumulative binomial probabilities for $p = 0.01(0.01)0.05, 0.06, 0.08,$ $0.10(0.05)0.50$ and $n = 2(1)20$. Larson (1966) developed a useful chart and Muench (1984) provides a convenient slide-rule for cumulative binomial distribution probabilities. References to extensive tabulations of the binomial distribution are given by Hahn and Chandra (1981). $B(x'; n, p)$ is tabled, for example, by the Harvard University Computations Laboratory (1955) for

$x' = 0(1)n,$
$n = 1(1)50(2)100(10)200(50)1000,$ and
$p = 0.01(0.01)0.50, \frac{1}{16}, \frac{1}{12}, \frac{1}{8}, \frac{1}{6}, \frac{3}{16}, \frac{5}{16}, \frac{1}{3}, \frac{3}{8}, \frac{5}{12}, \frac{7}{16}.$

If the sample size n is large, cumulative binomial distribution probabilities can be approximated by a normal distribution with mean $\mu = np$ and standard deviation $\sigma = [np(1 - p)]^{1/2}$. In particular,

$$B(x'; n, p) \approx \Phi\left(\frac{x' + 0.5 - \mu}{\sigma}\right)$$

where $\Phi(z)$ is the standard normal cumulative distribution function given in Table A.5 (i.e., the probability that a normally distributed random variable with mean 0 and standard deviation 1 is less than z). The $+0.5$ in this expression is a "continuity correction" that improves the approximation. With the continuity correction, the approximation is reasonably accurate as long as $np > 5$ and $n(1 - p) > 5$.

5.1.2.2 The Hypergeometric Distribution

The function

$$H(x'; n, D, N) = \Pr(x \le x') = \sum_{i=0}^{x'} \frac{\binom{D}{i}\binom{N - D}{n - i}}{\binom{N}{n}}$$

is the (cumulative hypergeometric) probability of obtaining x' or fewer nonconforming units in a random sample of size n (drawn without replacement) from a population of size N that contains exactly D nonconforming units. There are computer algorithms that will compute these probabilities [e.g., HYPDF in IMSL (1987)]. Also, $H(x'; n, D, N - D)$ is tabled in Lieberman and Owen (1961) for

$N = 2(1)100$, $n = 1(1)50$ for $x' = 0(1)n$,
$N = 1000$, $n = 500$, for $x = 0(1)n$, and
$N = 100(100)2000$, $n = 0.5N$, $x' = 1$ and $x' = n$.

5.1.2.3 The Inverse Hypergeometric Distribution

The function

$$H^*(L, U, D, n, N) = \sum_{i=L}^{U} h^*(i, D + i, n, N)$$

where

$$h^*(i, D + i, n, N) = \frac{\binom{D - 1}{i}\binom{N - D}{n - i}}{\binom{N}{n}}$$

Table 5.1 Ordered Measurements of the Amount of a Compound for 100 Batches from a Chemical Process

Batches	1	2	3	4	5	6	7	8	9	10
1–10	1.49	1.66	2.05	2.24	2.29	2.69	2.77	2.77	3.10	3.23
11–20	3.28	3.29	3.31	3.36	3.84	4.04	4.09	4.13	4.14	4.16
21–30	4.57	4.63	4.83	5.06	5.17	5.19	5.89	5.97	6.28	6.38
31–40	6.51	6.53	6.54	6.55	6.83	7.08	7.28	7.53	7.54	7.68
41–50	7.81	7.87	7.94	8.43	8.70	8.97	8.98	9.13	9.14	9.22
51–60	9.24	9.30	9.44	9.69	9.86	9.99	11.28	11.37	12.03	12.32
61–70	12.93	13.03	13.09	13.43	13.58	13.70	14.17	14.36	14.96	15.89
71–80	16.57	16.60	16.85	17.18	17.46	17.74	18.40	18.78	19.84	20.45
81–90	20.89	22.28	22.48	23.66	24.33	24.72	25.46	25.67	25.77	26.64
91–100	28.28	28.28	29.07	29.16	31.14	31.83	33.24	37.32	53.43	58.11

has been called the "inverse hypergeometric distribution." The properties of the inverse hypergeometric distribution are discussed by Guenther (1975). The listing of subroutine HSTAR to compute values of $H^*(\cdot)$ is given in Appendix C.

5.1.3 Example

The following example will be used to illustrate the intervals presented in this chapter. A production engineer wants to evaluate the capability of a chemical process to produce a particular compound. Measurements are available on the amount of the compound that was present in composite samples taken from each of $n = 100$ randomly selected batches from the process. Each batch was thoroughly mixed before sampling and can, therefore, be regarded as homogeneous. Measurement error was small enough to be ignored. (If measurement error had been a problem, it could have been reduced by taking a sufficient number of independent measurements from each batch and averaging the measurements).

Table 5.1 gives the readings, ordered from smallest to largest; these ordered observations will facilitate the application of the methods given in this chapter.

Whenever data are collected over time, it is important to check for possible time-related dependencies. For example, measuring instruments may drift with time or use, or readings may depend on ambient temperature which might change over the time that the readings are taken. A time-order plot of the readings, measured in parts per million, is given in Figure 5.1. The data do not exhibit any trend, cycle, or other departure from randomness. There are several large values, but these were known to be caused by random, uncontrollable shocks to the process. Otherwise, the process appears to be in statistical control.

Figure 5.2a gives a histogram of the data, while Figure 5.2b is the closely related stem-and-leaf display of the data. Both of these displays were

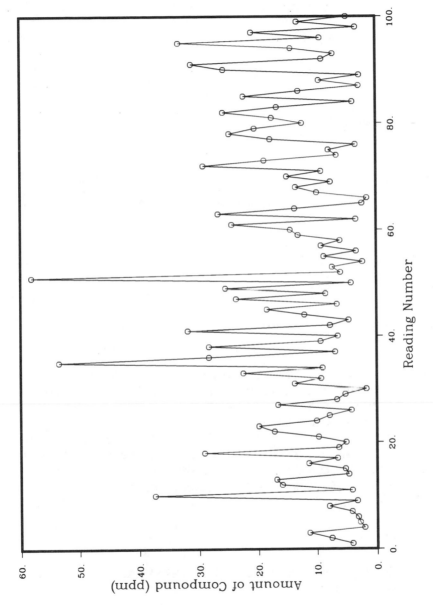

Figure 5.1 Time-order plot of chemical process data given in Table 5.1.

```
MIDDLE OF    NUMBER OF
INTERVAL     OBSERVATIONS
   0.           5      *****
   5.          32      ********************************
  10.          23      ***********************
  15.          15      ***************
  20.           8      ********
  25.           7      *******
  30.           6      ******
  35.           2      **
  40.           0
  45.           0
  50.           0
  55.           1      *
  60.           1      *
```

Figure 5.2a   Histogram of chemical process data given in Table 5.1.

```
Leaf Unit = 1.0

     23     0  1122222233333344444444
    (33)    0  55555666666677777777788889999999999
     44     1  1122233333444
     31     1  5666777889
     21     2  0022344
     14     2  55568899
      6     3  113
      3     3  7
      2     4
      2     4
      2     5  3
      1     5  8
```

Figure 5.2b   Stem-and-leaf display of chemical process data given in Table 5.1.

produced using the MINITAB® statistical analysis software package. In addition to the information on distribution shape and range provided by the more familiar histogram, the stem-and-leaf displays the distribution of the most significant digits of a data set.

Figures 5.2a and 5.2b show that the distribution of the amount of compound in the batches is skewed to the right and that the normal distribution clearly is not a good model for these data. Although a Box–Cox transformation (see Section 4.12) could be used so that the data might be approximated by a normal distribution, we will, in this chapter, use distribution-free intervals.

5.2   DISTRIBUTION-FREE CONFIDENCE INTERVALS FOR A PERCENTILE

5.2.1   Two-Sided Distribution-Free Confidence Intervals for a Percentile

A two-sided distribution-free conservative $100(1 - \alpha)\%$ confidence interval for Y_p, the $100p$th percentile of the sampled population, is obtained from a sample of size n as $[\underline{Y}_p, \tilde{Y}_p] = [x_{(\ell)}, x_{(u)}]$, where ℓ and u are given for

$p = 0.01, 0.05, 0.10, 0.20, 0.30, 0.40$, and 0.50 in Tables A.15a to A.15g, respectively. These tabulations provide integer values of ℓ and u that are symmetric or nearly symmetric, around $i = [np] + 1$, where $[r]$ is the integral part of r. This is done because the nonparametric estimate of the $100p$th percentile lies between $x_{([np])}$ and $x_{([np]+1)}$. One can also use Tables A.15a to A.15f to find distribution-free confidence intervals for the $100p$th percentile for $p = 0.99, 0.95, 0.90, 0.80, 0.70$, and 0.60. In this case, enter the table corresponding to $1 - p$ and use

$$\left[\underline{Y}_p, \tilde{Y}_p \right] = \left[x_{(n-u+1)}, x_{(n-\ell+1)} \right]$$

as the desired interval.

For situations that are not covered in the tabulations, one can choose integers ℓ and u symmetrically (or nearly symmetrically) around $p(n + 1)$ and as close together as possible subject to the requirements that

$$\text{EPYP}(n, \ell, u, p) = B(u - 1; n, p) - B(\ell - 1; n, p) \geq 1 - \alpha,$$
$$0 < \ell < u \geq n, \qquad 0 < p < 1, \quad (5.1)$$

which can be computed by using subroutine EPYP, listed in Appendix C. Determination of ℓ and u might require some iteration. For example, to find a distribution-free confidence interval for Y_p, one can start with the integers $\ell = i^-$ and $u = i^+$ that are closest to $p(n + 1)$, and then proceed to use $\ell = i^- - j$ and $u = i^+ + j$ for $j = 1, 2, \ldots$, incrementing j until the constraint in Equation (5.1) is met.

The preceding interval is conservative because the actual confidence level, given by the left-hand side of Equation (5.1), is greater than the specified value $1 - \alpha$. The actual confidence levels are also given in Tables A.15a to A.15g or can be calculated from $\text{EPYP}(n, \ell, u, p) = 1 - \alpha$.

If the underlying (unspecified) distribution can be assumed to be symmetric, somewhat shorter distribution-free confidence intervals for the median (i.e., $Y_{0.50}$) can be obtained; see Gibbons (1975).

As indicated in Section 5.1, with a limited sample size, it is sometimes impossible to construct a distribution-free statistical interval that has at least the desired confidence level. This problem is particularly acute when estimating percentiles in the tail of a distribution from a small sample. Situations where a desired confidence level cannot be met are indicated in Tables A.15a to A.15g by a *. In some cases, the analyst can cope with this problem by choosing ℓ and u nonsymmetrically. Another alternative may be to use a reduced confidence level.

5.2.2 Example

Assume that a distribution-free 95% confidence interval is needed for $Y_{0.50}$, the median of the population of batches described in Section 5.1.3. For

$n = 100$ and $1 - \alpha = 0.95$, Table A.15g gives $\ell = 41$ and $u = 61$. Thus, a distribution-free conservative 95% confidence interval for $Y_{0.50}$ is the interval enclosed by the values of the 41st and the 61st ordered observations:

$$\left[Y_{0.50}, \tilde{Y}_{0.50} \right] = [x_{(41)}, x_{(61)}] = [7.81, 12.93].$$

The actual confidence level, is also given in Table A.15g, or as EPYP(100, 41, 61, 0.50) = $B(60; 100, 0.50) - B(40; 100, 0.50) = 0.9540$, i.e., 95.40% confidence, instead of the desired 95%.

Confidence intervals for the other percentiles are found similarly. For example, for the 10th percentile with $n = 100$ and $1 - \alpha = 0.95$, Table A.15c gives $\ell = 4$ and $u = 16$. Thus,

$$\left[Y_{0.10}, \tilde{Y}_{0.10} \right] = [x_{(4)}, x_{(16)}] = [2.24, 4.04]$$

is a two-sided conservative distribution-free 95% confidence interval for $Y_{0.10}$. The actual confidence level is shown in Table A.15c to be 0.9523, or calculated as EPYP(100, 4, 16, 0.10) = $B(15; 100, 0.10) - B(3; 100, 0.10) = 0.9523$.

To find a distribution-free 95% confidence interval for $Y_{0.90}$, the 90th percentile, we again use Table A.15c with $n = 100$ and $1 - p = 1 - 0.90 = 0.10$. We again obtain $\ell = 4$ and $u = 16$, giving $n - u + 1 = 100 - 16 + 1 = 85$ and $n - \ell + 1 = 100 - 4 + 1 = 97$. Thus

$$\left[Y_{0.90}, \tilde{Y}_{0.90} \right] = [x_{(85)}, x_{(97)}] = [24.33, 33.24]$$

is a two-sided conservative distribution-free 95% confidence interval for $Y_{0.90}$. The actual confidence level is EPYP(100, 85, 97, 0.90) = $B(96; 100, 0.90) - B(84; 100, 0.90) = 0.9523$.

Note from Table A.15c that for $n = 50$ or less, a two-sided distribution-free 99% confidence interval for $Y_{0.10}$ cannot be obtained with symmetrically chosen order statistics. Then, if one wishes to choose ℓ and u symmetrically about $p(n + 1)$, one must settle for the highest achievable confidence level from the 1st and 11th order statistics. Thus, for $n = 50$, the highest achievable confidence level for a near-symmetric interval is 98.55%, corresponding to $\ell = 1$ and $u = 11$. Alternately, one may lengthen the interval on the upper side to use the 1st and 12th ordered observations. This would give a nonsymmetric distribution-free confidence interval of [1.49, 3.29] with a confidence level of 99.16% [i.e., EPYP(50, 1, 12, 0.1) = $B(11; 50, 0.1) - B(0; 50, 0.1) = 0.9916$].

5.2.3 One-Sided Distribution-Free Confidence Bounds for a Percentile

One can use Table A.16 to determine the order statistic to provide a one-sided distribution-free conservative *upper* $100(1 - \alpha)\%$ confidence bound for the $100p$th percentile of the sampled population for $p = 0.75, 0.90, 0.95$, and 0.99, and $(1 - \alpha) = 0.90, 0.95$, and 0.99. Alternately, one can use Figures 5.3a to 5.3d to determine such an order statistic for values of p ranging from

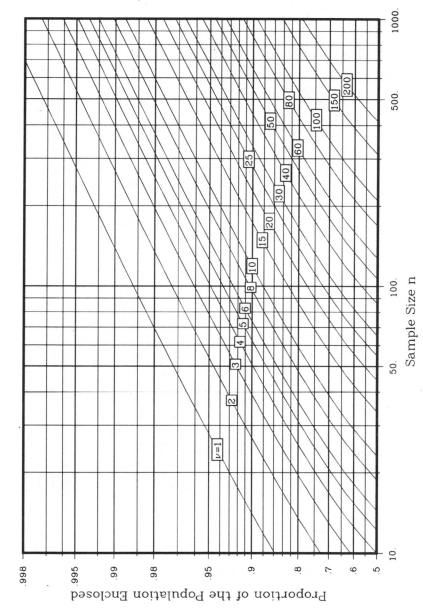

Figure 5.3a Proportion of the population enclosed by a two-sided distribution-free tolerance interval or a one-sided distribution-free tolerance bound with 90% confidence when ν extreme observations are excluded from the ends of a sample of size n. A similar figure first appeared in Murphy (1948). Adapted with permission of the Institute of Mathematical Statistics. Also see Table A.16.

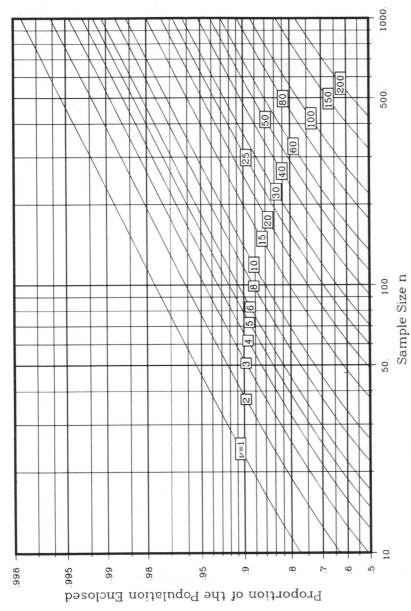

Figure 5.3b Proportion of the population enclosed by a two-sided distribution-free tolerance interval or a one-sided distribution-free tolerance bound with 90% confidence when ν extreme observations are excluded from the ends of a sample of size n. A similar figure first appeared in Murphy (1948). Adapted with permission of the Institute of Mathematical Statistics. Also see Table A.16.

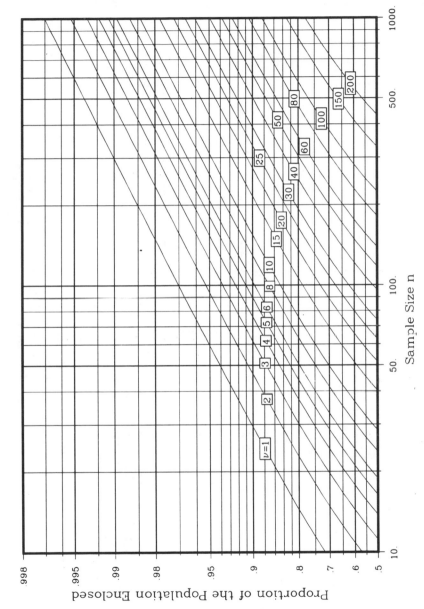

Figure 5.3c Proportion of the population enclosed by a two-sided distribution-free tolerance interval or a one-sided distribution-free tolerance bound with 95% confidence when ν extreme observations are excluded from the ends of a sample of size n. A similar figure first appeared in Murphy (1948). Adapted with permission of the Institute of Mathematical Statistics. Also see Table A.16.

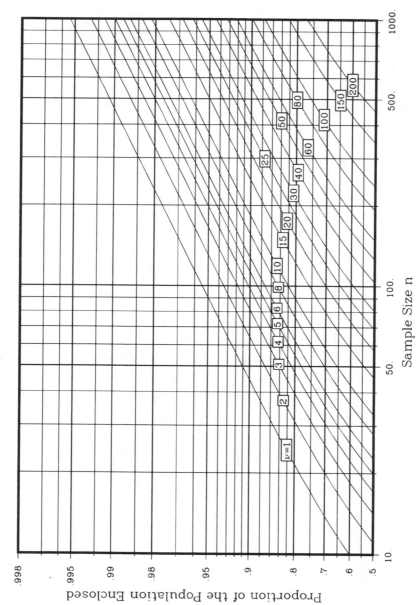

Figure 5.3d Proportion of the population enclosed by a two-sided distribution-free tolerance interval or a one-sided distribution-free tolerance bound with 99% confidence when ν extreme observations are excluded from the ends of a sample of size n. A similar figure first appeared in Murphy (1948). Adapted with permission of the Institute of Mathematical Statistics. Also see Table A.16.

0.5 to 0.998 for $(1 - \alpha) = 0.75, 0.90, 0.95,$ and 0.99. For *lower* confidence bounds on the $100p$th percentile one can use the same table for $p = 0.25, 0.10, 0.05, 0.01$, and the same figures for values of p from 0.002 to 0.5, in both cases using $1 - p$ in place of p. The table is more precise, but the figures give more insight. This table and these figures provide ν for values of n ranging from 10 to 1000. The integer ν is the number of extreme observations to be removed from the lower (upper) end of the sample of size n to obtain the order statistic that provides the desired lower (upper) one-sided confidence bound for the $100p$th percentile. For all entries in Table A.16, an actual confidence level is shown. When the actual confidence level is less than the desired confidence level, it is marked with a *. In using Figures 5.3a to 5.3d, one selects the value of ν corresponding to the next highest integer (interpolating, if necessary, for the given values of $n, p,$ and α). The desired lower confidence bound is $x_{(\nu)}$. The desired upper confidence bound is $x_{(n-\nu+1)}$.

To find a one-sided distribution-free conservative *lower* $100(1 - \alpha)\%$ confidence bound for situations that are not covered in the tabulations, ℓ is chosen as the largest integer such that

$$\text{EPYP}(n, \ell, n + 1, p) = 1 - B(\ell - 1; n, p) \geq 1 - \alpha,$$

$$0 \leq \ell < n + 1, \quad 0 < p < 1.$$

Again, the actual confidence level is given by the left-hand side of this inequality, which can be evaluated with the subroutine EPYP listed in Appendix C.

Similarly, a one-sided distribution-free conservative *upper* $100(1 - \alpha)\%$ confidence bound for the $100p$th percentile of the sampled population is obtained as $\tilde{Y}_p = x_{(u)}$, where u is chosen as the smallest integer such that

$$\text{EPYP}(n, u, n + 1, p) = B(u - 1; n, p) \geq 1 - \alpha,$$

$$0 < u \leq n + 1, \quad 0 < p < 1.$$

The actual confidence level is again given by the left-hand side of this inequality.

Sommerville (1958) gives tabulations that are similar to Table A.16, but without the actual confidence levels. Murphy (1948) first produced charts like those in Figures 5.3a to 5.3d.

5.2.4 Example

Assume that we need a one-sided distribution-free *lower* 95% confidence bound for the 10th percentile of the distribution for the example in Section 5.1. Entering Table A.16 or Figure 5.3c (and reading up to the next highest curve) with $n = 100$, $1 - p = 0.9$, and $1 - \alpha = 0.95$ gives $\ell = \nu = 5$. Thus, $\underset{\sim}{Y}_{0.10} = x_{(5)} = 2.29$ provides the desired lower confidence bound.

Similarly, an *upper* 95% confidence bound for the 90th percentile of the population is obtained by using the same entries in Table A.16 (i.e., $p = 0.9$ and $n = 100$), giving $\nu = 5$ and $u = 100 - \nu + 1 = 96$. Thus, $\tilde{Y}_{0.90} = x_{(96)} = 31.83$ provides the desired upper confidence bound. The actual confidence level for these bounds is seen from Table A.16 to be 97.63% [i.e., $EPYP(100, 5, 101, 0.1) = 1 - B(4; 100, 0.10) = B(95; 100, 0.90) = 0.9763$].

5.3 DISTRIBUTION-FREE TOLERANCE INTERVALS AND BOUNDS TO CONTAIN A SPECIFIED PERCENTAGE OF A POPULATION

5.3.1 Two-Sided Distribution-Free Tolerance Intervals

A two-sided distribution-free conservative $100(1 - \alpha)\%$ tolerance interval to contain (or cover) at least $100p\%$ of the sampled population from a sample of size n is $[\underline{T}_p, \tilde{T}_p] = [x_{(\ell)}, x_{(u)}]$. To determine the order statistics that will provide the desired confidence level, one generally chooses ℓ and u symmetrically or nearly symmetrically around the integers that are closest to $0.5(n + 1)$. Either Table A.16 or Figures 5.3a through 5.3d can be used to find appropriate values of ℓ and u. Specifically, from either the table or the figures, choose the value of ν that gives the desired level of confidence for enclosing $100p\%$ of the population (reading up to the next highest integer when using the figures). The integer ν is the total number of observations (or "blocks") to be removed from the extremes of the sample to obtain the order statistics that define the desired tolerance interval. To find the number of observations to be removed from *each* end of the sample, prior to looking at the data, we need to divide ν into two parts as follows:

- Let $\nu_1 = \nu_2 = \nu/2$ if ν is even.
- Let $\nu_1 = \nu/2 - 1/2$ and $\nu_2 = \nu_1 + 1$ or $\nu_1 = \nu/2 + 1/2$ and $\nu_2 = \nu_1 - 1$ if ν is odd (either will give a tolerance interval with the same level of confidence). Then use $\ell = \nu_1$ and $u = n - \nu_2 + 1$. The desired tolerance interval is then formed by the values of $x_{(\ell)}$ and $x_{(u)}$.

For situations that are not covered by the tabulations, ℓ and u are chosen symmetrically (or nearly symmetrically) and as close together as possible around the largest integer that is less than or equal to $(n + 1)/2$ such that

$$EPTI(n, \ell, u, p) = B(u - \ell - 1; n, p) \geq 1 - \alpha,$$

$$0 \leq \ell < u \leq n + 1, \quad 0 < p < 1. \quad (5.2)$$

The actual confidence level for the tolerance interval is given by the left-hand side of this inequality. It can be evaluated with the subroutine EPTI listed in Appendix C.

Table A.17 gives the sample size n needed to provide $100(1 - \alpha)\%$ confidence that the tolerance interval defined by the extreme values of the sample, i.e., $[x_{(1)}, x_{(n)}]$, will contain $100p\%$ of the sampled population for $(1 - \alpha) = 0.50, 0.75, 0.90, 0.95, 0.98, 0.99, 0.999,$ and $p = 0.50(0.05)0.95(0.01)0.99, 0.995, 0.999$. This table was obtained by substituting $u = n$ and $\ell = 1$ into Equation (5.2), yielding the expression

$$1 - \alpha = 1 - np^{n-1} + (n - 1)p^n, \tag{5.3}$$

which is solved for n using standard numerical techniques.

5.3.2 Example

The manufacturer needs a two-sided tolerance interval that will, with 95% confidence, contain the compound amounts for at least 90% of the batches from the sampled production process, based on the sample of 100 batches. From Table A.16, we find $v = 5$; we choose $v_1 = 2$ and $v_2 = 3$, yielding $\ell = 2$ and $u = 98$. ($v_1 = 3$ and $v_2 = 2$ might have been chosen, instead, yielding $\ell = 3$ and $u = 99$.) Thus,

$$\left[\underaccent{\tilde}{T}_{0.90}, \tilde{T}_{0.90} \right] = [x_{(2)}, x_{(98)}] = [1.66, 37.32]$$

is a two-sided tolerance interval to contain at least 90% of the population values with (at least) 95% confidence. The actual confidence level for this interval is $\text{EPTI}(100, 2, 98, 0.90) = B(98 - 2 - 1; 100, 0.90) = 0.9763$.

Note from Table A.16 that a sample of size $n = 100$ is not sufficient to obtain a two-sided distribution-free 95% tolerance interval to contain 99% of a population. In fact, using the extreme values of a given sample gives only a confidence level of 26.42% for containing 99% of the sampled population [i.e., $\text{EPTI}(100, 1, 100, 0.99) = B(100 - 1 - 1; 100, 0.99) = 0.2643$]. To obtain a 95% confidence interval to contain at least 99% of the population, we see from Table A.17 that a minimum of $n = 473$ observations are needed. The actual confidence level is $\text{EPTI}(473, 1, 473, 0.99) = B(473 - 1 - 1; 473, 0.99) = 0.9502$.

5.3.3 One-Sided Distribution-Free Tolerance Bounds

A one-sided distribution-free tolerance bound is equivalent to a one-sided distribution-free confidence bound for a percentile of that population. That is, a one-sided distribution-free lower (upper) $100(1 - \alpha)\%$ tolerance bound that will be exceeded by (that will exceed) at least $100p\%$ of the population is the same as a distribution-free lower (upper) $100(1 - \alpha)\%$ confidence bound for the $100p$th percentile of the population. Such bounds and examples were given in Section 5.2.3.

Table A.18 gives the smallest sample size needed to have $100(1 - \alpha)\%$ confidence that the largest (smallest) observation in the sample will exceed (be exceeded by) at least $100p\%$ of the population for $(1 - \alpha) =$ 0.50, 0.75, 0.90, 0.95, 0.98, 0.99, 0.999,   and $p = 0.50(0.05)0.95(0.01)0.99$, 0.995, 0.999. This table was obtained by substituting $\ell = 0$ and $u = n$ (or equivalently $\ell = 1$ and $u = n + 1$) into Equation (5.2), yielding the expression

$$1 - \alpha = 1 - p^n. \tag{5.4}$$

Solving for n gives $n = \log(\alpha)/\log(p)$.

5.3.4   Example

A test is planned to estimate the life of a newly designed engine bearing. The results are to provide a lower 95% tolerance bound to be exceeded by the lifetimes of 99% of the population of bearings. It is desired to determine the minimum number of bearings that will have to be tested to obtain the desired bound if no distributional assumptions are made. This would require using the first bearing failure time as the desired bound. If the bearings are placed on test simultaneously, the test can be terminated after the first failure. From a practical viewpoint, the usefulness of the results will depend on the actual value of the first failure time. From Table A.18, we note that 299 bearings are needed, or $n = \log(0.05)/\log(0.99) = 298.07$.

5.4   DISTRIBUTION-FREE PREDICTION INTERVALS TO CONTAIN AT LEAST k OF m FUTURE OBSERVATIONS

5.4.1   Two-Sided Distribution-Free Prediction Intervals

A two-sided conservative distribution-free $100(1 - \alpha)\%$ prediction interval to contain at least k of m future observations from a previously sampled population is obtained as

$$\left[\underline{y}_{k;m}, \tilde{y}_{k;m}\right] = \left[x_{(\ell)}, x_{(u)}\right],$$

where ℓ and u can be obtained by using Tables A.19a to A.19c, which was adapted from Danziger and Davis (1964). Using Tables A.19a to A.19c, with given n, m, and the desired confidence level $1 - \alpha$, one finds the largest value of ν that provides the desired value of k in the body of the table. Then ℓ and u are chosen symmetrically in the manner described for the two-sided nonparametric tolerance intervals in Section 5.3.1., and illustrated by the example in Section 5.3.2. Tables A.19a to A.19c contain entries for

$(1 - \alpha) = 0.50, 0.75, 0.90, 0.95, 0.99, \quad n = 5, 10, 25, 50, 75, 100, \quad m = 5, 10, 25, 50, 75, 100, \infty$, and $\nu = 1(1)12$. Also, Hall, Prairie, and Motlagh (1975) provide tables for the confidence level associated with one-sided and two-sided distribution-free prediction intervals for the special case where the extreme observation(s) from the given sample are used as the endpoints of the interval.

For situations that are not covered by the available tabulations, ℓ and u are chosen symmetrically (or nearly symmetrically) from the extremes of the sample and as close together as possible such that

$$\text{EPKM}(n, \ell, u, m, k) = H^*(k, n, u - \ell, n, n + m) \geq 1 - \alpha,$$
$$n > 0, \quad 0 \leq \ell < u \leq n + 1, \quad m > 0, \quad 0 \leq k \leq m; \quad (5.5)$$

the actual confidence level of the prediction interval is given by the left-hand side of this inequality and can be computed with subroutine EPKM listed in Appendix C.

For the special case where the prediction interval is formed by the extreme values of the given sample (i.e., $\ell = 1$ and $u = n$), the confidence level associated with the inclusion of all $k = m$ future observations is

$$1 - \alpha = \frac{n(n - 1)}{(n + m)(n + m - 1)}. \quad (5.6)$$

Tables A.20a through A.20c give the necessary sample size n so that a two-sided distribution-free prediction interval $[x_{(1)}, x_{(n)}]$ will enclose

- all m,
- at least $m - 1$ of m, and
- at least $m - 2$ of m

observations in a future sample of size m from a previously sampled population with $100(1 - \alpha)\%$ confidence for $(1 - \alpha) = 0.50, 0.75, 0.90, 0.95, 0.98, 0.99, 0.999$, and $m = 1(1)25(5)50(10)100$.

5.4.2 Example

Based on the previous sample of $n = 100$ batches, the manufacturer wants a distribution-free 90% prediction interval to contain all the observations from a future sample of five batches from the same population. We search Table A.19a with $n = 100$, $m = 5$, and $1 - \alpha = 90$ to find the largest ν giving $k = 5$ in the body of the table. In this case, we obtain $\nu = 2$. Proceeding as in Section 5.3.1, we divide ν into $\nu_1 = 1$ and $\nu_2 = 1$ to obtain $\ell = \nu_1 = 1$ and $u = n - \nu_2 + 1 = 100$. Thus the desired 90% two-sided prediction interval to contain all $k = 5$ of m future observations is the range formed by the smallest and largest observations; i.e.,

$$\left[\underline{y}_{5;5}, \tilde{y}_{5;5}\right] = \left[x_{(1)}, x_{(100)}\right] = [1.49, 58.11].$$

Because this interval uses the two extreme observations in the previous sample as the endpoints of the prediction interval, we note from Equation (5.6) that the actual confidence level is $100(100 - 1)/[(100 + 5)(100 + 5 - 1)] = 0.9066$. This number can also be read from Table 4 of Hall, Prairie, and Motlagh (1975).

Similarly, suppose we require a 95% prediction interval to contain at least $k = 4$ of $m = 5$ future observations. We enter Table A.19 with $n = 100$, $m = 5$, and $1 - \alpha = 0.95$ and find the largest ν such that $k = 4$. This is found under $\nu = 7$. Breaking down $\nu = 7$ into $\nu_1 = 3$ and $\nu_2 = 4$ gives $\ell = \nu_1 = 3$ and $u = n - \nu_2 + 1 = 97$, following the procedure in Section 5.3.1. Thus, the desired 95% prediction interval to contain at least four of five future observations is

$$\left[\underline{y}_{4;5}, \tilde{y}_{4;5}\right] = \left[x_{(3)}, x_{(97)}\right] = [2.05, 33.24].$$

The actual confidence level for this interval, computed using the subroutine EPKM, is 0.9545. If we had broken down $\nu = 7$ into $\nu_1 = 4$ and $\nu_2 = 3$ instead, we would have obtained $\ell = 4$ and $u = 98$, and

$$\left[\underline{y}_{4;5}, \tilde{y}_{4;5}\right] = \left[x_{(4)}, x_{(98)}\right] = [2.24, 37.32].$$

This provides the same level of confidence as before (i.e., 0.9545) for containing at least four of the five future observations.

In addition, Table 5.2 gives, for various symmetric ℓ and u, the confidence level for a two-sided prediction interval, based on a previous sample of size

Table 5.2   Confidence Levels for Various Symmetric Two-Sided Distribution-Free Prediction Intervals[†] To Contain at Least k of $m = 5$ Future Observations, Based upon $n = 100$ Past Observations

ℓ	u	k:	1	2	3	4	5
1	100		1.0000 −	1.0000 −	0.9998	0.9946	0.9066
2	99		1.0000 −	1.0000 −	0.9990	0.9827	0.8203
3	98		1.0000 −	0.9999	0.9972	0.9652	0.7407
4	97		1.0000 −	0.9997	0.9942	0.9426	0.6674
5	96		1.0000 −	0.9993	0.9898	0.9159	0.6000
6	95		1.0000 −	0.9987	0.9836	0.8855	0.5382
7	94		0.9999	0.9978	0.9756	0.8521	0.4816
8	93		0.9998	0.9964	0.9655	0.8163	0.4299
9	92		0.9997	0.9946	0.9534	0.7785	0.3827
10	91		0.9996	0.9921	0.9392	0.7393	0.3397

[†]Prediction interval is $[x_{(\ell)}, x_{(u)}]$; 1.0000 − indicates a number slightly less than 1.0.

$n = 100$, to contain at least a specified number $k = 1, 2, \ldots, 5$ of the $m = 5$ future observations. This table was obtained by using Equation (5.5) evaluated with the subroutine EPKM. For example,

$$\left[\underline{y}_{4;5}, \bar{y}_{4;5} \right] = \left[x_{(1)}, x_{(100)} \right] = [1.49, 58.11]$$

is an interval which, we can assert with 99.46% confidence, will contain at least four out of five future observations.

5.4.3   One-Sided Distribution-Free Prediction Bounds

A one-sided lower (upper) conservative distribution-free $100(1 - \alpha)\%$ prediction bound to be exceeded by (to exceed) at least k of m future observations from a previously sampled population is $\underline{y}_{k;m} = x_{(\ell)}$ $(\bar{y}_{k;m} = x_{(u)})$, where one can obtain ℓ and u from Tables A.19a to A.19c in the following manner. For given n, m, and $1 - \alpha$, one finds the largest ν that provides the desired value of k in the body of the table. Then $\ell = \nu$ (or $u = n - \nu + 1$). Alternatively, using Equation (5.5) with u replaced by $n + 1$ (with ℓ replaced by 0), one can obtain ℓ as the largest (u as the smallest) integer such that the inequality holds. Again, the actual confidence level is given by the left-hand side of this inequality.

Also, a one-sided prediction bound that uses the smallest (or largest) of n observations as the lower (or upper) limit, to enclose all of m future observations has associated confidence level

$$1 - \alpha = \frac{n}{n + m}. \tag{5.7}$$

Tables A.21a through A.21c give the necessary sample size so that a one-sided lower (upper) distribution-free prediction bound defined by the smallest (largest) observation from the previous sample will be exceeded by (will exceed)

- all m,
- at least $m - 1$ of m, and
- at least $m - 2$ of m

observations in a future sample of size m from the previously sampled population using a $100(1 - \alpha)\%$ confidence level for $1 - \alpha = 0.50, 0.75, 0.90, 0.95, 0.98, 0.99, 0.999$, and $m = 1(1)25(5)50(10)100$.

5.4.4   Example

Consider again the example in Section 5.4.2, but now assume that a one-sided distribution-free lower 99% prediction bound is desired. This bound is to be .

exceeded by the observed values for at least four of five future batches, using a 99% confidence level. We enter Tables A.19a to A.19c, with $n = 100$, $m = 5$, $1 - \alpha = 0.99$, and find the largest ν for which k is at least 4. The largest such value is $\nu = 2$. Thus, taking $\ell = \nu = 2$,

$$\underset{\sim}{y}_{4;5} = x_{(2)} = 1.66$$

is the desired bound. The actual confidence level for this interval (i.e., 1.66 to ∞), computed with subroutine EPKM, is 0.9946.

An upper 99% prediction bound is found in a similar manner. In this case, one uses $u = n - \nu + 1 = 99$, and obtains

$$\tilde{y}_{4;5} = x_{(99)} = 53.43$$

as an upper prediction bound that one can claim with approximately 99% confidence, will exceed at least four of five future observations randomly sampled from the same population.

Table 5.3 gives values of $1 - \alpha$ [again computed from Equation (5.5) with subroutine EPKM] for one-sided distribution-free lower (and upper) prediction bounds to be exceeded by (to exceed) at least k of $m = 5$ future observations for $k = 1, 2, \ldots, 5$. For example, for $k = 3$, the largest ℓ giving $1 - \alpha > 0.99$ is $\ell = 9$ and thus, $\tilde{y}_{3;5} = x_{(9)} = 3.10$ is a lower prediction bound that one can claim with 99.22% confidence will be exceeded by at least three of the five future observations. Similarly, using $u = 92$, one can claim, also with 99.22% confidence, that the upper prediction bound, $\tilde{y}_{3;5} = \tilde{x}_{(92)} = 28.28$, will exceed at least three of the next five observations.

Table 5.3 Confidence Levels for Various One-Sided Distribution-Free Prediction Bounds[†] to Contain at Least k of $m = 5$ Future Observations, Based upon $n = 100$ Past Observations

ℓ	u	k: 1	2	3	4	5
1	100	1.0000 −	1.0000 −	0.9999	0.9982	0.9524
2	99	1.0000 −	1.0000 −	0.9998	0.9946	0.9066
3	98	1.0000 −	1.0000 −	0.9995	0.9894	0.8626
4	97	1.0000 −	1.0000 −	0.9990	0.9827	0.8203
5	96	1.0000 −	0.9999	0.9982	0.9746	0.7797
6	95	1.0000 −	0.9999	0.9972	0.9652	0.7407
7	94	1.0000 −	0.9998	0.9959	0.9545	0.7033
8	93	1.0000 −	0.9997	0.9942	0.9426	0.6674
9	92	1.0000 −	0.9995	0.9922	0.9297	0.6330
10	91	1.0000 −	0.9993	0.9898	0.9159	0.6000

[†]Prediction bound is $x_{(\ell)}$ or $x_{(u)}$; 1.0000 − indicates a number slightly less than 1.0.

5.5 PREDICTION INTERVALS TO CONTAIN A SPECIFIED ORDERED OBSERVATION IN A FUTURE SAMPLE

The intervals in Section 5.4 are constructed to contain all or a specified number of observations from a future sample. In contrast, one might be interested in a distribution-free interval to contain a specified ordered observation, such as the median, in the future sample. For theory, tabulations, and other details for such prediction intervals, see Fligner and Wolfe (1976, 1979a, 1979b). Also, Guilbaud (1983) gives more general theory for the particular problem of setting prediction intervals for sample medians.

5.5.1 Two-Sided Distribution-Free Prediction Interval

A two-sided conservative distribution-free $100(1 - \alpha)\%$ prediction interval to contain the jth largest observation in a future sample of size m from a previously sampled population is obtained as $[x_{(\ell)}, x_{(u)}]$, where ℓ and u are generally chosen symmetrically or nearly symmetrically about the $100[j/(m + 1)]$th sample percentile from the previous sample of size n, and as close together as possible such that

$$\text{EPYJ}(n, \ell, u, m, j) = H^*(\ell, u - 1, j, n, n + m) \geq 1 - \alpha,$$

$$n > 0, \quad 1 \leq \ell < u \leq n, \quad 1 \leq j \leq m. \quad (5.8)$$

The left-hand side of this inequality gives the actual confidence level for the prediction interval and can be evaluated using subroutine EPYJ listed in Appendix C.

5.5.2 One-Sided Distribution-Free Prediction Bounds

A one-sided lower prediction bound to be exceeded by the jth largest of m future observations is the same as the one-sided lower prediction bound to be exceeded by j of m future observations, described in Section 5.4.3. Similarly, a one-sided upper prediction bound to exceed the jth largest of m future observations is the same as the one-sided upper prediction bound to exceed j of m future observations.

5.5.3 Example

The manufacturer desires a distribution-free prediction interval to contain the median $y_{(j)}$ of a future sample of m randomly selected batches, based upon the previous sample of 100 batches, from the production process. Table 5.4 gives the confidence level for the two-sided distribution-free prediction

Table 5.4   Confidence Levels for Various Symmetric Two-Sided Distribution-Free Prediction Intervals† for the Median of a Future Sample of Size m, Based upon $n = 100$ Past Observations

ℓ	u	m:	5	9	19	39	59
41	60		0.3382	0.4268	0.5598	0.6903	0.7561
40	61		0.3719	0.4673	0.6072	0.7392	0.8029
39	62		0.4051	0.5065	0.6516	0.7827	0.8429
38	63		0.4377	0.5443	0.6930	0.8209	0.8766
37	64		0.4696	0.5808	0.7312	0.8540	0.9044
36	65		0.5008	0.6157	0.7662	0.8824	0.9271
35	66		0.5313	0.6490	0.7981	0.9064	0.9453
35	67		0.5462	0.6649	0.8126	0.9165	0.9525
34	67		0.5611	0.6808	0.8270	0.9265	0.9597
33	68		0.5900	0.7109	0.8528	0.9430	0.9707
32	69		0.6180	0.7393	0.8759	0.9564	0.9791

†Prediction interval is $[x_{(\ell)}, x_{(u)}]$.

interval defined by

$$\left[\hat{Y}_{0.50}, \tilde{\tilde{Y}}_{0.50} \right] = \left[\underset{\sim}{y}_{(j)}, \tilde{y}_{(j)} \right] = [x_{(\ell)}, x_{(u)}]$$

for several values of ℓ and u and for $m = 5, 9, 19, 39,$ and 59. This table was obtained by using Equation (5.8) with $n = 100$ and $j = (m + 1)/2$ using subroutine EPYJ. For example,

$$\left[\hat{Y}_{0.50}, \tilde{\tilde{Y}}_{0.50} \right] = \left[\underset{\sim}{y}_{(20)}, \tilde{y}_{(20)} \right] = [x_{(32)}, x_{(69)}] = [6.53, 14.96]$$

is a 95.64% prediction interval for $y_{(20)}$ the median of a future sample of size $m = 39$. Also,

$$\left[\hat{Y}_{0.50}, \tilde{\tilde{Y}}_{0.50} \right] = \left[\underset{\sim}{y}_{(30)}, \tilde{y}_{(30)} \right] = [x_{(35)}, x_{(67)}] = [6.83, 14.17]$$

is a 95.25% prediction interval for $y_{(30)}$, the median of a future sample of size $m = 59$.

Table 5.5 contains similar confidence levels for one-sided lower and upper prediction bounds for the same sample sizes and values of ℓ and u. The confidence levels were computed with EPYJ with Equation (5.8) using $u = n + 1$ and $j = (m + 1)/2$. For example,

$$\hat{Y}_{0.50} = \underset{\sim}{y}_{(30)} = x_{(37)} = 7.28$$

is a lower prediction bound that one can claim, with 95.22% confidence, will

Table 5.5 Confidence Levels for Various One-Sided Distribution-Free Prediction Bounds[†] for the Median of a Future Sample of Size m, Based upon $n = 100$ Past Observations

ℓ	u	m: 5	9	19	39	59
41	60	0.6691	0.7134	0.7799	0.8452	0.8781
40	61	0.6860	0.7337	0.8036	0.8696	0.9015
39	62	0.7026	0.7533	0.8258	0.8913	0.9214
38	63	0.7188	0.7722	0.8465	0.9104	0.9383
37	64	0.7348	0.7904	0.8656	0.9270	0.9522
36	65	0.7504	0.8078	0.8831	0.9412	0.9636
35	66	0.7657	0.8245	0.8991	0.9532	0.9727
34	67	0.7805	0.8404	0.9135	0.9632	0.9798
33	68	0.7950	0.8554	0.9264	0.9715	0.9854
32	69	0.8090	0.8697	0.9379	0.9782	0.9896

[†]Prediction bounds are $x_{(\ell)}$ or $x_{(u)}$.

be exceeded by the median of a future sample of size 59. Similarly,

$$\tilde{\hat{Y}}_{0.50} = \tilde{y}_{(30)} = x_{(64)} = 13.43$$

is an upper prediction bound that one can claim with 95.22% confidence will exceed the median from the sample of size 59.

CHAPTER 6

Statistical Intervals
for Proportions and Percentages
(Binomial Distribution)

6.1 INTRODUCTION

This chapter describes statistical intervals for proportions and percentages. Such intervals are used, for example, when each observation is either a "conforming" or a "nonconforming" unit and the data consist of the number (or equivalently, the proportion or percentage) of nonconforming units, in a random sample of n units from a population (or process). Two examples are:

- An integrated circuit passes an operational test only if it successfully completes a specified set of operations after a 48-hour "burn-in" at 85 °C and 85% relative humidity. Thus, the given data consist of the proportion of the n units that failed (or passed) the test. The goal is to estimate the proportion of potentially failing (or passing) units in the sampled process.
- Federal regulations require that the level of a certain pollutant in the exhaust from an internal combustion engine be less than 10 ppm (parts per million). If an engine fails to meet this standard, it must undergo expensive rework. Management wishes to estimate the proportion of units from a specified process that will require such rework. The available data consist of the number of units that needed rework in a random sample of n engines from the process.

Our discussion will be mainly in terms of "conforming" and "nonconforming" units to suggest the most common quality control application. The applicability of the intervals is, however, much more general. For example, a conforming unit could be an individual indicating a preference for a particular candidate in a forthcoming election, the survival of an animal in a biological experiment, etc.

100

Problems involving the proportion of conforming (or nonconforming) units in a random sample of size n from a large population can be modeled with the binomial distribution. In particular, the probability of observing x' nonconforming units in a random sample of size n, assuming a proportion p of nonconforming units in the population, is given by the binomial probability function

$$\Pr(x = x') = f(x'; n, p) = \frac{n!}{x'!(n - x')!} p^{x'}(1 - p)^{n - x'}, \quad x' = 0, 1, \ldots, n.$$

For example, if the proportion of nonconforming units in a large population is known to be $p = 0.10$, the probability of obtaining 0 nonconforming units in a random sample of size $n = 5$ is

$$\Pr(x = 0) = f(0; 5, 0.10) = \frac{5!}{0!(5 - 0)!}(0.10)^0(1 - 0.10)^{5 - 0} = 0.5905.$$

The probability of exactly 1 nonconforming unit in the sample is

$$\Pr(x = 1) = f(1; 5, 0.10) = \frac{5!}{1!(5 - 1)!}(0.10)^1(1 - 0.10)^{5 - 1} = 0.3280.$$

Dixon and Massey (1969), Duncan (1974), Grant and Leavenworth (1988), Hahn and Shapiro (1967), Natrella (1963), and numerous other books provide further discussion of the binomial distribution.

For the binomial distribution problems considered here, the parameter p is unknown. Instead, all that is known is that a past random sample of size n resulted in x nonconforming units. We then seek a confidence interval for p.

The binomial is a discrete distribution. A binomial random variable can take on only the integer values $x = 0, 1, 2, \ldots, n$, but cannot take on values between these integers. Because of this, statistical intervals do not generally have exactly the desired confidence level. Thus, the statistical intervals given here are approximate and, like those in the Chapter 5, tend to be conservative (i.e., the actual confidence level is larger than the nominal value).

6.1.1 Overview

In this chapter we will show how to obtain the following:

- Confidence intervals for p, the (true) proportion nonconforming in the sampled population (or process).
- Confidence intervals for the probability that the number of nonconforming units in a future sample of size m will be less than or equal to (or greater than) a specified number.

- One-sided tolerance bounds for the distribution of the number of nonconforming units in future samples of m units.
- Prediction intervals for the number of nonconforming units in a future sample of m units.

6.1.2   Notation and Tabulations

In this chapter we use the normal and F distributions; these were discussed in Section 4.1.2. We also use the cumulative binomial and hypergeometric distributions; see Section 5.1.2 for references that give tabulations, approximations, and further information on these distributions.

6.1.3   Examples

For the first example, assume that a random sample of $n = 1000$ integrated circuits has been selected from production and that $x = 20$ of these units fail the test. From the data, an estimate for p, the proportion of failed units in the population, is $\hat{p} = x/n = 20/1000 = 0.02$. The resulting data can be used to make inferences or predictions about the sampled production process, assuming that it is in statistical control.

For the second example, the level of a pollutant was measured for each engine in a random sample of $n = 10$ engines. If we are willing to assume that the pollution measurements are adequately described by a normal distribution, the desired statistical intervals can be obtained by using the procedures in Chapter 4, based upon the sample mean and sample standard deviation of the pollutant measurements (in ppm). These were $\bar{x} = 8.05$ ppm and $s = 1.09$ ppm. The assumption of a normal distribution, however, might be questionable, and one might wish to make inferences concerning the population proportion outside specification limits without making such an assumption. In this case, one simply counts the number of observations outside the specified limits, defines this as the number of "nonconforming units," and then uses the techniques described in this chapter to construct the desired statistical intervals. In the example, a pollutant level above 10 ppm was deemed to be nonconforming, and one of the 10 engine exhaust measurements exceeded this value. Thus, there was $x = 1$ nonconforming unit in a sample of size $n = 10$. (The intervals apply even if the number of nonconforming units is $x = 0$.) Such dichotomizing of measurement data might be especially useful when there are multiple (possibly correlated) measurements, each of which must meet some stated specification limit. In that case, a nonconforming unit would simply be one that fails to meet the specification limit for at least one of the measurements.

As discussed in the previous chapter, the price that one has to pay for abandoning the assumption of a specific distribution, such as the normal distribution, is that some information is sacrificed by not using the exact measurements; thus the resulting intervals will generally be longer. As a result, in the second example, a confidence interval to contain the true proportion of nonconforming engines will tend to be longer if the interval is based only on a count of the number of such engines in the sample, rather than on the actual pollution level measurements, assuming a particular distribution. Such loss of precision, in some situations, is a serious concern, especially in evaluating a high reliability product. Thus, one should avoid dichotomizing measurement data, if possible. Of course, in some other examples, the data may be inherently of a go/no-go nature, e.g., the operation of a switch, the cure of a patient, and, possibly, the integrated circuit test in the first example, and one has no choice other than to use the methods of this chapter.

As in previous chapters, we reiterate the importance to assure valid inferences, in both examples, of the random selection of sample units from the population or process of interest. The analyst must consider the practical considerations for this to be the case; see Chapter 1.

6.2 CONFIDENCE INTERVALS FOR THE (TRUE) PROPORTION NONCONFORMING IN THE SAMPLED POPULATION (OR PROCESS)

The sample proportion $\hat{p} = x/n$ is a point estimate for p, the true population (or process) proportion. However, $\hat{p}$ differs from p due to sampling fluctuations. Thus, one frequently desires to compute, from the sample data, a two-sided confidence interval or a one-sided confidence bound for p.

We will present the most commonly used methods for computing these intervals. For a detailed description of the underlying theory, other methods, and a comparison of various approximate methods of computing confidence intervals for a population proportion, see Blyth (1986) and Blyth and Still (1983).

6.2.1 Graphical Method

Figures 6.1a to 6.1d can be used to obtain approximate two-sided 80, 90, 95, or 99% confidence intervals (or one-sided 90, 95, 97.5, or 99.5% confidence bounds) for p, the population or process proportion nonconforming. To use the charts, one computes $\hat{p} = x/n$ from the data, locates this value on the horizontal axis, draws a vertical line from this point to the curves corresponding to the sample size n, and then draws a horizontal line to the vertical axis to obtain the endpoint(s) of the desired interval (bound). Charts like these were first given by Clopper and Pearson (1934) and are based on the calculational method to be described in the next subsection.

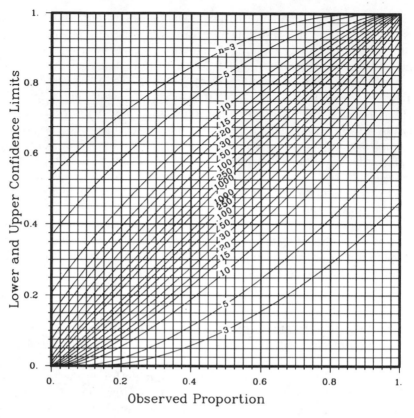

Figure 6.1a Two-sided 80% confidence intervals (one-sided 90% confidence bounds) for a binomial proportion. A similar figure first appeared in Clopper and Pearson (1934). Adapted with permission of the Biometrika Trustees.

6.2.2 Calculational Method

For x observed nonconforming units in a sample of size n, a conservative two-sided $100(1 - \alpha)\%$ confidence interval for p is

$$\left[\underset{\sim}{p}, \tilde{p}\right] = \left[\left\{1 + \frac{(n - x + 1)F_{(1 - \alpha/2; 2n - 2x + 2, 2x)}}{x}\right\}^{-1},\right.$$

$$\left.\left\{1 + \frac{n - x}{(x + 1)F_{(1 - \alpha/2; 2x + 2, 2n - 2x)}}\right\}^{-1}\right] \qquad (6.1)$$

where $F_{(\gamma; r_1, r_2)}$ is the 100γth percentile of the F distribution with r_1 and r_2

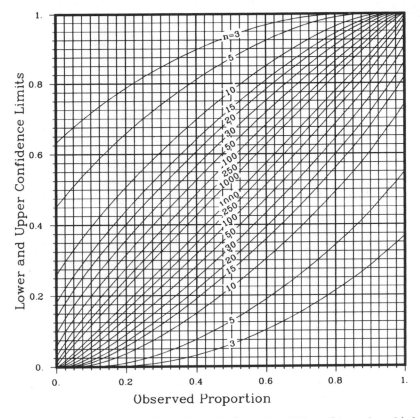

Figure 6.1b Two-sided 90% confidence intervals (one-sided 95% confidence bounds) for a binomial proportion. A similar figure first appeared in Clopper and Pearson (1934). Adapted with permission of the Biometrika Trustees.

degrees of freedom (see Section 4.1.2). The lower limit is defined to be $p = 0$ if $x = 0$, and the upper limit is $\tilde{p} = 1$ if $x = n$. One-sided upper and lower $100(1 - \alpha)\%$ confidence bounds are obtained by replacing $\alpha/2$ by α in Equation (6.1).

6.2.3 Small-Sample Tabular Method

Tables A.23a and A.23b can be used to obtain either a two-sided confidence interval or a one-sided confidence bound for p based upon x nonconforming units in a sample of size n for $n \le 20$ and $1 - \alpha = 0.80$, 0.90, and 0.98 for two-sided confidence intervals, and $1 - \alpha = 0.90$, 0.95, and 0.99 for one-sided confidence bounds. Odeh and Owen (1983) provide more extensive tabulations with values of n ranging from 20 to 1000.

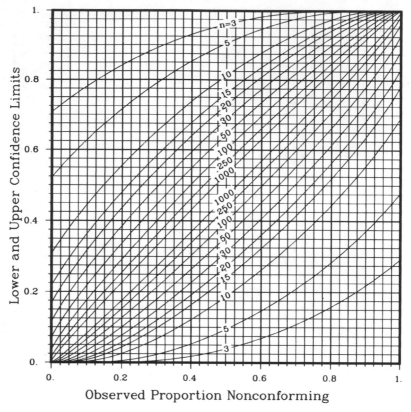

Figure 6.1c Two-sided 95% confidence intervals (one-sided 97.5% confidence bounds) for a binomial proportion. A similar figure first appeared in Clopper and Pearson (1934). Adapted with permission of the Biometrika Trustees.

6.2.4 Large-Sample Approximate Method

Tables A.23a and A.23b are limited to $n \leq 20$ and certain values of $1 - \alpha$. Also, sometimes the use of Equation (6.1) may not be convenient. For example, commonly available tables for the F distribution give percentiles for a limited number of values of r_1 and r_2, and interpolation may be necessary. Fortunately, an approximate expression which uses only tabulations of the normal distribution percentiles provides adequate accuracy when both $n\hat{p}$ and $n(1 - \hat{p})$ exceed 10. The approximate two-sided $100(1 - \alpha)\%$ confidence interval for p is

$$\left[\underset{\sim}{p}, \tilde{p} \right] = \hat{p} \pm z_{(1-\alpha/2)}\left[\frac{\hat{p}(1 - \hat{p})}{n} \right]^{1/2}, \qquad (6.2)$$

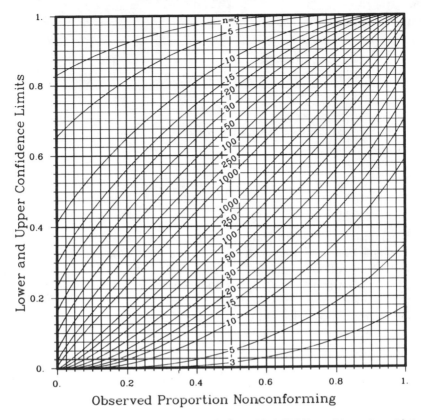

Figure 6.1d Two-sided 99% confidence intervals (one-sided 99.5% confidence bounds) for a binomial proportion. A similar figure first appeared in Clopper and Pearson (1934). Adapted with permission of the Biometrika Trustees.

where $z_{(\gamma)}$ is the 100γth percentile of the standard normal distribution. One-sided lower and upper $100(1 - \alpha)\%$ confidence bounds are obtained by replacing $\alpha/2$ by α in the lower and upper parts, respectively, of this expression. More refined approximations, based upon the normal distribution, are given by Dixon and Massey (1969), Fujino (1980), and Hahn and Chandra (1981).

6.2.5 Examples

For the engine exhaust pollutant example, management wants a confidence interval for p, the proportion of engines in the sampled population with pollutant levels greater than 10 ppm. Using the calculational method in Equation (6.1), with $n = 10$, $x = 1$, we obtain $F_{(0.975; 20, 2)} = 39.45$, $F_{(0.975; 4, 18)} = 3.608$, and $F_{(0.95; 4, 18)} = 2.928$ from Tables A.9d and A.9e. Thus, a 95%

confidence interval for p is

$$\left[\underset{\sim}{p}, \tilde{p}\right] = \left[\left\{1 + \frac{(10)39.45}{1}\right\}^{-1}, \left\{1 + \frac{9}{(2)3.608}\right\}^{-1}\right] = [0.0025, 0.44].$$

An upper 95% confidence bound for p is

$$\tilde{p} = \left\{1 + \frac{9}{(2)2.928}\right\}^{-1} = 0.39,$$

which can also be obtained from Table A.23a.

In comparison, the procedure in Section 4.5 (which assumes that the measurements follow a normal distribution), with $n = 10$, $\bar{x} = 8.05$, and $s = 1.09$, yields a two-sided 95% confidence interval for p of 0.0026 to 0.227, and a one-sided upper 95% confidence bound for p of 0.18. Such intervals are shorter than those given in this chapter, because the measured values are used directly, rather than being dichotomized.

For the other example, $x = 20$ nonconforming units were found in the sample of $n = 1000$ integrated circuits. Because $n\hat{p} = 1000(0.02) = 20$ and $n(1 - \hat{p}) = 1000(0.98) = 980$, we can use the large-sample approximate method in Equation (6.2). An approximate 95% confidence interval for p, using $\hat{p} = 0.02$ and $z_{(0.975)} = 1.96$, is

$$\left[\underset{\sim}{p}, \tilde{p}\right] = 0.02 \pm (1.96)\left\{\frac{(0.02)(0.98)}{1000}\right\}^{1/2} = [0.011, 0.029].$$

An upper approximate 95% confidence bound for p, using $z_{(0.95)} = 1.645$, is

$$\tilde{p} = 0.02 + (1.645)\left\{\frac{(0.02)(0.98)}{1000}\right\}^{1/2} = 0.027.$$

Thus, we can claim, with approximately 95% confidence, that the true proportion of nonconforming units in the sampled population is less than $\tilde{p} = 0.027$. For this example, the calculational method, based on Equation (6.1), gives $[\underset{\sim}{p}, \tilde{p}] = [0.012, 0.031]$ for the two-sided 95% confidence interval and $\tilde{p} = 0.029$ for the one-sided upper 95% confidence bound.

For these examples, confidence limits can also be read directly from Figures 6.1b and 6.1c.

6.3 CONFIDENCE INTERVALS FOR THE PROBABILITY THAT THE NUMBER OF NONCONFORMING UNITS IN A FUTURE SAMPLE IS LESS THAN OR EQUAL TO (OR GREATER THAN) A SPECIFIED NUMBER

Some applications require inferences concerning the probability that the number of nonconforming units in a future sample is less than or equal to (or greater than) some specified number J. For example, units may be selected at random from a production process and packaged in lots of size m. Based on the information in a previous random sample of size n, the manufacturer wants to find a confidence interval on the probability p_{LE} that the number of nonconforming units in a future sample of size m is less than or equal to J, where J is a prespecified number (which could be any integer from 0 to m). These intervals were described previously in Chandra and Hahn (1981).

6.3.1 Method

If the proportion of nonconforming units in the population were *known* to equal p, the probability p_L that y, the number of nonconforming units in a sample of size m, will be less than or equal to a prespecified number J is computed from the binomial cumulative distribution function as

$$p_{LE} = \Pr(y \leq J) = B(J; m, p), \qquad (6.3)$$

where $B(J; m, p)$ was defined in Section 5.1.2.1. Usually, p is unknown and only sample data on the number of nonconforming units x in the previous sample of size n are available. Because p_{LE} is a decreasing function of p, the following two-step procedure is used to find a two-sided confidence interval for p_{LE}:

1. Obtain a two-sided confidence interval for p, based on the data, using one of the methods given in Section 6.2.
2. Substitute these values for p into Equation (6.3) to obtain the desired two-sided confidence interval on p_{LE}.

For example, if $[\underline{p}, \tilde{p}]$ is a two-sided 95% confidence interval for p, a two-sided 95% confidence interval for p_{LE} is

$$\left[\underline{p}_{LE}, \tilde{p}_{LE} \right] = \left[B(J; m, \tilde{p}), B(J; m, \underline{p}) \right].$$

A one-sided lower (upper) $100(1 - \alpha)\%$ confidence bound for p_{LE} is found by substituting a one sided upper (lower) $100(1 - \alpha)\%$ confidence bound for p into Equation (6.3).

Similarly, if the proportion of nonconforming units in the population is known to equal p, the probability p_{GT} that y, the number of nonconforming units in a future sample of size m, will be greater than J is the complement of the binomial cumulative distribution function

$$p_{GT} = 1 - p_{LE} = \Pr(y > J) = 1 - B(J; m, p). \qquad (6.4)$$

When p is unknown, because p_{GT} is an increasing function of p, a $100(1 - \alpha)\%$ confidence interval for p_{GT} is

$$\left[\underset{\sim}{p}_{GT}, \tilde{p}_{GT} \right] = \left[1 - \tilde{p}_{LE}, 1 - \underset{\sim}{p}_{LE} \right] = \left[1 - B(J; m, \underset{\sim}{p}), 1 - B(J; m, \tilde{p}) \right].$$

A one-sided lower (upper) $100(1 - \alpha)\%$ confidence bound for p_{GT} is found by substituting a one-sided lower (upper) $100(1 - \alpha)\%$ confidence bound for p into Equation (6.4).

6.3.2   Example

Assume now that the manufacturer of integrated circuits in the first example ships packages of 50 units. A 95% confidence interval is desired on the probability that a package will have two or fewer nonconforming units, or, equivalently, on the proportion of packages that will have two or fewer nonconforming units. Using the 95% confidence interval $[\underset{\sim}{p}, \tilde{p}] =$ [0.0123, 0.0307], obtained in Section 6.2.5 (carrying another digit for the intermediate calculation), the desired 95% confidence interval for $p_{LE} = \Pr(y \le J)$ is

$$\left[\underset{\sim}{p}_{LE}, \tilde{p}_{LE} \right] = \left[B(2; 50, 0.0307), B(2; 50, 0.0123) \right] = [0.80, 0.98].$$

Thus, we are 95% confident that between 80 and 98% of such packages will have no more than two nonconforming units. A one-sided lower 95% confidence bound for p_{LE}, using the previously obtained upper confidence bound $\tilde{p} = 0.0289$, is

$$\underset{\sim}{p}_{LE} = B(2; 50, 0.0289) = 0.82.$$

Thus, we are 95% confident that at least 82% of the packages of 50 units will have no more than two defective units. The binomial cumulative distribution function $B(\cdot)$ was evaluated for the examples with the IMSL (1987) subroutine BINDF.

This example depends heavily on the assumption that p, the population proportion nonconforming, is constant, i.e., that the production process is "in control." If the process produces varying proportions of nonconforming units during differing periods of time this assumption is not met, and the preceding confidence interval or bound could be misleading.

6.4 ONE-SIDED TOLERANCE BOUNDS FOR THE DISTRIBUTION OF NUMBER OF NONCONFORMING UNITS IN FUTURE SAMPLES OF *m* UNITS

Some applications require one-sided lower or upper tolerance bounds for the distribution of y, the number of nonconforming units in future samples of m units. For example, assume units from a production process are packaged in groups of size m. Suppose that it is desired to state, with a specified degree of confidence, that at least $100p_{LE}\%$ of such packages contain J or fewer (or $100p_{GE}\%$ contain J or more) nonconforming units; here $p_{LE}(p_{GE})$ is specified and a "safe" value of J is desired. This problem is the inverse of the one in Section 6.3; there J was specified and a confidence interval for the probability p_{LE} of J or fewer nonconforming units (or p_{GT} of more than J nonconforming units) in a future sample was desired.

The following one-sided binomial tolerance bounds were first given by Hahn and Chandra (1981).

6.4.1 Upper Binomial Tolerance Bound

If the population (or process) proportion p of nonconforming units is *known*, the smallest integer J such that

$$\Pr(y \le J) = B(J; m, p) \ge p_{LE} \qquad (6.5)$$

is an upper *probability* bound on the number of nonconforming units that will be in $100p_{LE}\%$ of the future samples of size m from the sampled population (or process). In our problem, p is unknown and only sample data, consisting of x nonconforming units in a random sample of n units, are available. Thus, one can only obtain a *statistical* upper tolerance bound for the distribution of y. Because the smallest integer J satisfying Equation (6.5) is an increasing function of p, an upper tolerance bound for the distribution of y can be found by the following two-step procedure:

1. Use one of the methods in Section 6.2 to compute $\tilde{p}$, an upper $100(1 - \alpha)\%$ confidence bound for p, and

2. Substitute $\tilde{p}$ for p, and find the smallest integer J such that the inequality given by Equation (6.5) holds. This integer is the desired upper tolerance bound $\tilde{T}_{pLE}$.

Thus, one can say with $100(1 - \alpha)\%$ confidence that $\Pr(y \le \tilde{T}_{pLE}) \ge p_{LE}$.

That is, we are $100(1 - \alpha)\%$ confident that at least $100p_{LE}\%$ of the future samples of size m will contain $\tilde{T}_{pLE}$ or fewer nonconforming units.

6.4.2   Lower Binomial Tolerance Bound

If the population or process proportion p of nonconforming units is *known*, the largest integer J such that

$$\Pr(y \geq J) = 1 - B(J - 1; m, p) \geq p_{GE} \qquad (6.6)$$

is a lower *probability* bound on the number of nonconforming units that will be in $100p_{GE}\%$ of the future samples of size m from the sampled population (or process). When p is unknown, and only sample data (consisting of x nonconforming units in a random sample of n units) are available, a lower *statistical* tolerance bound on y can be found by the following two-step procedure:

1. Use one of the methods in Section 6.2 to compute $\underset{\sim}{p}$, which is a lower $100(1 - \alpha)\%$ confidence bound for p.
2. Substitute $\underset{\sim}{p}$ for p, and find the largest integer J such that the inequality given by Equation (6.6) holds. This integer is the desired lower tolerance bound $\underset{\sim}{T}_{pGE}$.

Thus, one can say with $100(1 - \alpha)\%$ confidence that $\Pr(y \geq \underset{\sim}{T}_{pGE}) \geq p_{GE}$. That is, we are $100(1 - \alpha)\%$ confident that at least $100p_{GE}\%$ of the future samples of size m will contain at least $\underset{\sim}{T}_{pGE}$ nonconforming units.

6.4.3   Example

The manufacturer of the integrated circuits ships packages of $m = 50$ units. An upper tolerance bound $\tilde{T}_{0.90}$ is desired, such that one can claim, with 95% confidence, that the number of defective units will be less than or equal to $\tilde{T}_{0.90}$ in 90% of such packages. As in Section 6.3, one must assume that the units in each package are a random sample from the production process, and that the production process is in "statistical control."

For the example, $p_{LE} = 0.90$, and $\tilde{p} = 0.0289$ is the previously computed one-sided upper 95% confidence bound for p. Because $\Pr(y \leq 2) = B(2; 50, 0.0289) = 0.825$ and $\Pr(y \leq 3) = B(3; 50, 0.0289) = 0.944$, the desired tolerance bound is $\tilde{T}_{0.90} = 3$. Thus, we are 95% confident that at least 90% (more precisely, at least 94.4%) of the packages will have three or fewer defective units. The binomial cumulative probability $B(\cdot)$ was evaluated for this example with the IMSL (1987) subroutine BINDF.

6.5 PREDICTION INTERVALS FOR THE NUMBER NONCONFORMING IN A FUTURE SAMPLE OF m UNITS

Assume, as before, that x nonconforming units have been observed in a random sample of size n from a population (or process). From these data, it is desired to find a prediction interval that will, with some specified degree of confidence, contain y, the number of nonconforming units in a future random sample of size m from the same population.

6.5.1 Calculational Method

The following exact procedure, based on the hypergeometric distribution, was suggested by Thatcher (1964). Consider the subpopulation defined by the combined samples of size n and m. This subpopulation contains $D = x + y$ nonconforming units out of a total of $N = n + m$ units. Because x follows a hypergeometric distribution one can use standard inferential procedures for that distribution to obtain a confidence interval to contain D (see Section 11.3.17 for references). Moreover, because x is known, these intervals can be translated directly into an interval for y, the number of nonconforming units in the future sample. (This procedure does not involve direct estimation of the proportion of nonconforming units in the population.)

Operationally, a $100(1 - \alpha)\%$ prediction interval $[\underset{\sim}{y}, \tilde{y}]$ for y can, thus, be obtained by finding the largest value $\underset{\sim}{y}$ and the smallest value $\tilde{y}$ such that

$$1 - H(x - 1; n, x + \underset{\sim}{y}, n + m) > \alpha/2 \quad \text{and} \quad H(x; n, x + \tilde{y}, n + m) > \alpha/2 \tag{6.7}$$

where $\underset{\sim}{y} = 0$ if $x = 0$ and $\tilde{y} = m$ if $x = n$ and $H(\cdot)$ is the hypergeometric cumulative distribution function, defined in Section 5.1.2.2. One-sided lower and upper $100(1 - \alpha)\%$ prediction bounds for y are obtained by replacing $\alpha/2$ by α in the lower and upper parts, respectively, of this expression.

6.5.2 Large-Sample Approximate Method

A large-sample approximate $100(1 - \alpha)\%$ prediction interval for y (see Hahn and Nelson 1973) is

$$\left[\underset{\sim}{y}, \tilde{y}\right] = m\hat{p} \pm z_{(1 - \alpha/2)}\left[m\hat{p}(1 - \hat{p})\frac{m + n}{n}\right]^{1/2}. \tag{6.8}$$

This approximate interval is easier to obtain than the one by the previous method, especially if a computer program or extensive tables of the hypergeometric distribution are not available. The approximate intervals will likely be sufficient for practical purposes if $x, n - x, y$, and $m - y$ are all reasonably large; specifically, Nelson (1982) suggests that each be at least 10. Because y is unknown, one might require that the estimated expected value of y and $m - y$ be larger than 10 [i.e., that $m\hat{p}$ and $m(1 - \hat{p})$ be both larger than 10].

Again, one-sided lower and upper $100(1 - \alpha)\%$ prediction bounds for y are obtained by replacing $\alpha/2$ by α in the lower and upper parts, respectively, of this expression.

6.5.3   Example

Based on the $x = 20$ nonconforming integrated circuits in the test of $n = 1000$ randomly selected units, the manufacturer requires a 95% prediction interval to contain the number of nonconforming units in a future lot of 1000 units that are randomly sampled from the same production process.

The calculational method of Section 6.5.1 yields $[y, \tilde{y}] = [9, 35]$ for a 95% prediction interval and $\tilde{y} = 32$ for an upper 95% prediction bound for y. Thus, based on $x = 20$ nonconforming units from the $n = 1000$ sample units, one can assert, with 95% confidence, that the number of nonconforming units in the future sample of $m = 1000$ will not exceed 32 units. The hypergeometric probabilities needed for these intervals can be computed with the IMSL (1987) subroutine HYPDF.

The corresponding approximate 95% prediction interval for y using (6.8) is

$$\left[y, \tilde{y} \right] = (1000)0.02 \pm (1.96)\left\{ (1000)(0.02)(0.98)\frac{2000}{1000} \right\}^{1/2} = [8, 32].$$

This agrees reasonably well with the previously computed interval. One-sided prediction bounds can be obtained similarly.

CHAPTER 7

Statistical Intervals
for the Number of Occurrences
(Poisson Distribution)

7.1 INTRODUCTION

This chapter describes statistical intervals for the number of occurrences over some interval of time or region of space, assuming independent occurrences and a constant mean occurrence rate. Such situations are modeled by the Poisson distribution. For example, the Poisson distribution may be used to represent the distribution of the number of flaws on the surface of a product. Specifically, suppose that units of the product are all of the same size and that flaws occur at random and independently of each other at a constant rate λ. Then the number of flaws per unit follows a Poisson distribution. Similarly, the number of unscheduled shutdowns of a complex system over some specified period of time might follow a Poisson distribution. This would be the case if shutdowns occur independently of one another and if their mean occurrence rate is constant over time and from one system to another. This assumption would not be correct if, for example, (1) environmental factors that cause failure (e.g., lightning) simultaneously affect more than one system or (2) if the failure rate changes with time (this might be the case if the system parts are subject to wearout). In this chapter, our discussion will frequently be in terms of x, the number of occurrences (e.g., unscheduled shutdowns of a computer system) in a given time period of length n. We could similarly have discussed variables that involve events over constant length, area, or volume, such as the number of defects per foot of wire or the number of flaws per square meter of a finished surface.

As indicated, problems involving the number of occurrences of independent, randomly occurring events per unit of space or time can be modeled with the Poisson distribution. The probability function for the Poisson

115

distribution, i.e., the probability of exactly x' occurrences, is

$$\Pr(x = x') = f(x'; \lambda) = \frac{\exp(-\lambda)\lambda^{x'}}{x'!}$$

where the parameter λ is the mean occurrence rate. For example, if the number of unscheduled shutdowns per year follows a Poisson distribution with a mean occurrence rate of $\lambda = 5$ per year, the probability of zero shutdowns in a one-year period is

$$\Pr(x = 0) = f(0; 5) = \frac{\exp(-5)5^0}{0!} = 0.006738.$$

The probability of exactly two shutdowns is

$$\Pr(x = 2) = f(2; 5) = \frac{\exp(-5)5^2}{2!} = 0.08422.$$

Also, if the number of occurrences per unit is Poisson distributed with occurrence rate λ, the number of occurrences in n units of time (where n is not necessarily an integer) is Poisson distributed with occurrence rate $n\lambda$. Duncan (1974), Grant and Leavenworth (1988), and Hahn and Shapiro (1967) discuss the Poisson distribution further.

In the problems considered here, the Poisson distribution parameter, or mean occurrence rate λ is unknown. Instead, all that is known is that over a past period of length (or duration) of n units of time, there were x occurrences of the event of interest. One of our problems, then, is to obtain a confidence interval for λ.

The Poisson, like the binomial, is a discrete distribution. A Poisson random variable can take on the integer values $x = 0, 1, 2, \ldots$ (with no theoretical upper limit). Because the distribution of x is discrete, statistical intervals do not generally have exactly the desired confidence level. Thus, the statistical intervals given here are approximate and, like those in the previous two chapters, tend to be conservative (i.e., the actual confidence level is larger than the nominal value).

7.1.1 Overview

In this chapter we will show how to obtain:

- Confidence intervals for λ, the (true) mean occurrence rate of the sampled population (or process).
- Confidence intervals for the probability that the number of occurrences in a specified period will be less than or equal to (or greater than) a specified value.

- One-sided tolerance bounds for the number of occurrences in a specified period.
- Prediction intervals for the number of occurrences in a future period.

These are analogous to the four topics that were considered in Chapter 6 for statistical intervals for proportions and percentages (binomial distribution), and the discussion that follows resembles that of the previous chapter.

7.1.2 Notation and Tabulations

The function

$$P(x'; \lambda) = \Pr(x \le x') = \sum_{i=0}^{x'} \frac{\exp(-\lambda)\lambda^i}{i!}$$

denotes the cumulative Poisson probability of observing x' or fewer occurrences in a unit of time or space, where λ is the constant occurrence rate per unit of time or space.

Table A.24 gives $P(x'; \lambda)$ for $x' = 0(1)31$ and $\lambda = 0.01(0.01)0.06(0.02)0.12$, 0.15, 1.0(0.1)2.0(0.2)7.0(1.0)9.0. $P(x'; \lambda)$ is tabled for many values of λ between 10^{-8} and 205 in the book of tables published by the General Electric Company, Defense Systems Department (1962). Other tabulations are referenced and approximations are given by Hahn and Chandra (1981). A useful chart for finding cumulative Poisson probabilities is given by Dodge and Romig (1944). Also, computer programs compute Poisson probabilities [e.g., POIDF in IMSL (1987)].

The Poisson cumulative probability for the number of occurrences in n units of time or space, when the mean occurrence rate is λ for each unit of time or space, can be approximated by a normal distribution with a mean $\mu = n\lambda$ and a standard deviation $\sigma = (n\lambda)^{1/2}$. In particular,

$$P(x'; n\lambda) \approx \Phi\left(\frac{x' + 0.5 - \mu}{\sigma}\right)$$

where $\Phi(z)$ is the standard normal cumulative distribution function given in Table A.6. (The $+0.5$ in this expression is a "continuity correction" that improves the approximation.) The approximation is reasonably accurate as long as $n\lambda > 10$.

7.1.3 Example

For the examples that follow, assume that the data consist of $x = 24$ unscheduled shutdowns in $n = 5.0$ system-years of operation for a computing system. Furthermore, assume that unscheduled shutdowns occur independently of one another at a constant mean rate from one system to the next and from one year to the next. [This assumption can be assessed from the observed shutdown times; see Ascher and Feingold (1984)]. Under such assumptions, the number of unscheduled shutdowns follows a Poisson distribution. The mean number of shutdowns per year (or shutdown rate) is then estimated from the given data as $\hat{\lambda} = x/n = 24/5.0 = 4.8$. We note that in this example a system-year was taken to be the unit time interval.

7.2 CONFIDENCE INTERVALS FOR THE MEAN OCCURRENCE RATE OF A POISSON DISTRIBUTION

The observed occurrence rate $\hat{\lambda} = x/n$ is a point estimate of the true occurrence rate λ. However $\hat{\lambda}$ differs from λ due to sampling fluctuations. Thus, one frequently desires to compute, from the sample data, a two-sided confidence interval or a one-sided confidence bound for λ.

7.2.1 Graphical Method for Obtaining an Upper Confidence Bound

Figures 7.1a to 7.1d allow one to easily find upper 60, 90, 95, and 99% confidence bounds for λ, the Poisson mean occurrence rate. (These figures do not provide lower confidence bounds or two-sided confidence intervals.) The horizontal axes of these figures run from $n' = 1$ to 10, where n' is a standardized unit of time. To use these charts, one first finds both the integer r that satisfies $n' = n/10^r$ so that n' is between 1 and 10 and the associated value n'. For example, if the data consist of the number of occurrences in $n = 27.3$ units of time (or time intervals), we obtain $r = 1$ and $n' = 27.3/10^1 = 2.73$. We then enter the horizontal axis of the figure at n', go vertically to meet the line corresponding to x, the observed number of occurrences (interpolating for x, if necessary). Next, we go horizontally to the left and read $\tilde{\lambda}'$ from the vertical axis. The desired upper confidence bound on λ is

$$\tilde{\lambda} = \tilde{\lambda}'/10^r.$$

7.2.2 Tabular Method

A two-sided $100(1 - \alpha)\%$ confidence interval for λ, given x occurrences in n units of time, is

$$\left[\underset{\sim}{\lambda}, \tilde{\lambda}\right] = \left[\frac{G_{L(1-\alpha;\, x)}}{n}, \frac{G_{U(1-\alpha;\, x)}}{n}\right].$$

Table A.25 contains the convenient factors $G_{L(\gamma;\, x)}$ and $G_{U(\gamma;\, x)}$ for $x = 0(1)$

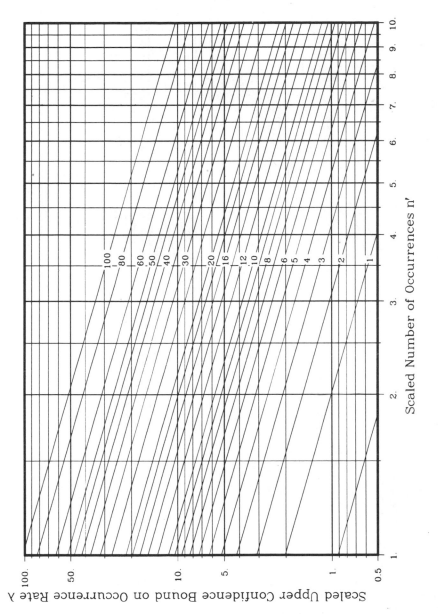

Figure 7.1a Factors for upper 60% confidence bounds for a Poisson occurrence rate. A similar figure appeared previously in Nelson (1972b). Adapted with permission of the American Society for Quality Control.

119

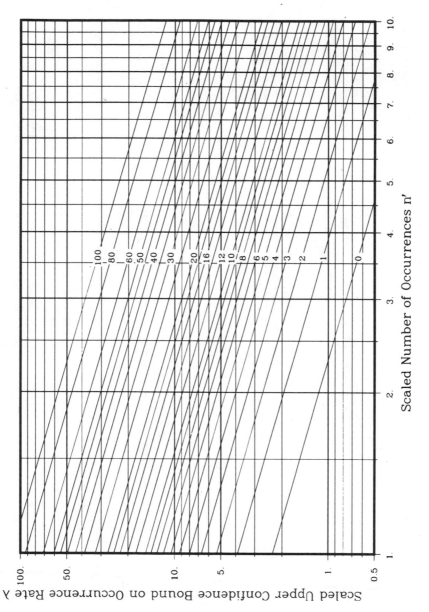

Figure 7.1b Factors for upper 90% confidence bounds for a Poisson occurrence rate. A similar figure appeared previously in Nelson (1972b). Adapted with permission of the American Society for Quality Control.

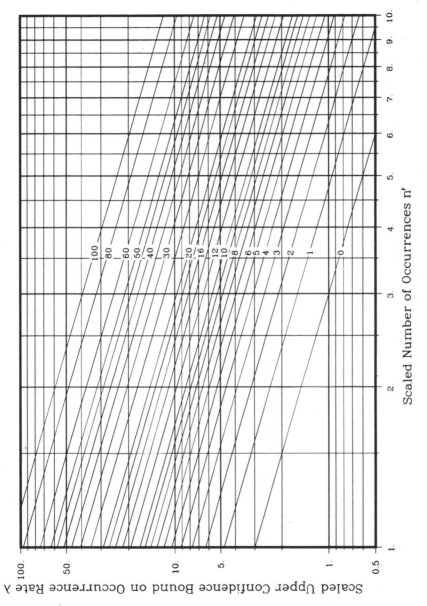

Figure 7.1c Factors for upper 95% confidence bounds for a Poisson occurrence rate. A similar figure appeared previously in Nelson (1972b). Adapted with permission of the American Society for Quality Control.

121

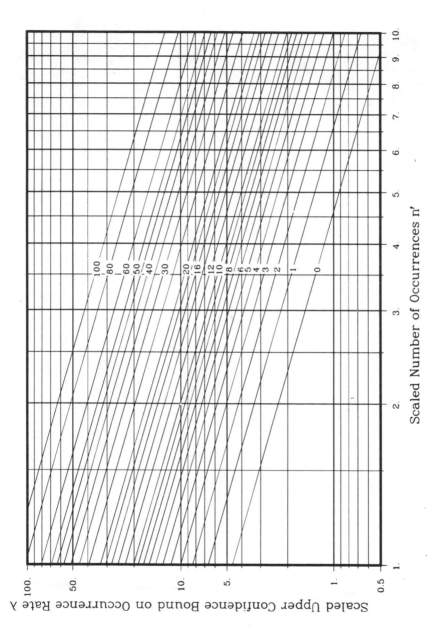

Figure 7.1d Factors for upper 99% confidence bounds for a Poisson occurrence rate. A similar figure appeared previously in Nelson (1972b). Adapted with permission of the American Society for Quality Control.

30(10)60, 80, 100 and $\gamma = 0.998$, 0.99, 0.98, 0.95, 0.90, 0.80, and 0.50 for two-sided confidence intervals and $\gamma = 0.999$, 0.995, 0.99, 9.975, 0.95, 0.90 and 0.75 for one-sided confidence bounds.

Equivalently,

$$\left[\underset{\sim}{\lambda}, \tilde{\lambda} \right] = \left[\frac{0.5\chi^2_{(\alpha/2;\, 2x)}}{n}, \frac{0.5\chi^2_{(1-\alpha/2;\, 2x+2)}}{n} \right] \tag{7.1}$$

where $\chi^2_{(\gamma;\, r)}$ is the 100γth percentile of a chi-square distribution with r degrees of freedom. This formula can also be used for values not tabulated in Table A.25. Table A.8 provides chi-square percentiles $\chi^2_{(\gamma;\, r)}$. One-sided lower and upper $100(1 - \alpha)\%$ confidence bounds for λ are obtained by replacing $\alpha/2$ by α in the first and second parts, respectively, of the preceding expressions.

7.2.3 Large-Sample Approximate Method

For large values of x, an approximate $100(1 - \alpha)\%$ confidence interval for λ is

$$\left[\underset{\sim}{\lambda}, \tilde{\lambda} \right] = \hat{\lambda} \pm z_{(1-\alpha/2)} \left(\frac{\hat{\lambda}}{n} \right)^{1/2} \tag{7.2}$$

where $\hat{\lambda} = x/n$. This approximation is quite good for $x > 20$. Again, one-sided lower (upper) $100(1 - \alpha)\%$ confidence bounds for λ are obtained by replacing $\alpha/2$ by α in the lower (upper) part of Equation (7.2).

7.2.4 Example

For the example, $x = 24$ failures were observed in $n = 5.0$ years. From Table A.25, $G_{L(0.975;\, 24)} = 15.38$, $G_{U(0.975;\, 24)} = 35.71$, and $G_{U(0.95;\, 24)} = 33.75$. A two-sided 95% confidence interval for the yearly failure rate λ is

$$\left[\underset{\sim}{\lambda}, \tilde{\lambda} \right] = \left[\frac{15.38}{5.0}, \frac{35.71}{5.0} \right] = [3.08, 7.14].$$

An upper 95% confidence bound for λ is

$$\tilde{\lambda} = \frac{33.75}{5} = 6.75.$$

This value could also have been obtained from reading Figure 7.1c using $n' = 5$, with $r = 0$, and interpolating between the appropriate lines for $x = 24$. Equivalently, using Equation (7.1) with $\chi^2_{(0.025;\, 48)} = 30.76$, $\chi^2_{(0.975;\, 50)} =$

71.42, and $\chi^2_{(0.95;\,50)} = 67.50$, we again obtain the two-sided 95% confidence interval $[\underline{\lambda}, \tilde{\lambda}] = [3.08, 7.14]$ and the one-sided upper 95% confidence bound $\tilde{\lambda} = 6.75$. Using the large sample approximation in Equation (7.2), with $\hat{\lambda} = 24/5.0 = 4.8$, $z_{(0.975)} = 1.96$ for a two-sided 95% confidence interval, gives

$$\left[\underline{\lambda}, \tilde{\lambda}\right] = 4.8 \pm 1.96\left(\frac{4.8}{5.0}\right)^{1/2} = [2.88, 6.72].$$

This compares reasonably well with the interval obtained above.

7.3 CONFIDENCE INTERVALS FOR THE PROBABILITY THAT THE NUMBER OF OCCURRENCES IN A SPECIFIED PERIOD IS LESS THAN OR EQUAL TO (OR GREATER THAN) A SPECIFIED NUMBER

In some applications, it is desired to estimate the probability p_{LE} (p_{GT}) that the number of occurrences in a (future) period of specified length will be less than or equal to (greater than) some specified value J. For example, a manufacturer might require a confidence interval for the probability p_{GT} of more than J unscheduled shutdowns during some future time period of specified length, based on the data for the number of such shutdowns in a previous period of length n. These intervals were described previously by Chandra and Hahn (1981).

7.3.1 Method

If the true occurrence rate were known to equal λ, the probability p_{LE} that the future number of occurrences y will be less than or equal to a specified value J in a period of length m is computed from the Poisson cumulative distribution function as

$$p_{LE} = \Pr(y \le J) = P(J; m\lambda), \qquad (7.3)$$

where $P(J; m\lambda)$ is defined in Section 7.1.2. In practice, λ is unknown and only sample data on the number of occurrences x in a previous period of length n units are available. Because p_{LE} is a decreasing function of λ, the following two-step procedure is used to find a two-sided confidence interval for p_{LE}:

1. Obtain a two-sided confidence interval for λ based on the data, using one of the methods in Section 7.2.
2. Substitute these values for λ into Equation (7.3) to obtain the desired two-sided confidence interval for p_{LE}.

For example, if $[\underline{\lambda}, \tilde{\lambda}]$ is a two-sided 95% confidence interval for λ, a two-sided 95% confidence interval for p_{LE} is

$$\left[\underline{p}_{LE}, \tilde{p}_{LE}\right] = \left[P(J; m\tilde{\lambda}), P(J; m\underline{\lambda})\right].$$

A one-sided lower (upper) $100(1 - \alpha)\%$ confidence bound for p_{LE} is found by substituting a one-sided upper (lower) $100(1 - \alpha)\%$ confidence bound for λ into Equation (7.3).

Similarly, if the true occurrence rate in the population is known to equal λ, the probability p_{GT} that the number of occurrences y in a period of length m will be greater than J is the complement of the Poisson cumulative distribution function

$$p_{GT} = 1 - p_{LE} = \Pr(y > J) = 1 - P(J; m\lambda). \qquad (7.4)$$

When λ is unknown, a $100(1 - \alpha)\%$ confidence interval for p_{GT}, because p_{GT} is an increasing function of λ, is

$$\left[\underline{p}_{GT}, \tilde{p}_{GT}\right] = \left[1 - \tilde{p}_{LE}, 1 - \underline{p}_{LE}\right] = \left[1 - P(J; m\underline{\lambda}), 1 - P(J; m\tilde{\lambda})\right].$$

A one-sided lower (upper) $100(1 - \alpha)\%$ confidence bound for p_{GT} is found by substituting a one-sided lower (upper) $100(1 - \alpha)\%$ confidence bound for λ into Equation (7.4).

7.3.2 Example

The manufacturer in the previous example would like to compute a 95% confidence interval for the probability that $J = 5$ or fewer unscheduled shutdowns will occur in the next $m = \frac{1}{2}$ year of system operation. Using the previously obtained confidence interval $[\underline{\lambda}, \tilde{\lambda}] = [3.08, 7.14]$ for λ, a 95% confidence interval for $p_{LE} = \Pr(y \leq 5)$ is

$$\left[\underline{p}_{LE}, \tilde{p}_{LE}\right] = \left[P(5; (0.50)7.14), P(5; (0.50)3.08)\right] = [0.848, 0.995].$$

Thus, we are 95% confident that the probability of five or fewer unscheduled shutdowns in the next half year of operation is between 0.848 and 0.995. Similarly, a one-sided lower 95% confidence bound on p_{LE} is

$$\underline{p}_{LE} = P[5; (0.50)6.75] = 0.874.$$

This example depends heavily on the assumption that the true Poisson mean occurrence rate λ does not change in the next one-half year from what it was during the previous 5.0 years.

7.4 ONE-SIDED TOLERANCE BOUNDS FOR THE NUMBER OF OCCURRENCES IN A SPECIFIED PERIOD

Some applications require one-sided lower or upper tolerance bounds for the distribution of y, the number of occurrences in a specified future time period. For example, the manufacturer of a system might wish to claim, with a specified degree of confidence, that at least $100p_{LE}\%$ of its systems will experience J or fewer unscheduled shutdowns (or $100p_{GE}\%$ will experience J or more shutdowns) during m units of time, where m and p_{LE} (or p_{GE}) are specified and a "safe" value for J is desired. This problem is the inverse of the one in Section 7.3; there J was specified and we computed a confidence interval for the probability p_{LE} of J or fewer occurrences (or p_{GT} of more than J occurrences).

The following one-sided Poisson tolerance bounds were first given by Hahn and Chandra (1981).

7.4.1 Upper Poisson Tolerance Bound

If the population (or process) occurrence rate λ is *known*, the smallest integer J such that

$$\Pr(y \le J) = P(J; m\lambda) \ge p_{LE} \qquad (7.5)$$

is an upper *probability* bound on the number of occurrences in $100p_{LE}\%$ of the future periods of length m from the sampled population (or process). When λ is unknown and only sample data (consisting of x occurrences in a period of length n) are available, one can obtain an upper tolerance bound for the distribution of y. Because the smallest integer J satisfying Equation (7.5) is an increasing function of λ, one can obtain an upper tolerance bound for the distribution of y by the following two-step procedure:

1. Use one of the methods in Section 7.2 to compute $\tilde{\lambda}$, an upper $100(1 - \alpha)\%$ confidence bound for λ.
2. Substitute $\tilde{\lambda}$ for λ and find the smallest integer J such that the inequality in Equation (7.5) holds. This integer is the desired upper tolerance bound $\tilde{T}_{pLE}$.

Thus, one can say with $100(1 - \alpha)\%$ confidence that $\Pr(y \le \tilde{T}_{pLE}) \ge p_{LE}$. That is, we are $100(1 - \alpha)\%$ confident that at least $100p_{LE}\%$ of the future periods of length m will result in at most $\tilde{T}_{pLE}$ occurrences.

7.4.2 Lower Poisson Tolerance Bound

If the occurrence rate λ is *known*, the largest integer J such that

$$\Pr(y \ge J) = 1 - P(J - 1; m\lambda) \ge p_{GE} \qquad (7.6)$$

is a lower *probability* bound on the number of occurrences in $100p_{GE}\%$ of the future periods of length m from the sampled population (or process). When λ is unknown and only sample data (consisting of x occurrences in a period of length n) are available, a lower statistical tolerance bound can be found by using the following two-step procedure:

1. Use one of the methods in Section 7.2 to compute λ, which is a lower $100(1 - \alpha)\%$ confidence bound for λ.
2. Substitute λ for λ and find the largest integer J such that the inequality given by Equation (7.6) holds. This integer is the desired lower tolerance bound T_{pGE}.

Thus, one can say with $100(1 - \alpha)\%$ confidence that $\Pr(y \geq T_{pGE}) \geq p_{GE}$. That is, we are $100(1 - \alpha)\%$ confident that at least $100p_{GE}\%$ of the future periods of length m will contain at least T_{pGE} occurrences.

7.4.3 Example

The manufacturer in the previous example needs an upper tolerance bound $\tilde{T}_{0.90}$ on the maximum number of unscheduled shutdowns that can be expected, with 95% confidence, to occur in 90% of the systems in one year of operation (i.e., $m = 1$). In the example, with $p_{LE} = 0.90$, the previously computed one-sided upper 95% confidence bound for λ (Section 7.2) is $\tilde{\lambda} = 6.75$, giving $\Pr(y \leq 9) = P[9; 1.0(6.75)] = 0.855$, and $\Pr(y \leq 10) = P[10; 1.0(6.75)] = 0.918$. Thus, the desired upper tolerance bound is $\tilde{T}_{0.90} = 10$ and we are 95% confident that at least 90% (more precisely, at least 91.8%) of the systems will have no more than 10 unscheduled shutdowns during a year of operation. The Poisson cumulative probability was evaluated for the examples with the IMSL (1987) subroutine POISDF. Similar results could be obtained by interpolating in Table A.24.

7.5 PREDICTION INTERVALS FOR THE NUMBER OF OCCURRENCES IN A FUTURE PERIOD

Some applications require prediction bounds on y, the number of occurrences in a specified future number of units of time or space, based on previous sample data. For example, a system user may wish a prediction interval to bound the number of unscheduled shutdowns that will occur over some future specified period of time.

7.5.1 Graphical Method

A two-sided $100(1 - \alpha)\%$ prediction interval $[y, \tilde{y}]$ for y, the number of occurrences in a future period of length m, based on x occurrences in a previous period of length n, can be obtained from Figures 7.2a to 7.2c. These

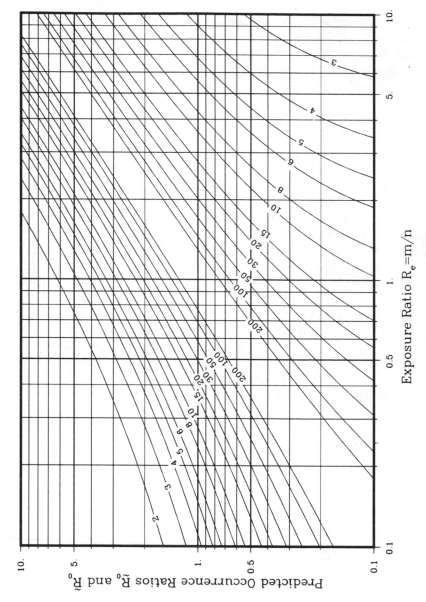

Figure 7.2a Factors for Poisson two-sided 90% prediction intervals (one-sided 95% prediction bounds). A similar figure first appeared in Nelson (1970). Adapted with permission of IEEE.

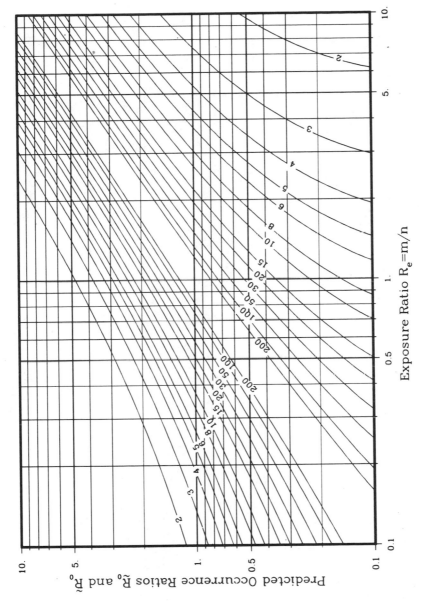

Figure 7.2b Factors for Poisson two-sided 95% prediction intervals (one-sided 97.5% prediction bounds). A similar figure first appeared in Nelson (1970). Adapted with permission of IEEE.

129

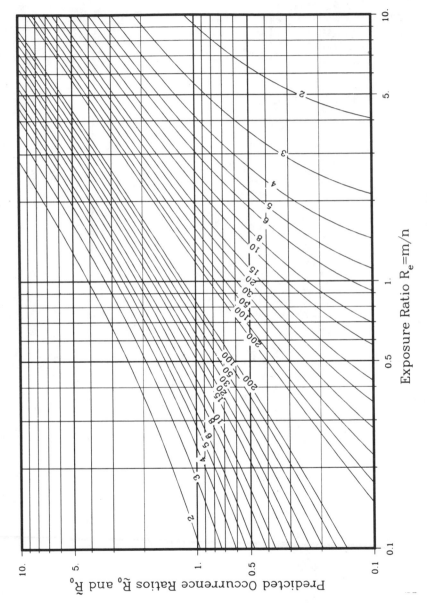

Figure 7.2c Factors for Poisson two-sided 99% prediction intervals (one-sided 99.5% prediction bounds). A similar figure first appeared in Nelson (1970). Adapted with permission of IEEE.

figures, for $\alpha = 90$, 95, and 99%, respectively, were developed by Nelson (1970). First compute the exposure ratio $R_e = m/n$, and use it to enter the horizontal axis of the appropriate figure. Then move up to the lower curve labeled with the observed number of occurrences x (interpolation may be necessary) and read across to the vertical axis. Do the same with the upper curve for the observed value of x. Denote the occurrence ratio values read from the vertical axis as R_0 and $\tilde{R}_0$. Then a $100(1 - \alpha)$% prediction interval to contain y, the number of future occurrences in a specified period of length m, is

$$\left[\underset{\sim}{y}, \tilde{y} \right] = \left[x\underset{\sim}{R}_0, x\tilde{R}_0 \right].$$

Figures 7.2a to 7.2c can also be used, in a similar manner, to obtain one-sided 95, 97.5 and 99.5% prediction bounds, respectively.

7.5.2 Calculational Method

A two-sided $100(1 - \alpha)$% prediction interval $[\underset{\sim}{y}, \tilde{y}]$ for the future number of occurrences in a period of length m, based upon x past occurrences in a period of length n, can be obtained by finding the largest value $\underset{\sim}{y}$ and the smallest value $\tilde{y}$ such that

$$\frac{m}{\underset{\sim}{y} + 1} \leq \left(\frac{n}{x} \right) F_{(1 - \alpha/2;\, 2\underset{\sim}{y} + 2,\, 2x)} \quad \text{and} \quad \frac{\tilde{y}}{m} \geq \left(\frac{x + 1}{n} \right) F_{(1 - \alpha/2;\, 2x + 2,\, 2\tilde{y})}.$$

$$(7.7)$$

One-sided lower and upper $100(1 - \alpha)$% prediction bounds for y are obtained by replacing $\alpha/2$ by α in the lower and upper parts, respectively, of the preceding expressions. This interval was first given by Nelson (1970).

7.5.3 Large-Sample Approximate Method

A large-sample approximate $100(1 - \alpha)$% prediction interval for y is

$$\left[\underset{\sim}{y}, \tilde{y} \right] = m\hat{\lambda} \pm z_{(1 - \alpha/2)} m \left\{ \hat{\lambda} \left(\frac{1}{n} + \frac{1}{m} \right) \right\}^{1/2}.$$

$$(7.8)$$

One-sided lower and upper $100(1 - \alpha)$% prediction bounds for y are obtained by replacing $\alpha/2$ by α and using the lower and upper limits, respectively. Nelson (1982) suggests that this approximation will be reasonably accurate if x and y are both larger than 10. Because y is unknown, one might require that the estimated expected value of y be at least 10 (i.e., $m\hat{\lambda} > 10$).

7.5.4   Example

A consumer who is installing one of the computer systems described in our example would like to use the manufacturer's data to obtain an upper 95% prediction bound for y, the number of unscheduled shutdowns during a future half year of operation (i.e., $m = 0.5$).

To find the upper prediction bound, enter the horizontal axis of Figure 7.2a with the exposure ratio $R_e = m/n = 0.50/5.0 = 0.10$. Moving up to the upper line with $x = 24$ (using interpolation) and reading across to the vertical axis gives the occurrence ratio $\tilde{R}_0 = 0.26$. Then we compute $\tilde{y} = x\tilde{R}_0 = 24(0.26) \simeq 6$. Thus, we are 95% confident that the number of shutdowns in the next half year will be less than or equal to six.

Say now that a two-sided 95% prediction interval is required for the number of unscheduled shutdowns in $m = 4.0$ years of operation. Proceeding as before, $R_e = m/n = 4.0/5.0 = 0.80$, which, using $x = 24$ in Figure 7.2b, gives $\underline{R}_0 = 0.36$ and $\tilde{R}_0 = 1.4$, and we obtain the prediction interval

$$\left[\underline{y}, \tilde{y}\right] = \left[x\underline{R}_0, x\tilde{R}_0\right] = [9, 34].$$

Thus, we are 95% confident that the number of shutdowns during 4 years will be between 9 and 34. These are also the limits that one gets by using the calculational method in Section 7.5.2.

Because the estimated expected value for y is $m\hat{\lambda} = (4)(4.8) = 19.2$, the large-sample approximation can also be used. A 95% two-sided approximate prediction interval for y, using Equation (7.8), is

$$\left[\underline{y}, \tilde{y}\right] = 4(4.8) \pm (1.96)(4.0)\{4.8(\tfrac{1}{5} + \tfrac{1}{4})\}^{1/2} = [8, 31].$$

This compares reasonably well with the previous interval.

CHAPTER 8

Sample Size Requirements for Confidence Intervals on Population Parameters

8.1 INTRODUCTION

This chapter addresses the frequently asked question "How large a sample do I need to obtain a confidence interval?" It deals with sample size requirements for estimating the mean and standard deviation of a normally distributed population, for estimating a binomial proportion, and for estimating an occurrence rate. To determine sample size requirements, one can start with a statement of the desired precision (e.g., in terms of interval length) and then use the procedures for constructing statistical intervals described in the previous chapters "in reverse."

This and the following two chapters are concerned with data *quantity* (sample size). We wish, however, to reiterate that the issue of data quantity is often secondary to that of the *quality* of the data. In particular, in making a statistical estimate or constructing a statistical interval, one assumes that the available data were obtained by using a random sample from a defined population or process of interest. As stated previously, when this is not the case, "all bets are off." Just increasing the sample size—without broadening the scope of the investigation—does not compensate for lack of randomness; all it does is allow one to obtain a (possibly) biased estimate with greater precision. Putting it another way, increasing the sample size per se usually improves the precision of an estimate, but not necessarily its accuracy.

Mace (1964) and Brush (1988) give comprehensive collections of methods for choosing sample sizes for different problems. In addition, the books by Beyer (1968), Natrella (1963), Odeh and Fox (1975), Odeh and Owen (1980, 1983), Odeh, Owen, Birnbaum, and Fisher (1977), and Dixon and Massey (1969) contain tables, charts, and figures that can be used to determine the necessary sample size for various problems. In this chapter and the following two chapters we present and adapt methods from these and other sources (and, where needed, develop new ones) for choosing the sample size for many situations described in this book. Sometimes, these methods will lead to the finding that to attain the desired degree of precision, one requires a larger sample than is practical. Discouraging as this may be, it is better to know before starting an investigation than at its conclusion.

8.2   BASIC REQUIREMENTS IN SAMPLE SIZE DETERMINATION

To determine the required sample size, one generally requires:

* A specification of the objectives of the investigation.
* A statement of desired precision.
* A decision of what statistical distribution, if any, is to be assumed, and, frequently initial guesses or estimates, to be referred to as "planning values," for one or more parameters of that distribution.

These requirements are discussed further below.

8.2.1   The Objectives of the Investigation

Before one can determine how large a sample is needed, one must specify what is to be computed from the resulting data. This could, for example, be a confidence interval for a specified parameter, such as the mean or standard deviation of a normal population, a binomial proportion, a Poisson occurrence rate, a tolerance interval to contain a specified proportion of the population, a prediction interval to contain one or more future observations. One should also state whether the resulting interval is to be one-sided or two-sided.

8.2.2   Statement of Desired Precision

Statisticians are often asked how many observations are needed to estimate some quantity (e.g., the mean of a sampled population) with "95% confidence." Such a question, however, is generally insufficient per se. In fact, a literal answer is often one or two observations, depending upon the specific quantity being estimated. Unfortunately, the resulting precision is often so poor (reflected by a very long interval) as to make the estimate of little value.

For example, suppose a vendor wants to provide a customer an interval that contains, with 95% confidence, the percentage of conforming units in a large manufacturing lot. Such an interval can be computed even if one has only a single randomly selected unit. If the unit is in conformance, the 95% confidence interval for the percentage of conforming units in the lot is 2.5 to 100% (see Section 6.2). Similarly, if two units had been randomly selected and both found to be conforming, the calculated 95% confidence interval is 16 to 100%. In these cases, a statistical confidence interval has been found, but it is of little practical value.

As a second example, assume that the mean of a normal population is to be estimated from a sample of two observations. A random sample of two units has resulted in readings of 15.13 and 15.25. A 95% confidence interval to include the population mean is 14.43 to 15.95 (see Section 4.2). The 99% confidence interval is much longer still: 11.37 to 19.01!

These examples illustrate that it is often possible to compute confidence intervals from a sample of two, and in some cases, even a single observation. The resulting intervals, however, are, because of their immense length, generally of little value (except to demonstrate how little the data tell us about the characteristic of interest).

Thus, to determine the sample size required to obtain a more *useful* interval, one must specify not only the desired confidence level, but also the desired precision. Such precision is measured, for example, by the allowable error in the resulting estimate or the length of the statistical interval that is to be constructed from the data. For example, for a product packaged in jars labeled to contain "one pound net weight," one might desire a sufficiently large sample to be able to estimate the true mean content weight within 0.1 ounce with 95% confidence.

8.2.3 Assumed Statistical Distribution and Initial Parameter Estimates

Chapters 4, 6, and 7 describe statistical intervals for a normal distribution, a binomial distribution, and a Poisson distribution, respectively. The problem context should make clear which of these models apply. Of course, there are also many other distributions that might be appropriate in a particular situation (see Chapter 11). Also, Chapter 5 describes distribution-free intervals as an alternative to those for the normal, or some other specific, distribution.

The formulas for computing statistical intervals depend on the assumed distribution. Thus, the assumed distribution, if any, must be specified before the required size of the sample can be determined. Moreover, if a particular distribution is assumed, the sample size determination often requires a "planning value" for an unknown distribution parameter. For example, to determine the sample size:

- To estimate the mean of a normal distribution will require a "planning value" of the population standard deviation.
- To estimate the proportion of nonconforming units in a binomial population requires a planning value of the proportion to be estimated.

In general, such information is unknown before the investigation. If it were known, the investigation would be unnecessary in many cases, such as the second example. One can, however, usually provide conservative planning values. These, in turn, will usually result in conservative (i.e., larger than needed) sample sizes.

8.3 SAMPLE SIZE FOR A CONFIDENCE INTERVAL
FOR A NORMAL DISTRIBUTION MEAN

8.3.1 Introduction

This section shows how to choose a sample size large enough to estimate, with a specified precision, the mean of a normal distribution. The length (or half-length) of the resulting confidence interval (described in Section 4.2) is a convenient way to specify the desired precision. We will show how to choose the sample size n such that the resulting confidence interval for μ has the form $\bar{x} \pm d$, where d is the desired confidence interval half-length.

The first few methods described here require a planning value for the population standard deviation. Alternative methods use a two-stage sampling procedure to avoid having the length of the confidence interval depend on a planning value.

8.3.2 Example

An experiment to estimate the mean tensile strength of a new alloy is to be conducted. The experimenters must decide how many specimens to test so that the 95% confidence interval for μ will have a length of ± 1500 kilograms (i.e., $d = 1500$). Experience with similar alloys suggests that variability in specimen strength can be adequately modeled by a normal distribution. A conservative (high) guess for the standard deviation of the distribution of tensile strength is $\sigma^* = 2500$ kilograms.

8.3.3 Tabulations and Simple Formula for the Case When σ Is Assumed to Be Known

Table A.26a gives the sample size needed to estimate μ within $\pm k\sigma$ with probability $1 - \alpha$, where $\sigma^* = \sigma$ is assumed to be *known*. Specifically, the table provides the sample size needed to obtain a confidence interval that has a half-length of $d = k\sigma^*$, as a function of k and $1 - \alpha$. The table provides entries for values of $1 - \alpha$ from 0.50 to 0.999 and for k from 0.01 to 2.00.

The quantities in Table A.26a were computed from the simple approximate formula

$$n = \left(\frac{z_{(1-\alpha/2)}\sigma^*}{d} \right)^2 \qquad (8.1)$$

and then rounding n to the next largest integer.

Of course, Equation (8.1) can also be used directly in place of the tabulations or for untabulated values of k and α. For example, assume σ is known to be $\sigma^* = 2500$, and that the desired half-length for the 95% confidence interval is $d = k\sigma^* = 1500$. Then $k = d/\sigma^* = 1500/2500 = 0.60$ and the necessary sample size from Table A.26a is 11. Equation (8.1) gives

$$n = \left(\frac{1.96(2500)}{1500} \right)^2 = 10.67$$

which, when rounded up to 11, agrees with the value obtained from Table A.26a.

8.3.4   Tabulations for the Case When σ Is Unknown

When σ is unknown, the situation is more complicated because the confidence interval half-length is now itself a random variable, and can, therefore, not be determined exactly ahead of time. Many elementary textbooks give Equation (8.1) as a simple way to determine the approximate sample size needed to obtain a confidence interval that will be close to the specified length. In this case, a planning value σ^* is used in place of σ as a *prediction* of s, the estimate of σ that will be obtained from the sample. At the time the study is planned, neither σ nor s is known. However, to be on the safe side, one can use a conservatively large planning value σ^* for σ.

Kupper and Hafner (1989) show, however, that even if the planning value σ^* were exactly the same as the true value σ the chances are (a probability between 0.53 and 0.87 for the cases considered) that in using Equation (8.1) the half-length of the resulting confidence interval will be larger than the desired value d. The main reason for this is that the sample standard deviation s obtained from the sample, and used in Equation (4.1) for computing the confidence interval for μ in Section 4.2.1, is likely to be larger than σ.

As we have indicated, the length of the confidence interval is a random variable. Thus it is appropriate to select the sample size so that, with a prespecified probability, the resulting interval half-length is not more than d, assuming that the planning value σ^* is, indeed, equal to σ. Table A.26b, reproduced from Kupper and Hafner (1989), provides a means for doing this. The table is used as follows. Assume that we wish the future sample to yield a $100(1 - \alpha)\%$ confidence interval that has a half-length that is no larger than d with $100(1 - \gamma)\%$ probability if $\sigma = \sigma^*$. First one uses Table A.26a or Equation (8.1) to get an initial value n, based again upon a planning value σ^* for σ. Then one enters Table A.26b with this value and the values of $1 - \alpha$ and $1 - \gamma$ and reads the adjusted sample size from the body of the table. The γ' column in Table A.26b gives the probability that the confidence interval half-length is less than d when the initial sample size from the simple formula in Equation (8.1) is used if indeed $\sigma^* = \sigma$, but the sample estimate s, subsequently obtained from the data, is used for σ^* in Equation (8.1).

For the example, from Section 8.3.2, $n \approx 11$. Entering Table A.26b with $1 - \alpha = 0.95$ and $1 - \gamma = 0.90$, and interpolating between $n = 10$ and 15 for $n = 11$ gives a final sample size of about 19. Note that the price one pays to be 90% sure that the half-length of the confidence interval does not exceed $d = 1500$, as compared to using the simpler, less conservative method (that does not provide such assurance), is an increase in the sample size from 11 to 19. We note, moreover, from Table A.26b that if σ indeed were equal to σ^*, then the probability is only 0.34 that the half-length of the confidence interval will be less than d if a sample of size $n = 11$ were used.

8.3.5 Iterative Formula for the Case When σ Is Assumed to Be Unknown

For situations not covered in Table A.26b, and to explain how the table works, we give the following iterative method of finding the sample size for unknown σ.

We wish to find the smallest sample size that allows us to be $100(1 - \gamma)\%$ sure that the resulting $100(1 - \alpha)\%$ confidence interval will have a half-length less than d (assuming that the planning value $\sigma^* > \sigma$). A suitably modified version of Equation (8.1) can be obtained by substituting $t_{(1-\alpha/2;\, n-1)}$ for $z_{(1-\alpha/2)}$ and $\tilde{s}$ for σ^*, where $\tilde{s}$ is the upper $100(1 - \gamma)\%$ prediction bound for s, the standard deviation of the future sample. Thus, we start with the expression

$$n \geq \left(\frac{t_{(1-\alpha/2;\, n-1)}\tilde{s}}{d} \right)^2 \tag{8.2}$$

The upper $100(1 - \gamma)\%$ prediction bound for s (assuming that σ^* is the *true* value of σ) is

$$\tilde{s} = \sigma^* \left(\frac{\chi^2_{(1-\gamma;\, n-1)}}{n - 1} \right)^{1/2}.$$

Substituting this for $\tilde{s}$ in Equation (8.2) yields

$$n \geq \left(\frac{t_{(1-\alpha/2;\, n-1)}\sigma^*}{d} \right)^2 \left(\frac{\chi^2_{(1-\gamma;\, n-1)}}{n - 1} \right) \tag{8.3}$$

This is equivalent to an equation given by Kupper and Hafner (1989). Because n appears on both sides of Equation (8.3), iteration is required to find a solution. In particular, we need to find the smallest value of n such that the left-hand side is greater than the right-hand side. It is easy to write a computer program to do this with standard numerical methods. Working manually, one can use the following steps:

1. Use Equation (8.1) to get a starting value n_1 for n, based upon the planning value σ^*.
2. Use the formula

$$n_2 = \left(\frac{t_{(1-\alpha/2; n_1-1)}\sigma^*}{d} \right)^2 \left(\frac{\chi^2_{(1-\gamma; n_1-1)}}{n_1 - 1} \right)$$

and round to the next higher integer to obtain an adjusted sample size value n_2.
3. Obtain successive additional values for n by using the recursion formula

$$n_i = \left(\frac{t_{(1-\alpha/2; n_{i-1}-1)}\sigma^*}{d} \right)^2 \left(\frac{\chi^2_{(1-\gamma; n_{i-1}-1)}}{n_{i-1} - 1} \right) \qquad (8.4)$$

and round up at each stage until $n_i = n_{i-1}$ or $n_i = n_{i-1} - 1$. Generally no more than about five to seven iterations are required.

Using the previous example, assume now that we wish to be 90% sure that the half-length of the 95% confidence interval to contain μ does not exceed 1500, and our planning value for σ is 2500. Thus, as before, $\sigma^* = 2500$, $d = 1500$, $1 - \alpha = 0.95$, and $1 - \gamma = 0.90$. Our first guess is $n_1 = 11$—the solution for the case where σ is known. Then, using $t_{(0.975; 10)} = 2.228$ and $\chi^2_{(0.90, 10)} = 15.99$ in Equation (8.3), we obtain

$$n_2 = \left(\frac{2.228(2500)}{1500} \right)^2 \left(\frac{15.99}{10} \right) = 22.047.$$

Rounding up to $n_2 = 23$ and using $t_{(0.975; 22)} = 2.074$ and $\chi^2_{(0.90; 22)} = 30.81$ gives

$$n_3 = \left(\frac{2.074(2500)}{1500} \right)^2 \left(\frac{30.81}{22} \right) = 16.73.$$

The next three iterations give, after rounding up, $n_4 = 19$, and $n_5 = 18$, and $n_6 = 19$, allowing us to terminate the iterations.

Thus $n = 19$ is the smallest integer such that the left-hand side of Equation (8.3) is greater than the right-hand side. This also agrees with the value obtained from Table A.26b.

If one has an estimate of σ from a previous random sample from the same population, one can obtain a more appropriate modified sample size formula by using a *statistical* prediction bound for s, instead of using the planning value σ^*; see Section 4.9. In this case, Table A.26b cannot be used. However the iterative scheme that we have just described again applies.

In particular, if σ_ℓ^* is a planning value (in this case, the sample standard deviation) based on a previous sample with $\ell - 1$ degrees of freedom, one needs to solve

$$n \geq \left(\frac{t_{(1-\alpha/2; n-1)}\sigma_\ell^*}{d} \right)^2 F_{(1-\gamma; n-1, \ell-1)} \tag{8.5}$$

for the smallest value of n such that this inequality holds. This can be done by trial and error or with a simple computer program. As $\ell \to \infty$, (and thus $\sigma_\ell^* \to \sigma$) this approach is equivalent to that proposed by Kupper and Hafner (1989). It is also possible to develop correction tables like those given in Kupper and Hafner (1989), but a separate table would be required for each value of ℓ. A simple (large-sample) approximation for the required sample size is

$$n \approx n_1 F_{(1-\gamma; n_1-1, \ell-1)}$$

where n_1 is obtained from Equation (8.1) or (8.2), using σ_ℓ^* for σ^*.

8.3.6 A Double Sampling Method

When an investigation can be conducted in two stages, it is possible to obtain a confidence interval with exactly the desired half-length d, even if σ is unknown. This is done by using the sample standard deviation of the first stage to compute a confidence interval, and ignoring the sample standard deviation of the second stage in constructing the confidence interval. The sample size n_1 for the first stage should be chosen as large as possible, but smaller than the anticipated total sample size. Specifically, obtain n_1, say, from Equation (8.1) using a planning value for σ^*, which now need not be conservative. The sample size for the second stage of the investigation is then

$$n_2 = \left(\frac{t_{(1-\alpha/2; n_1-1)}s_1}{d} \right)^2 - n_1$$

where s_1 is the sample standard deviation from the first stage. The resulting $100(1 - \alpha)\%$ confidence interval for μ is

$$\bar{x} \pm d$$

where $\bar{x}$ is the mean of the $n_1 + n_2$ observations from both stages. If n_2

turns out to be negative, then the data from the first stage will be sufficient and, in fact, will give an interval with a half-length that is less than d. In this case the second sample is not needed.

We continue with the previous example, but now use $d = 500$. Assume that an initial sample of 20 units yielded $s_1 = 2500$. Then, using $t_{(0.975; 19)} = 2.093$, the investigation would require a total of

$$n_1 + n_2 = \left(\frac{2.093(2500)}{500} \right)^2 \approx 110$$

observations, i.e., $n_2 = 110 - 20 = 90$ observations in the second stage.

If an initial sample of size 10, instead of size 20, had been taken, and had again given $s_1 = 2500$, the total sample size requirements would have been estimated, using $t_{(0.975; 9)} = 2.262$, to be

$$n_1 + n_2 = \left(\frac{2.262(2500)}{500} \right)^2 \approx 128$$

instead of 110.

This double sampling procedure, first suggested by Stein (1945), is subject to criticism because, though exact, it does not, as previously indicated, use the information from the second stage to estimate σ. However, this will not be a serious practical concern if the sample size in the first stage is sufficiently large (e.g., $n_1 > 30$).

8.4 SAMPLE SIZE TO ESTIMATE A NORMAL DISTRIBUTION STANDARD DEVIATION

This section gives easy-to-use methods for choosing the sample size needed to estimate a normal distribution standard deviation σ.

8.4.1 Example

An experiment is to be conducted to estimate σ, the standard deviation of the measurement error of a chemical assay procedure. From previous experience with similar procedures, one can assume that the measurement error is normally distributed. The experiment will be conducted with specimens that are known to contain exactly the same amount of a chemical and that are destroyed during the assay. The experimenters need to know how many specimens to prepare so that, with 95% probability, the sample standard deviation s will underestimate the true measurement error standard deviation σ by no more than 20%.

8.4.2 Graphical Method

Figure 8.1 gives the sample size needed so that, with a specified probability $1 - \alpha$, σ will be underestimated by no more than $100P\%$. Curves are provided for $1 - \alpha = 0.70, 0.80, 0.90, 0.95, 0.99,$ and 0.999.

For the example, entering the "bound on percent error" axis with $100P\% = 20\%$ and reading down from the 95% curve gives $n \cong 36$. As a check, we note that the exact probability bound on the percent error is computed as

$$100\left\{ 1 - \left[\frac{\chi^2_{(\alpha; n-1)}}{n - 1} \right]^{1/2} \right\}.$$

In the example, $n - 1 = 35$; thus, using $\chi^2_{(0.05; 35)} = 22.46$, the 95% upper bound on the the percent error for estimating σ is

$$100\left\{ 1 - \left(\frac{22.46}{35} \right)^{1/2} \right\} = 19.8\%,$$

or approximately 20%.

8.4.3 Tabular Method

Table A.27 provides the minimum sample size n to assure that $\Pr(s < k\sigma) \geq 1 - \alpha$ for $1 - \alpha = 0.80, 0.85, 0.90, 0.95, 0.98, 0.99,$ and 0.999 and values of k from 1.01 to 3.0. For the example, taking $k = 1.20$ and $1 - \alpha = 0.95$ gives $n \cong 34$.

8.4.4 Large-Sample Approximate Formula

Nelson (1982) suggests using

$$n \cong 1.0 + 0.5\left[\frac{z_{(1-\alpha/2)}}{\log_e(k)} \right]^2$$

to find the approximate sample size needed to estimate σ to within $\pm 100(k - 1)\%$ of the true value with $100(1 - \alpha)\%$ probability. For one-sided bounds, replace $\alpha/2$ by α. This formula is based on the fact that, in large samples, the sampling distribution of $\log_e(s)$ is approximately normal. The approximation is adequate for two-sided intervals for most practical purposes when the resulting n exceeds 15, but may be quite coarse for one-sided bounds.

In the example, the sample size needed so that the estimate of σ is no more than 20% less than the true value (i.e., $k = 1.20$) with 95% probability

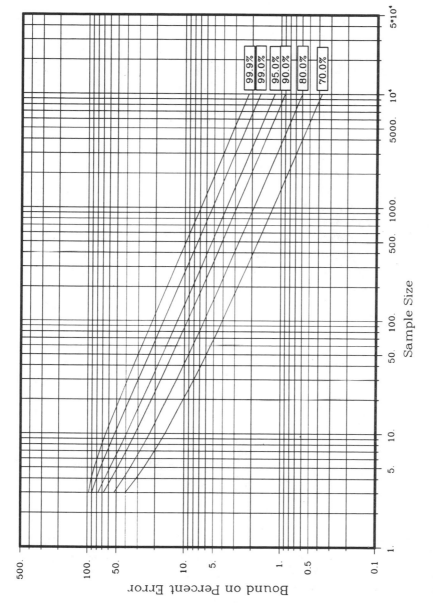

Figure 8.1 Sample size needed to estimate a normal distribution standard deviation for various probability levels. This figure is based on methodology described by Greenwood and Sandomire (1950).

[using $z_{(0.95)} = 1.645$] is approximately

$$n \cong 1.0 + 0.5 \left[\frac{1.645}{\log_e(1.20)} \right]^2 \cong 42,$$

as compared to $n = 36$ and $n = 34$ obtained previously.

8.5   SAMPLE SIZE TO ESTIMATE A BINOMIAL PROPORTION

8.5.1   Introduction

This section shows how to determine the approximate sample size needed to estimate a population proportion, with specified precision. More specifically, we wish the resulting confidence interval to be no larger than $\hat{p} \pm d$, where d is the confidence interval half-length. To make this determination, one must provide a "planning value," to be denoted by p^*, for the sample estimate $\hat{p}$ of p that will be obtained from the data. A conservative planning value for p^* (in the sense that, if incorrect, it will tend to overestimate the required sample size) is the value closest to 0.50 that still appears plausible.

8.5.2   Example

Some unknown proportion p of a large number of installed devices of a particular type were assembled incorrectly and have to be repaired. To assess the magnitude of the problem, the manufacturer of the devices wishes to estimate p, the proportion of incorrectly assembled devices in the field. In particular, the manufacturer needs to know how many units to sample at random so that $\hat{p}$ will be within ± 0.08 of p with 90% confidence, i.e., the 90% confidence interval around the estimate $\hat{p}$ is no larger than ± 0.08. It is conceivable that p could be near 0.50, and, therefore, $p^* = 0.50$ will be taken as the planning value for p.

8.5.3   Graphical Procedure

As explained in Section 6.2, Figures 6.1a to 6.1d or Table A.23 and A.23b can be used to obtain confidence limits for a population proportion p. The figures can be used in reverse to determine a sample size for estimating p. To do this one must first choose a planning value p^*. One then uses Figure 6.1a to 6.1d (depending upon the desired confidence level) and sets $\hat{p} = p^*$ on the horizontal axis. The resulting confidence intervals, for various sample sizes are read on the (vertical) p axis. Thus, at a glance, one can readily assess the effect on the confidence interval length of using different sample sizes.

For example, using $p^* = 0.50$ we see that a sample of size 10 gives a 90% confidence interval of [0.22, 0.78], a sample of size 30 gives a 90% confidence interval of [0.34, 0.66], a sample size of 100 gives a 90% confidence interval of [0.41, 0.59], and a sample size of 200 gives a confidence interval of [0.44, 0.56]. Thus, to estimate p within ± 0.08 with 90% confidence, requires a sample size somewhat above 100 if p^* is taken to be 0.50. In practice, one would, of course, use the resulting sample value $\hat{p}$, in place of p^* in obtaining the desired 90% confidence interval to contain p, and this interval would be smaller in length than ± 0.08 unless $\hat{p} = 0.50$. Note also that if p^* is taken either smaller or larger than 0.50, the required sample size to achieve an interval of the same length would be somewhat smaller.

8.5.4   A Simple Approximate Expression for Sample Size

The expression

$$n = \left[p^*(1 - p^*) \right] \left[\frac{z_{(1-\alpha/2)}}{d} \right]^2$$

rounded to the next largest integer gives the approximate sample size needed to obtain a $100(1 - \alpha)\%$ confidence interval for p with length $\pm d$. This normal distribution large sample size approximation is generally satisfactory if both np and $n(1 - p)$ exceed 10. Like the graphical procedure, use of this expression requires a planning value p^*. The most conservative (largest) n is again obtained by choosing $p^* = 0.50$.

In the example, $d = 0.08$, $1 - \alpha = 0.90$ (or $1 - \alpha/2 = 0.95$), and $p^* = 0.50$. Thus, using $z_{(0.95)} = 1.645$, the required sample size is

$$n = \left[0.50(1 - 0.50) \right] \left[\frac{1.645}{0.08} \right]^2 \cong 106.$$

Suppose, on the other hand, one expects $\hat{p}$ either to be less than 0.20 (or to be greater than 0.80). Then using the planning value $p^* = 0.20$, the required sample size to obtain a 90% confidence interval on p with length $\pm d$ is only

$$n = \left[0.20(1 - 0.20) \right] \left[\frac{1.645}{0.08} \right]^2 \cong 68.$$

8.6   SAMPLE SIZE TO ESTIMATE A POISSON OCCURRENCE RATE

This section shows how to choose the approximate sample size to estimate a Poisson occurrence rate λ with a specified precision. That is, we require the sample to be sufficiently large so that the upper or lower confidence bound does not differ from the sample estimate by more than a prespecified percentage. It will be assumed that events occur independently at a constant occurrence rate λ, and, thus, that the distribution of the number of occurrences is Poisson. It is necessary to provide a "planning value" λ^* for $\hat{\lambda}$, the sample estimate of λ. A conservative approach (tending to result in a larger sample size than actually needed) is to take λ^* to be the largest value expected for $\hat{\lambda}$.

8.6.1 Example

Flaws on the painted surface of an appliance occur independently of one another at a constant rate λ. This implies a Poisson distribution for the number of flaws per constant surface area. A new process that is believed to have a flaw rate λ of not more than 0.10 flaws per appliance has been developed. An experiment is to be conducted to estimate this mean flaw rate precisely enough so that the one-sided upper 95% confidence bound on λ will exceed the sample estimate $\hat{\lambda}$ by not more than 20%. It is desired to determine how many appliances must be selected at random from the process to meet this criterion.

8.6.2 Graphical Method

For a specified confidence level, $100(1 - \alpha)\%$, Figure 8.2a shows the percent by which the lower confidence bound for a Poisson occurrence rate is less than the sample estimate as a function of the number of occurrences in the sample. In particular, curves are given for $1 - \alpha = 0.70, 0.80, 0.90, 0.95, 0.99$, and 0.999. Similarly, Figure 8.2b provides the percent by which the upper one-sided confidence bound for a Poisson occurrence rate exceeds the sample estimate. In Figure 8.2a, the percent error was computed as $100[1 - (\lambda/\hat{\lambda})]$ and for Figure 8.2b, the percent error rate was computed as $100[(\tilde{\lambda}/\hat{\lambda}) - 1]$. Which one of these figures one should use depends on the problem context. In our example Figure 8.2b appears appropriate, because after the data are obtained, one would generally calculate an *upper* confidence bound to be "safe."

   To use these figures to determine the needed sample size, enter the graph on the vertical axis at the point corresponding to the desired bound on the percent error in the estimate of the occurrence rate. Then draw a line horizontally to intersect the line corresponding to the desired degree of confidence. Now move down from the point of intersection to read x, the "needed" number of occurrences, from the horizontal axis. Using a planning value λ^* for the occurrence rate λ, the approximate required sample size is $n \cong x/\lambda^*$.

   For the example, from Figure 8.2b, the one-sided upper 95% confidence bound $\tilde{\lambda}$ will be about 20% more than $\hat{\lambda}$ if we observe 82 flaws. Using the

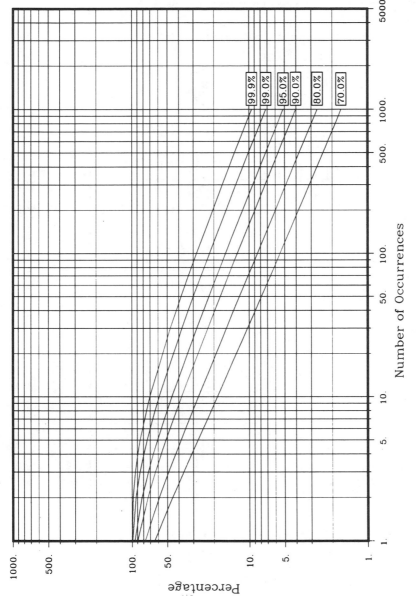

Figure 8.2a Percentage by which the lower confidence bound for the Poisson parameter λ is less than λ̂ for various confidence levels.

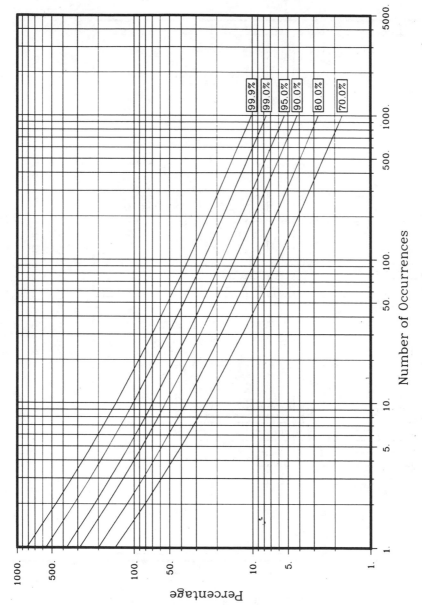

Figure 8.2b Percentage by which the upper confidence bound for the Poisson parameter λ exceeds $\hat{\lambda}$ for various confidence levels.

148

planning value $\lambda^* = 0.10$, the experiment will require approximately $n = x/\lambda^* = 82/0.10 = 820$ test units (again assuming flaws occur independently of one another, in accordance with the Poisson distribution assumption.) If this problem had called for a lower 95% confidence bound λ, we note, from Figure 8.2a, that λ will be about 20% less than $\hat{\lambda}$ if we observe 63 flaws, and thus the experiment would require approximately 630 appliances.

8.6.3 A Simple Approximate Expression for Sample Size

The expression

$$n = \lambda^* \left[\frac{z_{(1-\alpha/2)}}{d} \right]^2$$

rounded to the next larger integer gives the approximate sample size needed to obtain a $100(1 - \alpha)\%$ confidence interval of length $\pm d$ for the Poisson occurrence rate. For a one-sided bound, replace $1 - \alpha/2$ with $1 - \alpha$. This approximation is based on the fact that $\hat{\lambda}$ follows approximately a normal distribution in large samples and is often adequate, for practical purposes for problems requiring two-sided confidence intervals, when the sample size n is large enough to yield at least 10 occurrences.

For the example, $\lambda^* = 0.10$, $z_{(0.95)} = 1.645$, and $d = 0.02$ (20% of λ^*). Thus the experiment will require

$$n = 0.10 \left[\frac{1.645}{0.02} \right]^2 \cong 677$$

appliances. The difference between this result and the previous approximate sample size of 630, even though $n\lambda^* = (677)(0.10) \cong 68$, is attributed to the fact that this example deals with a one-sided confidence bound, rather than a two-sided confidence interval.

CHAPTER 9

Sample Size Requirements for Tolerance Intervals, Tolerance Bounds, and Demonstration Tests

9.1 INTRODUCTION

This chapter shows how to determine sample size requirements for tolerance intervals and for demonstration tests concerning the proportion of product that exceeds (or is exceeded by) a specified value. Sections 9.2 and 9.3 treat situations where the normal distribution is an appropriate model. Sections 9.4, 9.5, and 9.6 deal with distribution-free situations.

9.2 SAMPLE SIZE FOR NORMAL DISTRIBUTION TOLERANCE BOUNDS AND INTERVALS

This section provides simple methods for finding the sample size needed to achieve a specified level of precision when the data are to be used to compute a two-sided tolerance interval or a one-sided tolerance bound to contain at least a specified proportion of a sampled normal population. As the sample size increases, the computed tolerance interval will approach the probability interval that *actually* contains the specified population proportion. Small sample sizes can, however, result in a tolerance interval that is much longer than this limiting probability interval.

9.2.1 Criterion for the Precision of a Tolerance Interval

Faulkenberry and Weeks (1968) suggested the following criterion for finding the sample size to control the size of a tolerance interval. Choose the sample size to be large enough such that both (1) the probability is $1 - \alpha$ (large) that at least $100p'\%$ of the population will be included within the tolerance interval, and (2) the probability is δ (small) that more than $100p^*\%$ of the population will be included, where p' and p^* are specified proportions and p^* is larger than p'. The idea is that, with fixed $p^* > p'$, and probability

$$1 - \alpha = \text{Pr(interval will contain at least } 100p'\% \text{ of the population)},$$

the probability

$$\delta = \text{Pr(interval will contain at least } 100p^*\% \text{ of the population)}$$

is a decreasing function of the sample size n. That is, δ, the probability that the interval is so long that it will contain $100p^*\%$ of the population, will decrease to zero as n increases. This criterion can be used for both two-sided tolerance intervals and one-sided tolerance bounds.

9.2.2   Tabulation for Tolerance Interval Sample Sizes

Tables A.28 and A.29 give, respectively, the necessary sample sizes for two-sided tolerance intervals and one-sided tolerance bounds for a normal distribution for $p' = 0.50, 0.75, 0.90, 0.95, 0.99, 1 - \alpha = 0.80, 0.90, 0.95, 0.99,$ $\delta = 0.20, 0.10, 0.05, 0.01$, and several values of p^*, depending on p'. Similar tables were first presented by Faulkenberry and Daly (1970).

9.2.3   Example for a Two-Sided Tolerance Interval

The engineers responsible for a machined part want to establish limits for a critical dimension so that, for markcting purposes, they can claim, with 95% confidence, that the interval contains the dimension for a large proportion of the parts. Based on experience with the process control scheme and with similar processes, the engineers feel that the dimensions can be adequately modeled by a normal distribution. The measurements on the dimensions for a random sample of the parts will be used to compute a two-sided tolerance interval to contain 90% of the population of parts produced from the process with 95% confidence, and this will provide the desired limits.

   If the sample size is too small, it is possible that the tolerance interval providing the desired coverage with the specified level of confidence will be so long that it will appreciably overestimate the scatter in the distribution of the dimensions. Thus, in addition to the requirement that the tolerance interval contain at least 90% of the population ($p' = 0.90$) with 95% confidence ($1 - \alpha = 0.95$), the manufacturer wants to choose the sample size n sufficiently large so that the probability is only $\delta = 0.10$ that the interval will actually contain 96% or more ($p^* = 0.96$) of the dimensions of the manufactured parts. From Table A.28, the sample size needed to accomplish this is $n = 91$ units. After the sample has been obtained, the desired tolerance interval is calculated using the methods of Section 4.6.2.

9.2.4   Example for a One-Sided Tolerance Bound

The designers of a system want a one-sided lower 95% tolerance bound for the strength of a critical component. Again, the normal distribution model is felt to adequately describe the distribution of strengths. An experiment is to be conducted to obtain data to compute a lower tolerance bound that will, with 95% confidence, be exceeded by the strengths of at least 99% of the components in the sampled product population.

   If the chosen sample size is too small, the resulting lower tolerance bound for strength will be unduly conservative. Thus, in addition to the requirement that the lower tolerance bound be exceeded by at least 99% of the components in the population ($p' = 0.99$), with 95% confidence ($1 - \alpha = 0.95$), the engineers want to choose a sample large enough so that the probability is only 0.01 ($\delta = 0.01$) that the resulting lower tolerance bound will be exceeded by the strengths of 99.7% or more of the components in the population ($p^* = 0.997$). From Table A.29, the necessary sample size is $n = 370$. After the sample has been obtained the desired tolerance bound is found using the methods of Section 4.6.3.

9.3   SAMPLE SIZE TO PASS A ONE-SIDED DEMONSTRATION TEST BASED ON NORMALLY DISTRIBUTED MEASUREMENTS

This section shows how to find the sample size needed to conduct a test to demonstrate, with $100(1 - \alpha)\%$ confidence, that the $100p$th percentile of a normal distribution, denoted by Y_p, is less than a specified value Y_p^S. The value Y_p^S is often an upper specification limit that most of the product values (denoted by y) should not exceed, as in the example to follow. This problem is equivalent to finding the sample size needed to demonstrate that $p = \Pr(y < Y_p^S)$ is greater than p' for specified Y_p^S and p'. We want the demonstration to be successful with specified high probability p_{DEM} when the actual probability p is equal to a specified value $p^* > p'$.

   In reliability demonstrations, it is often necessary to demonstrate that $Y_p > Y_p^S$, where p is typically small, corresponding to a low percentile of the life distribution. In this case, Y_p^S is a lower, rather than an upper, specification limit, e.g., minimum life. This situation can be handled by the same methods as those to be described for an upper specification limit. This is because the sample size needed to demonstrate $q = 1 - p = \Pr(y > Y_p^S) < q'$ is the same as that needed to demonstrate $p = \Pr(y < Y_p^S) > p'$.

9.3.1   Example

A proposed federal standard requires that the measured noise level of a particular type of machinery not exceed 60 decibels at a distance of 20 meters from the source for at least 95% of the units; i.e., $Y_{0.95} < Y_{0.95}^S = 60$ decibels.

A manufacturer of such machinery needs to test a random sample of the many units in the field to demonstrate compliance. The data will be used to find an upper 90% confidence bound $(1 - \alpha = 0.90)$ on $Y_{0.95}$, the 95th percentile $(p' = 0.95)$ of the distribution of noise emitted from the population of units in the field. The demonstration will be successful if the upper 90% confidence bound on the 95th percentile of the population, i.e., $\tilde{Y}_{0.95}$, does not exceed 60 decibels. Equivalently, the demonstration requires $p > p' = 0.95$, where $\underset{\sim}{p}$ is a lower 90% confidence bound for p, the proportion of units with noise levels less than 60 decibels. The manufacturer wishes the sample size to be sufficiently large so that the probability of a successful demonstration is $p_{\text{DEM}} = \text{Pr}(\underset{\sim}{p} > p') \geq 1 - \delta = 0.95$ (i.e., $\delta = 0.05$) when the true proportion conforming is $p^* = 0.98$.

9.3.2 Graphical Method

Figures 9.1a to 9.1n show p_{DEM}, the probability of successfully demonstrating that the proportion conforming in the population is greater than p' at a $100(1 - \alpha)\%$ confidence level. This probability is a function of the true proportion conforming p^*, the sample size n, and the probability to be demonstrated p', assuming a normal distribution. These charts, which were developed from theory based on the noncentral t distribution, described by Odeh and Owen (1980, page 269), cover all combinations of $1 - \alpha = 0.90$ and 0.95, and $p' = 0.50$, 0.70, 0.80, 0.90, 0.95, 0.97, and 0.99.

For the example, we use Figure 9.1e, designed for determining the required sample size for demonstrating with $100(1 - \alpha)\% = 90\%$ confidence that $p > p' = 0.95$. In particular, we enter the horizontal scale at $p^* = 0.98$ and move up, and simultaneously, enter the vertical scale at the desired $p_{\text{DEM}} = 0.95$ and move to the right. After finding the point of intersection, we interpolate for the needed value of n. In this case, we interpolate between 126 and 148 to get $n \approx 137$.

9.3.3 Tabular Method

The tabular method to determine the required sample size for such problems uses Table A.29. This provides the required sample size for all combinations of $1 - \alpha = 0.80$, 0.90, 0.95, and 0.99, $p' = 0.500$, 0.750, 0.900, 0.950, and 0.990, $\delta = 1 - p_{\text{DEM}} = 0.01$, 0.05, 0.10, and 0.20, and various values of p^*, depending on p'. In our example we want to show, with 90% confidence $(1 - \alpha = 0.90)$, that $p > p' = 0.95$ and to have a success probability of $p_{\text{DEM}} = 0.95$ $(\delta = 0.05)$ when the actual conforming proportion is $p^* = 0.98$. From Table A.29, we find the necessary sample size to be $n = 138$. The demonstration will be successful if the resulting upper confidence bound for $Y_{0.95}$, $\tilde{Y}_{0.95} = \bar{x} + 1.83s$, is less than 60. Here $\tilde{Y}_{0.95}$ is calculated as shown in Section 4.4, using interpolation to obtain $g'_{(0.90; 0.95, 138)} = 1.83$.

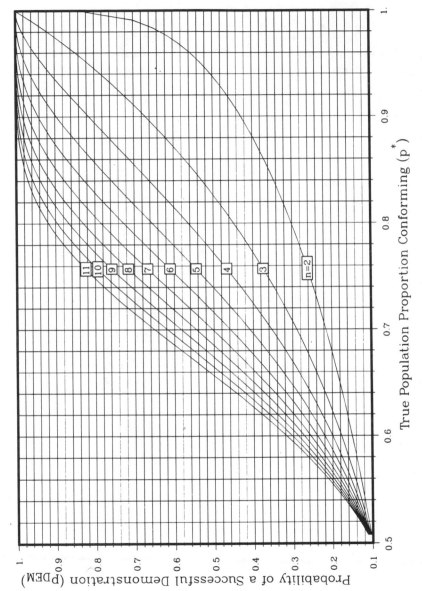

Figure 9.1a Probability of successfully demonstrating that $p > 0.50$ with 90% confidence (normal distribution).

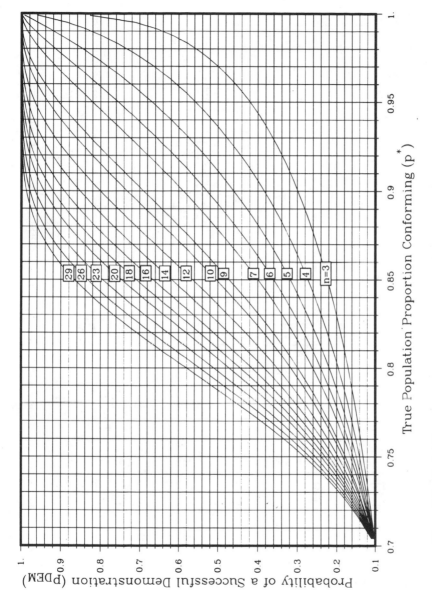

Figure 9.1b Probability of successfully demonstrating that $p > 0.70$ with 90% confidence (normal distribution).

155

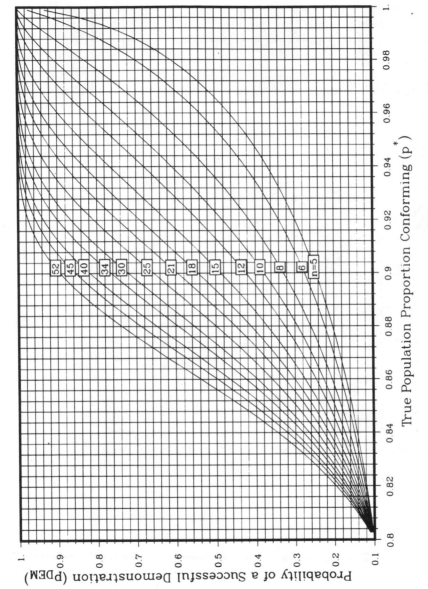

Figure 9.1c Probability of successfully demonstrating that $p > 0.80$ with 90% confidence (normal distribution).

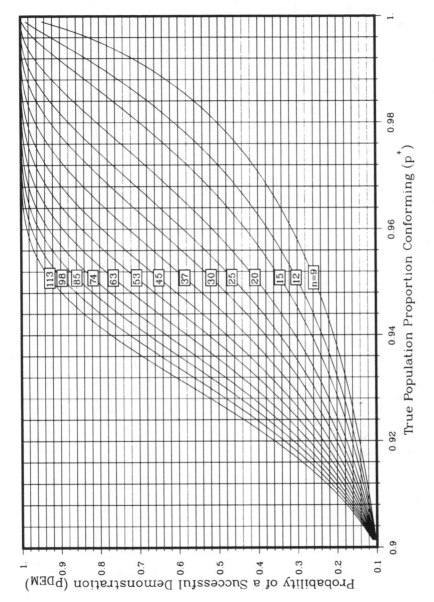

Figure 9.1d Probability of successfully demonstrating that $p > 0.90$ with 90% confidence (normal distribution).

157

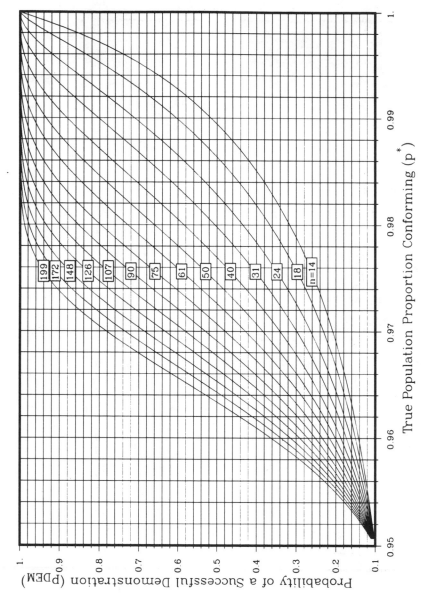

Figure 9.1e Probability of successfully demonstrating that $p > 0.95$ with 90% confidence (normal distribution).

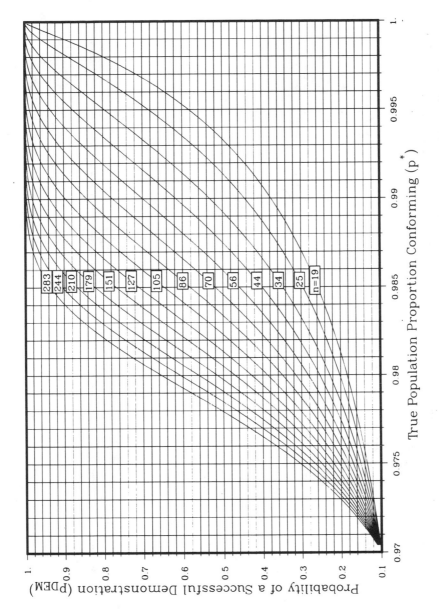

Figure 9.1f Probability of successfully demonstrating that $p > 0.97$ with 90% confidence (normal distribution).

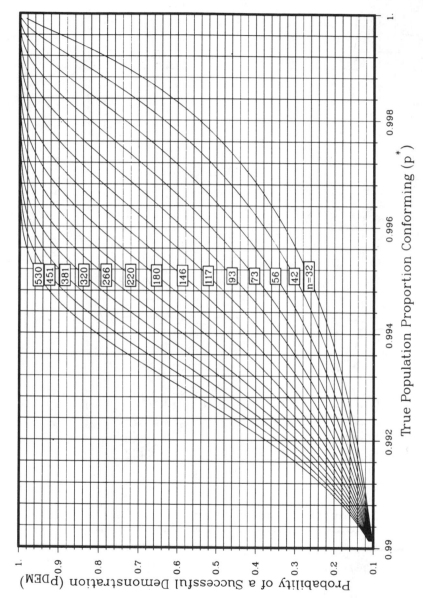

Figure 9.1g Probability of successfully demonstrating that $p > 0.99$ with 90% confidence (normal distribution).

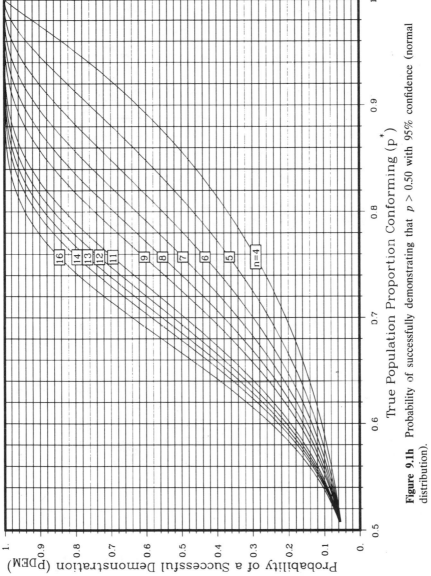

Figure 9.1h Probability of successfully demonstrating that $p > 0.50$ with 95% confidence (normal distribution).

161

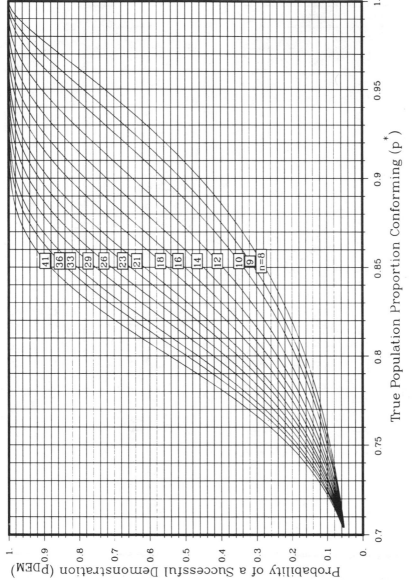

Figure 9.1i Probability of successfully demonstrating that $p > 0.70$ with 95% confidence (normal distribution).

162

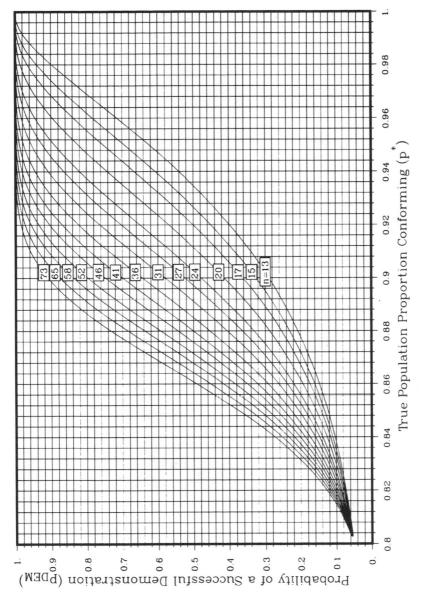

Figure 9.1j Probability of successfully demonstrating that $p > 0.80$ with 95% confidence (normal distribution).

163

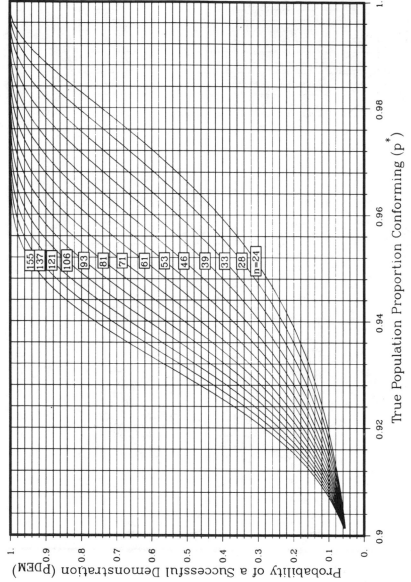

Figure 9.1k Probability of successfully demonstrating that $p > 0.90$ with 95% confidence (normal distribution).

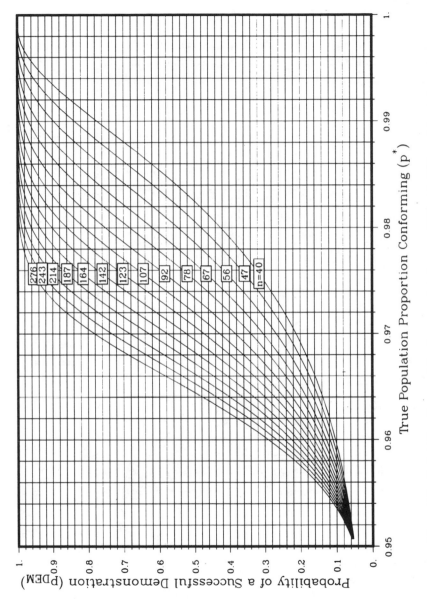

Figure 9.11 Probability of successfully demonstrating that $p > 0.95$ with 95% confidence (normal distribution).

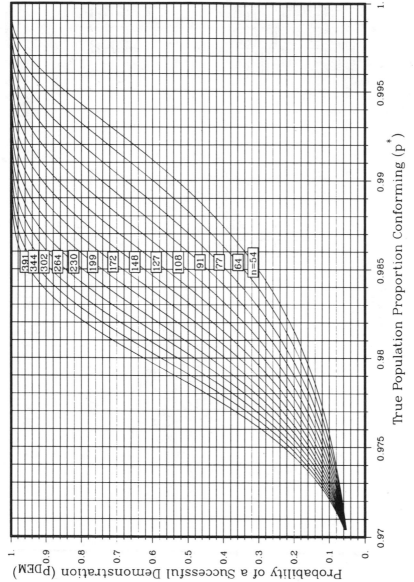

Figure 9.1m Probability of successfully demonstrating that $p > 0.97$ with 95% confidence (normal distribution).

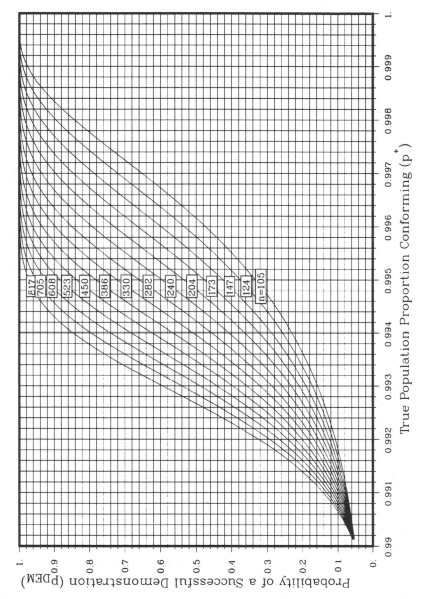

Figure 9.1n Probability of successfully demonstrating that $p > 0.99$ with 95% confidence (normal distribution).

167

9.3.4 Relationship to Odeh and Owen (1980) Tables for Normal Distribution Sampling Plans

Our Table A.29 and Odeh and Owen's (1980) Table 2 (which gives one-sided normal distribution sampling plans controlling producer's and consumer's risk) are closely related. The corresponding entries are Odeh and Owen's (O & O's) $p_2 = $ our $1 - p'$, O & O's $p_1 = $ our $1 - p^*$, O & O's $\alpha = $ our δ, and O & O's $\beta = $ our α. Odeh and Owen (1980) provide entries for all combinations of $p_1 = 0.005(0.005)0.05, 0.075, 0.10$; $p_2 = 2p_1(0.005)0.10, 0.15$, 0.20, 0.30, with $\alpha = 0.01, 0.025, 0.05$, and $\beta = 0.05, 0.10, 0.20$. Some additional combinations of p_1 and p_2 are provided for $\alpha = 0.05$ and $\beta = 0.10$.

9.4 MINIMUM SAMPLE SIZE FOR DISTRIBUTION-FREE TWO-SIDED TOLERANCE INTERVALS AND ONE-SIDED TOLERANCE BOUNDS

9.4.1 Two-Sided Tolerance Intervals

Section 5.3.1 describes how to compute two-sided distribution-free tolerance intervals. As indicated there, Table A.17 gives the smallest sample size needed to provide $100(1 - \alpha)\%$ confidence that the interval defined by the range of sample observations will contain at least $100p\%$ of the sampled population for $1 - \alpha = 0.50, 0.75, 0.90, 0.95, 0.98, 0.99, 0.999$, and $p = 0.50(0.05)0.95(0.01)0.99, 0.995, 0.999$.

9.4.2 Example

Assume that in the example in Section 9.2.3, the manufacturer now wants a distribution-free tolerance interval that does not require the assumption that the dimensions follow a normal distribution. We still, however, must assume that we are dealing with a random sample from the population of interest. Now the manufacturer wants to find the smallest sample size that will use the minimum and the maximum observations for a two-sided distribution-free tolerance interval to contain, with 95% confidence, the critical dimension for at least 90% of the units in the sampled population. From Table A.17 we see that a minimum random sample of size $n = 46$ units is needed.

One might be surprised that the required sample size for a distribution-free tolerance interval ($n = 46$) is smaller than that when normality is assumed ($n = 91$). However, the two intervals are not comparable. In particular, the normal distribution-based interval is more demanding, because, in addition to specifying that we wished to include 90% of the population with 95% confidence, we also required the probability to be no more than $\delta = 0.10$ that the interval will cover 96% or more of the population values. On the other hand, the sample size determination for the distribution-free case did *not* have this second requirement. In fact, it called for the *smallest* possible sample size to include 90% of the population with 95% confidence, irrespective of the coverage achieved (see Section 9.5 for sample size determination that controls the precision of a distribution-free tolerance interval).

9.4.3 One-Sided Tolerance Bounds

A distribution-free one-sided tolerance bound is equivalent to a one-sided distribution-free confidence bound for a percentile of that population. That is, a one-sided distribution-free lower (upper) $100(1 - \alpha)\%$ tolerance bound that will be exceeded by (that will exceed) at least $100p\%$ of the population is the same as a distribution-free lower (upper) $100(1 - \alpha)\%$ confidence bound for the $100p$th percentile of the population. Methods for constructing such bounds are given in Sections 5.2.3 and 5.3.3. As indicated there, Table A.18 gives the smallest sample size needed to obtain a one-sided tolerance bound that will, with $100(1 - \alpha)\%$ confidence, be exceeded by (exceed) at least $100p\%$ of the population. This table is based on the expression $n = \log(\alpha)/\log(p)$, which can, of course, also be used directly—and has to be—for nontabulated values.

9.4.4 Example

A group of reliability engineers is planning a test to estimate the life of a newly designed engine bearing. They wish to use the results to compute a lower tolerance bound that they can claim, with 95% confidence, will be exceeded by the lifetimes of at least 99% of the population of bearings, i.e., $1 - \alpha = 0.95$ and $p = 0.99$. The engineers want to know the minimum required sample size to obtain the desired lower bound with no distributional assumptions. This requires using the first bearing failure time as the bound. In this case, if the bearings are placed on test simultaneously, the test can be terminated after the first failure. The practical usefulness of the results will depend on the magnitude of the first failure. In particular, if the first failure occurs very early, all that we know is how little we know. From Table A.18, we note that 299 bearings need be tested. Equivalently, we find that $n = \log(0.05)/\log(0.99) \cong 299$.

9.5 SAMPLE SIZE FOR CONTROLLING THE PRECISION OF TWO-SIDED DISTRIBUTION-FREE TOLERANCE INTERVALS AND ONE-SIDED DISTRIBUTION-FREE TOLERANCE BOUNDS

Distribution-free tolerance intervals based on the smallest possible sample size will often be too long for the intended application. For example, in Section 9.4.2, a 95% tolerance interval to contain 90% of the population, based on the smallest possible sample size ($n = 46$), results in an interval that is so long that the probability is 0.554 [from Equation (5.2)] that the tolerance interval will, in fact, contain more than 96% of the population.

Table A.30 is similar to Tables A.28 and A.29 for the normal distribution tolerance intervals and bounds. It provides the sample size needed to control the precision of distribution-free two-sided tolerance intervals and one-sided tolerance bounds, using the Faulkenberry and Weeks criterion described in Section 9.2.1. For the distribution-free case, the criterion results in the same sample size for both two-sided tolerance intervals and one-sided tolerance bounds. This is because the level of confidence for the distribution-free tolerance intervals and bounds depend on the number of "blocks" that are removed from the end(s) of the sample and *not* the end(s) from which they come.

9.5.1 Two-Sided Tolerance Interval Example

Consider the example in Sections 9.2.3 and 9.4.2. Assume the distribution-free tolerance interval to contain 90% of the population ($p' = 0.90$) with 95% confidence ($1 - \alpha = 0.95$) should, in addition, be sufficiently short so that the probability is only 0.10 ($\delta = 0.10$) that the interval will contain more than 96% of the population ($p^* = 0.96$). From Table A.30, the necessary sample size is $n = 154$. Section 5.3.1 shows how to obtain the interval. In particular, we note from Equation (5.2) that increasing the sample size to $n = 154$ from $n = 46$ (Section 9.4.2), i.e., a 235% increase, reduces the probability that the distribution-free tolerance interval will contain more than 96% of the population from $\delta = 0.554$ to (the exact calculated value) $\delta = 0.091$.

We also note that the sample size of $n = 154$ is about 70% larger than the sample of $n = 91$ which was required to achieve the same coverage probabilities under the assumption that the measured dimension is normally distributed (see Section 9.2.3).

9.5.2 One-Sided Tolerance Bound Example

Continuing with the example in Section 9.4.4, assume that, in addition to requiring that the tolerance bound be exceeded by the lifetimes of at least 99% of the population of bearings ($p' = 0.99$) with 95% confidence ($1 - \alpha = 0.95$), we now require that the sample size also should be large enough so that the probability is only 0.10 ($\delta = 0.10$) that the bound is exceeded by more than 99.7% of the population ($p^* = 0.997$). From Table A.30, we note that the required sample size is $n = 1050$. Section 5.3.3 shows how to obtain the resulting interval. Moreover, using Equation (5.2), we find that increasing the sample size from $n = 299$ (Section 9.4.4) to $n = 1050$ reduces the probability that the bound will be exceeded by 99.7% of the population from $\delta = 0.593$ to $\delta = 0.0995$.

Returning to the example in Section 9.2.4—which assumed a normal distribution—we note that a sample size of $n = 370$ was required to obtain a tolerance interval with the desired coverage probabilities. If we enter Table A.30 instead, with $p' = 0.99$, $1 - \alpha = 0.95$, $\delta = 0.01$, and $p^* = 0.997$, we find that the required sample size is $n = 1941$ if no distributional assumption is made. The substantial increase in the required sample is the price paid for dropping the normality assumption and is due to the high degree of precision required in the tail of the distribution (which, unfortunately, is the part of the distribution where the normality assumption is most likely to be in doubt).

9.6 SAMPLE SIZE TO DEMONSTRATE THAT A PROPORTION EXCEEDS (IS EXCEEDED BY) A SPECIFIED VALUE

This section shows how to find the sample size needed to demonstrate, with a specified confidence $100(1 - \alpha)\%$, that the population proportion conforming (to a specified requirement), denoted by p, is larger (smaller) than or equal to a specified value, which will be denoted by p'. Assuming that nonconforming units occur independently of each other and with a constant probability implies that the binomial distribution is an appropriate model. The demonstration requires that a one-sided lower (upper) confidence bound for p be larger (smaller) than p'. We want to choose the sample size so that there is a specified high probability p_{DEM} of a successful demonstration when $p = p^*$, where $p^* > p'$. Note that demonstrating that $q > q'$ is equivalent to demonstrating that $p < p'$ where $q = 1 - p$. Thus the methods in this section can be used for both problems.

9.6.1 Example

A new electronic integrated circuit chip must pass a battery of diagnostic tests to conform to specifications. Assume that the producer of the chips must "demonstrate," with 90% confidence, that at least 99% of the manufactured units conform to specifications (i.e., show that $p > p' = 0.99$). This is done by obtaining p, a lower 90% confidence bound for p, based upon the results of a random sample of chips. The demonstration will be successful if $p > 0.99$. Assuming that 99.9% of the units in the sampled population are actually in conformance, how large does the sample size n have to be so that the probability of passing the demonstration is 0.95? In our notation, $p' = 0.99$, $1 - \alpha = 0.90$, and we want the probability of a successful demonstration to be $p_{DEM} = 0.95$ when $p^* = 0.999$.

9.6.2 Graphical Method

Figures 9.2a to 9.2n show p_{DEM}, the probability of a successful demonstration that $p > p'$ at the $100(1 - \alpha)\%$ confidence level as a function of the true proportion conforming p^*, and the sample size n for all combinations of $1 - \alpha = 0.90$ and 0.95 and $p' = 0.50, 0.70, 0.80, 0.90, 0.95, 0.97$, and 0.99.

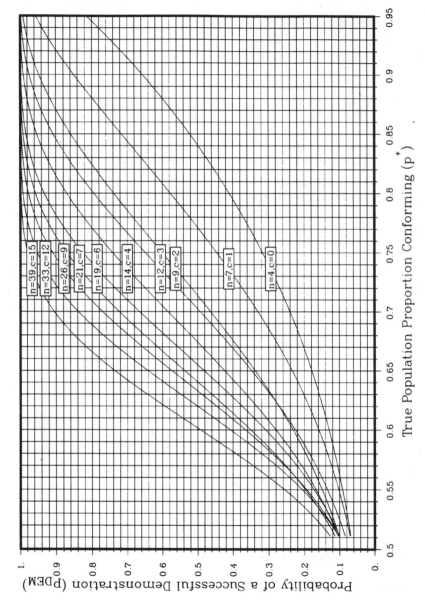

Figure 9.2a Probability of successfully demonstrating that $p > 0.50$ with 90% confidence (binomial distribution).

172

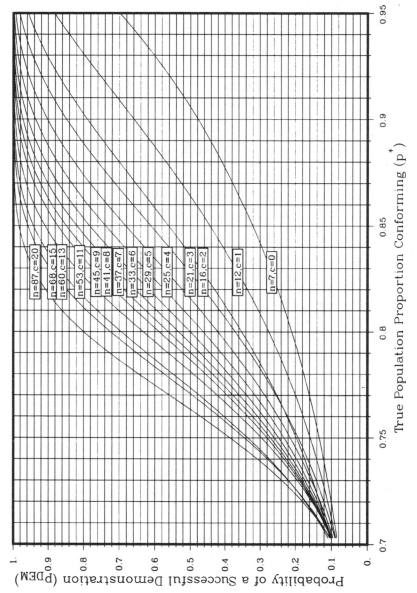

Figure 9.2b Probability of successfully demonstrating that $p > 0.70$ with 90% confidence (binomial distribution).

173

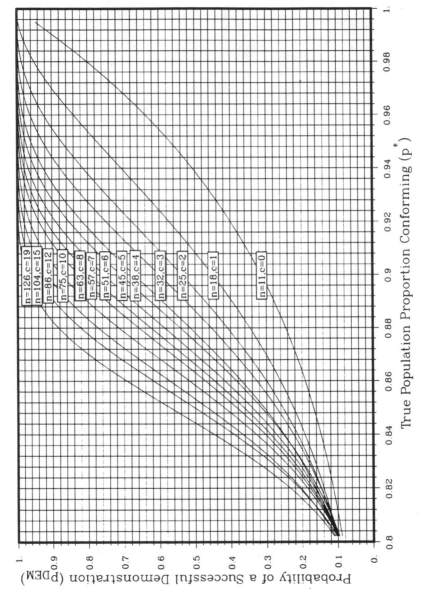

Figure 9.2c Probability of successfully demonstrating that $p > 0.80$ with 90% confidence (binomial distribution).

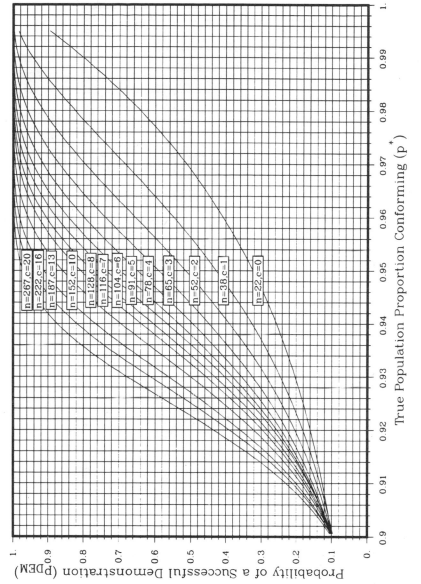

Figure 9.2d Probability of successfully demonstrating that $p > 0.90$ with 90% confidence (binomial distribution).

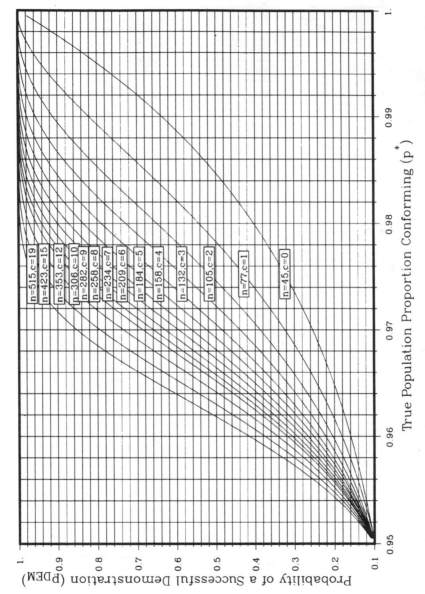

Figure 9.2e Probability of successfully demonstrating that $p > 0.95$ with 90% confidence (binomial distribution).

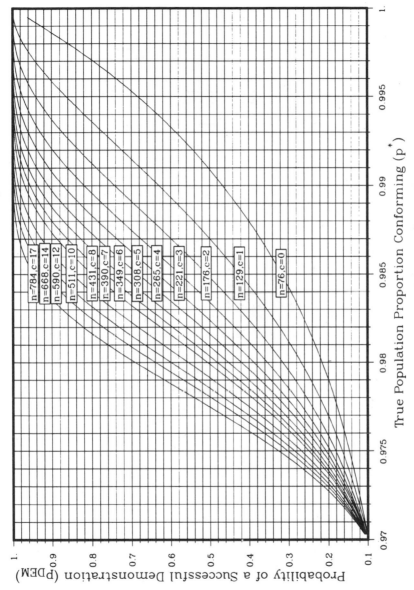

Figure 9.2f Probability of successfully demonstrating that $p > 0.97$ with 90% confidence (binomial distribution).

177

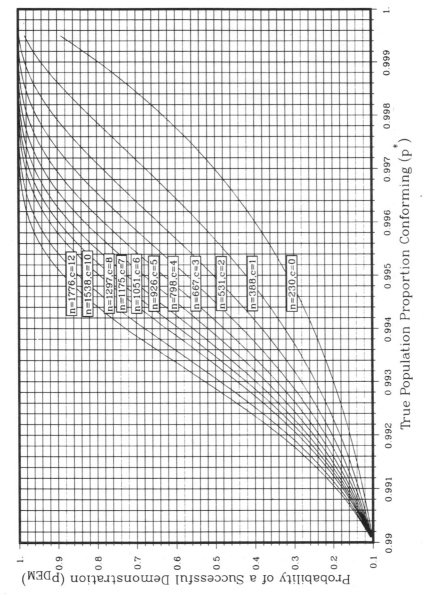

Figure 9.2g Probability of successfully demonstrating that $p > 0.99$ with 90% confidence (binomial distribution).

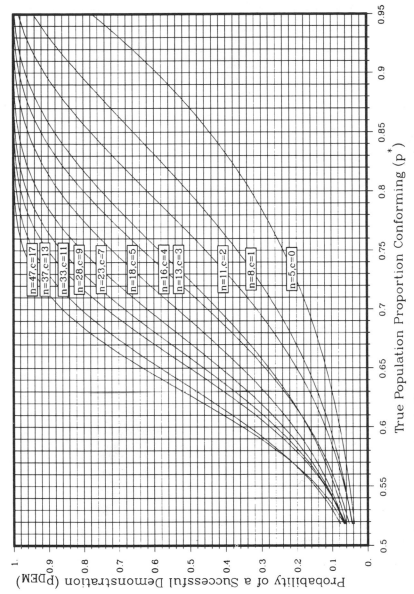

Figure 9.2h Probability of successfully demonstrating that $p > 0.50$ with 95% confidence (binomial distribution).

179

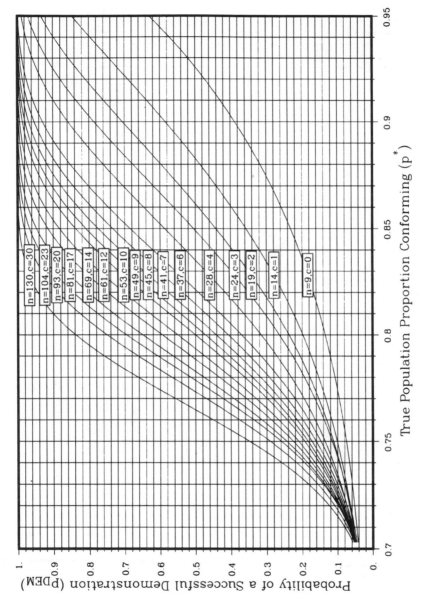

Figure 9.2i Probability of successfully demonstrating that $p > 0.70$ with 95% confidence (binomial distribution).

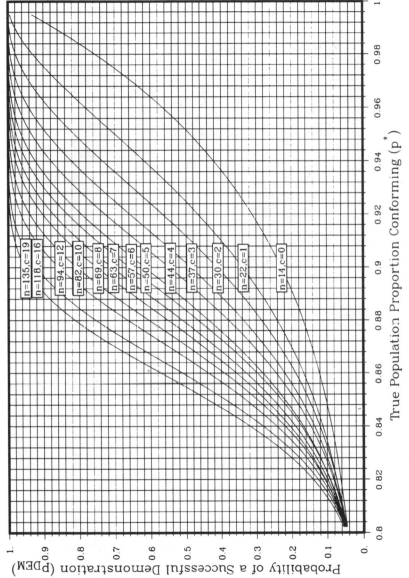

Figure 9.2j Probability of successfully demonstrating that $p > 0.80$ with 95% confidence (binomial distribution).

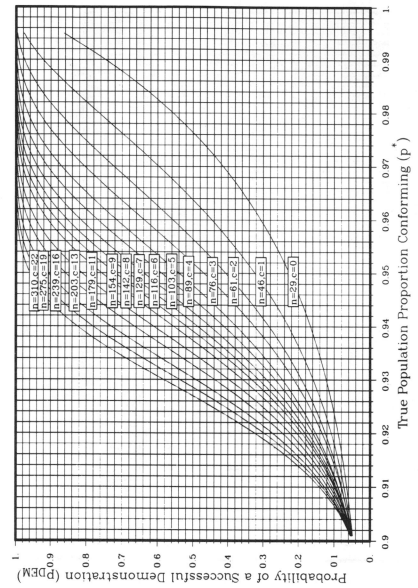

Figure 9.2k Probability of successfully demonstrating that $p > 0.90$ with 95% confidence (binomial distribution).

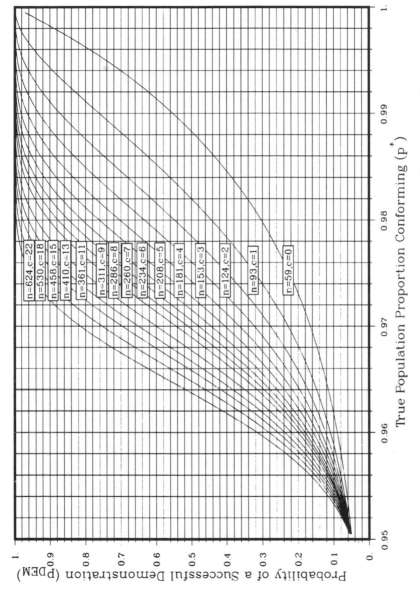

Figure 9.21 Probability of successfully demonstrating that $p > 0.95$ with 95% confidence (binomial distribution).

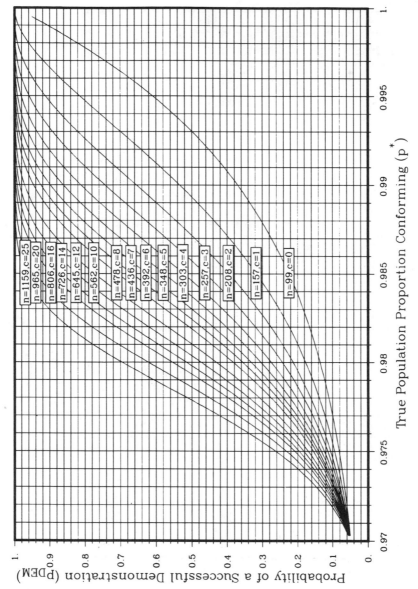

Figure 9.2m Probability of successfully demonstrating that $p > 0.97$ with 95% confidence (binomial distribution).

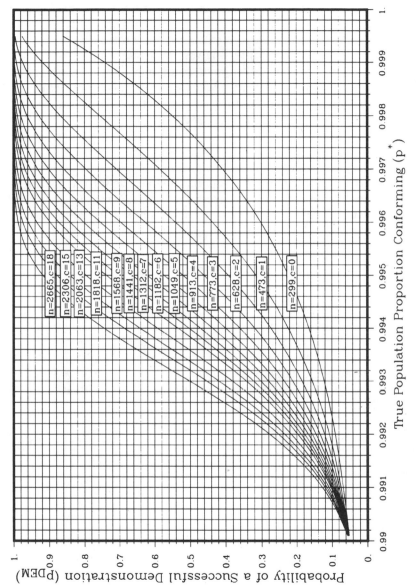

Figure 9.2n Probability of successfully demonstrating that $p > 0.99$ with 95% confidence (binomial distribution).

185

These can be used to determine n for specified values of p', p^*, p_{DEM}, and α. The figure also shows c, the maximum number of nonconforming units in the resulting random sample that are allowable for the demonstration to be successful. For our example, we use Figure 9.2g which was designed specifically for demonstrating with $100(1 - \alpha)\% = 90\%$ confidence that $p > p' = 0.99$. Enter the horizontal scale at the value $p^* = 0.999$ and move up from that point. Then, simultaneously enter the vertical scale at $p_{DEM} = 0.95$ and move to the right. After finding the point of intersection, read the values $n = 531$ and $c = 2$ from the line immediately above this point. Thus, the required sample size is $n = 531$, and the demonstration will be successful if there are $c = 2$ or fewer nonconforming items.

9.6.3 Tabular Method

The tabular method to determine the required sample size for such problems uses Table A.30. In our example, enter Table A.30 with $1 - \alpha = 0.90$, $p' = 0.99$, $p^* = 0.999$, and $\delta = 1 - p_{DEM} = 1 - 0.95 = 0.05$, and read the necessary sample size to be $n = 531$.

9.6.4 Relationship to Odeh and Owen (1983) Tables for Attribute Sampling Plans

Our Table A.30 and Odeh and Owen's (1983) Table 2 (which gives one-sided attribute sampling plans controlling producer's and consumer's risk) are closely related. The corresponding entries are O & O's $p_2 =$ our $1 - p'$, O & O's $p_1 =$ our $1 - p^*$, O & O's $\alpha =$ our δ, and O & O's $\beta =$ our α. Odeh and Owen provide entries for all combinations of $p_1 = 0.005(0.005)0.05$, 0.075, 0.10, $p_2 = 2p_1 (0.005)0.10$, 0.15, 0.20, 0.30, with $\alpha = 0.01$, 0.025, 0.05, and $\beta = 0.05$, 0.10, 0.20.

CHAPTER 10

Sample Size Requirements for Prediction Intervals

10.1 INTRODUCTION

This chapter provides guidelines for choosing the sample size required to obtain a prediction interval to contain a future single observation, a specified number of future observations, or some other quantity to be calculated from a future sample from a previously sampled population.

There are two sources of imprecision in statistical prediction: First, because the given data are limited, there is uncertainty with respect to the characteristics (e.g., parameters) of the previously sampled population. Second, there is the random variation in the future sample. Say, for example, that the results of an initial sample of size n from a normal population with unknown mean μ and unknown standard deviation σ are to be used to predict the value of a single future randomly selected observation from the same population. The mean $\bar{x}$ of the initial sample is used to predict the future observation. Now $\bar{x} = \mu + \varepsilon_1$, where ε_1, the random variation associated with the mean of the given sample, is itself normally distributed with mean 0 and variance σ^2/n. The future observation to be predicted is $y = \mu + \varepsilon_2$, where ε_2 is the random variation associated with the future observation, and is normally distributed with mean 0 and variance σ^2, independently of ε_1. Thus, the prediction error is $y - \bar{x} = \varepsilon_2 - \varepsilon_1$, and has variance $\sigma^2 + (\sigma^2/n)$. The length of a normal-theory prediction interval to contain y will be proportional to the square root of the estimate of this quantity (Section 4.7.1). Increasing the size of the initial sample will reduce the imprecision associated with the sample mean $\bar{x}$ (i.e., σ^2/n), but it will reduce only the sampling error in the estimate of the variation (σ^2) associated with the future sample. Thus, an increase in the size of the initial sample beyond the point where the inherent variation in the future sample tends to dominate will not materially reduce the length of the prediction interval.

The following discussion deals with selecting the size of the initial sample that will be used to construct a prediction interval to contain the mean of a *future* sample of size m from the same normal population. Figures and tables for the frequently encountered special case where the future sample size is $m = 1$, and also for $m = 10$, are provided, followed by a numerical example. Finally, it is shown how these ideas can be applied to assessing sample size requirements for some other prediction intervals.

10.2   SAMPLE SIZE FOR A NORMAL DISTRIBUTION PREDICTION INTERVAL LENGTH

Unlike a confidence interval to contain a population parameter, which converges to a point as the sample size increases, a prediction interval converges to an interval. It is thus not possible to obtain a prediction interval consistently shorter than this limiting interval, irrespective of how large an initial sample is taken. Thus, we suggest that the criterion for assessing the effect of sample size on prediction interval length be expressed in terms of this limiting interval. More specifically, because the length of the calculated prediction interval is an observed value of a random variable, we propose that one decide on the initial sample size based on either:

- The ratio of the expected length of the prediction interval to the length of the limiting interval or
- The ratio of an appropriate upper prediction bound on the prediction interval length to the length of the limiting interval.

10.2.1   Relative Length of the Prediction Interval

The ratio of the length of the two-sided $100(1 - \alpha)\%$ prediction interval for the mean of a future sample of size m from an initial sample of size n [Equation (4.2)] to that of the probability interval assuming an infinite initial sample size (i.e., $\mu \pm z_{(1-\alpha/2)}\sigma/\sqrt{m}$) is given by

$$L = \left(\frac{t_{(1-\alpha/2;\, n-1)}s_n}{z_{(1-\alpha/2)}\sigma} \right) \left(1 + \frac{m}{n} \right)^{1/2} \tag{10.1}$$

where $t_{(\gamma;\, r)}$ is the 100γ percentile of the Student t distribution with r degrees of freedom.

The relative length L involves the random variable s_n—the estimated standard deviation of the initial sample—and, therefore, is itself a random variable. The expected value of L is obtained by substituting the expected value of s_n [see Johnson and Kotz (1970a), page 62] for s_n in Equation (10.1) to obtain

$$E(L) = \left(\frac{t_{(1-\alpha/2;\, n-1)}}{z_{(1-\alpha/2)}}\right)\left(\frac{2}{n-1}\right)^{1/2}\left(\frac{\Gamma\left(\dfrac{n}{2}\right)}{\Gamma\left(\dfrac{n-1}{2}\right)}\right)\left(1 + \frac{m}{n}\right)^{1/2} \quad (10.2)$$

where $\Gamma(\cdot)$ is the gamma function [see, for example, page 534 of Mood, Graybill, and Boes (1974) for the definition of this function]. We propose $E(L)$, the expected value of the relative length of the prediction interval, as one criterion for selecting the initial sample size n.

One can also obtain a prediction bound on the relative length. In particular, an upper $100\gamma\%$ prediction bound on L is obtained by substituting an upper $100\gamma\%$ prediction bound for s_n/σ in Equation (10.1), as described in Section 4.9. Thus

$$\tilde{L}_U = \left(\frac{t_{(1-\alpha/2;\, n-1)}}{z_{(1-\alpha/2)}}\right)\left(\frac{\chi^2_{(\gamma;\, n-1)}}{n-1}\right)^{1/2}\left(1 + \frac{m}{n}\right)^{1/2} \quad (10.3)$$

where $\chi^2_{(\gamma;\, r)}$ is the 100γ percentile of the chi-square distribution with r degrees of freedom. The interpretation of $\tilde{L}_U$ is that, in repeated constructions of a $100(1-\alpha)\%$ prediction interval for $\bar{y}_m$ using Equation (10.1), from independent samples of size n, the relative length of the interval L will exceed this bound only $100(1-\gamma)\%$ of the time. We propose $\tilde{L}_U$ as an alternative criterion for choosing the initial sample size n.

10.2.2 Description of Figures and Tables for Two-Sided Prediction Intervals

Figure 10.1 gives the expected relative length of a two-sided prediction interval for a single future observation (i.e., $m = 1$) from a normal population as a function of the initial sample size for the confidence levels $1 - \alpha = 0.5$, 0.8, 0.9, 0.95, and 0.99 associated with the prediction interval. Figure 10.2 gives upper 95% prediction bounds on the relative lengths for the same values of $1 - \alpha$. Figures 10.3 and 10.4 provide information similar to that in Figures 10.1 and 10.2 for two-sided prediction intervals to contain the mean of $m = 10$ future observations. These figures were calculated from Equations (10.2) and (10.3). They can be used to assess the effect of the initial sample size on (1) the expected relative length and (2) on the upper prediction bound for the relative length, of prediction intervals. Thus they are useful in selecting the size of the initial sample.

Tables 10.1, 10.2, 10.5, and 10.6 provide information similar to that in Figures 10.1 through 10.4, respectively, for selected values of n. Tables 10.3

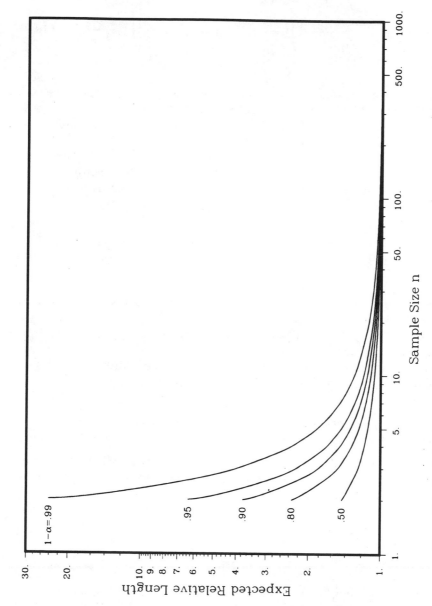

Figure 10.1 Prediction interval expected relative length ($m = 1$). A similar figure first appeared in Meeker and Hahn (1982). Adapted with permission of the American Society for Quality Control.

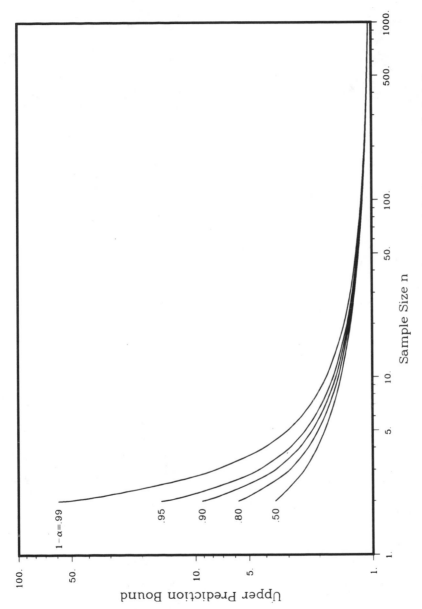

Figure 10.2 Upper 95% prediction bound on prediction interval relative length ($m = 1$). A similar figure first appeared in Meeker and Hahn (1982). Adapted with permission of the American Society for Quality Control.

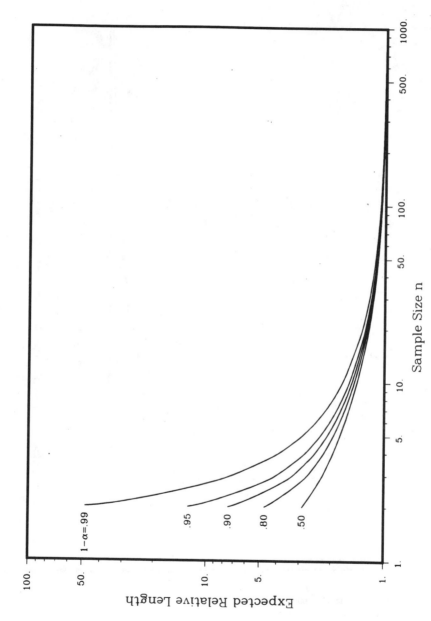

Figure 10.3 Prediction interval expected relative length ($m = 10$). A similar figure first appeared in Meeker and Hahn (1982). Adapted with permission of the American Society for Quality Control.

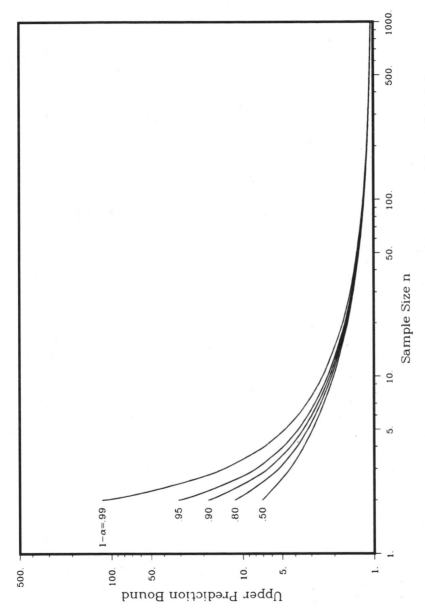

Figure 10.4 Upper 95% prediction bound on prediction interval relative length ($m = 10$). A similar figure first appeared in Meeker and Hahn (1982). Adapted with permission of the American Society for Quality Control.

Table 10.1   Expected Relative Length of Two-Sided Prediction Interval for $m = 1$

n	Confidence Level $(1 - \alpha)$				
	0.50	0.80	0.90	0.95	0.99
2	1.449	2.346	3.750	6.334	12.146
5	1.131	1.232	1.334	1.458	1.840
10	1.063	1.101	1.137	1.177	1.287
15	1.042	1.065	1.086	1.110	1.172
30	1.021	1.031	1.041	1.051	1.078
60	1.011	1.015	1.020	1.025	1.037
120	1.006	1.008	1.010	1.012	1.018
1000	1.001	1.001	1.001	1.001	1.002

A similar table first appeared in Mecker and Hahn (1982). Adapted with permission of the American Society for Quality Control.

Table 10.2   Upper 95% Prediction Bound on Relative Length of Two-Sided Prediction Interval for $m = 1$

n	Confidence Level $(1 - \alpha)$				
	0.50	0.80	0.90	0.95	0.99
2	3.561	5.764	9.212	15.558	59.315
5	1.853	2.018	2.186	2.389	3.015
10	1.499	1.552	1.602	1.659	1.814
15	1.338	1.410	1.438	1.470	1.552
30	1.248	1.260	1.272	1.285	1.318
60	1.167	1.172	1.177	1.183	1.197
120	1.114	1.116	1.119	1.121	1.128
1000	1.038	1.038	1.038	1.038	1.039

A similar table first appeared in Mecker and Hahn (1982). Adapted with permission of the American Society for Quality Control.

Table 10.3   Expected Relative Length of Two-Sided Prediction Interval for $m = 5$

n	Confidence Level $(1 - \alpha)$				
	0.50	0.80	0.90	0.95	0.99
2	2.214	3.584	5.728	9.675	36.884
5	1.460	1.590	1.723	1.883	2.376
10	1.242	1.285	1.327	1.375	1.503
15	1.165	1.190	1.214	1.241	1.311
30	1.085	1.096	1.106	1.117	1.146
60	1.043	1.048	1.053	1.058	1.071
120	1.022	1.024	1.026	1.029	1.035
1000	1.003	1.003	1.003	1.003	1.004

A similar table first appeared in Mecker and Hahn (1982). Adapted with permission of the American Society for Quality Control.

Table 10.4   Upper 95% Prediction Bound on Relative Length of Two-Sided Prediction Interval for $m = 5$

n	Confidence Level $(1 - \alpha)$				
	0.50	0.80	0.90	0.95	0.99
2	5.439	8.805	14.072	23.766	90.603
5	2.393	2.605	2.822	3.085	3.892
10	1.750	1.812	1.871	1.938	2.118
15	1.543	1.576	1.608	1.643	1.735
30	1.326	1.339	1.351	1.365	1.400
60	1.204	1.210	1.215	1.221	1.236
120	1.132	1.135	1.137	1.140	1.147
1000	1.040	1.040	1.040	1.040	1.040

A similar table first appeared in Mecker and Hahn (1982). Adapted with permission of the American Society for Quality Control.

Table 10.5   Expected Relative Length of Two-Sided Prediction Interval for $m = 10$

n	Confidence Level $(1 - \alpha)$				
	0.50	0.80	0.90	0.95	0.99
2	2.899	4.693	7.500	12.667	48.292
5	1.789	1.948	2.110	2.306	2.910
10	1.434	1.484	1.533	1.587	1.735
15	1.302	1.331	1.358	1.387	1.465
30	1.160	1.171	1.182	1.194	1.225
60	1.083	1.088	1.092	1.098	1.111
120	1.042	1.044	1.047	1.049	1.055
1000	1.005	1.005	1.005	1.006	1.007

A similar table first appeared in Mecker and Hahn (1982). Adapted with permission of the American Society for Quality Control.

and 10.4 give similar information for prediction intervals to contain the mean of $m = 5$ future observations. For $n > 15$, three-point harmonic interpolation (i.e., interpolation in $1/n$) in these tables provides excellent accuracy (three to four significant figures). The tables make clear the limited usefulness of increasing the initial sample size beyond a certain point if the eventual objective is to obtain a prediction interval for a future sample. In addition, the tables can be used directly to determine approximate sample size requirements for $m = 1$, i.e., a single observation—the case encountered most frequently in practice—and also for $m = 5$ and $m = 10$.

10.2.3   Example

A rocket engine is to be used in a critical, self-destructive operation. An experiment is being planned on a random sample of engines of a particular

Table 10.6   Upper 95% Prediction Bound on Relative Length
of Two-Sided Prediction Interval for $m = 10$

	Confidence Level $(1 - \alpha)$				
n	0.50	0.80	0.90	0.95	0.99
2	7.121	11.528	18.424	31.117	118.631
5	2.931	3.191	3.457	3.778	4.767
10	2.021	2.092	2.161	2.238	2.446
15	1.624	1.762	1.798	1.837	1.940
30	1.417	1.431	1.445	1.459	1.497
60	1.250	1.255	1.261	1.267	1.283
120	1.155	1.157	1.160	1.162	1.169
1000	1.042	1.043	1.043	1.043	1.044

A similar table first appeared in Meeker and Hahn (1982). Adapted with permission of the American Society for Quality Control.

type to determine how long the engine can deliver a certain amount of thrust for a specified amount of fuel. The delivery time is assumed to follow a normal distribution with unknown mean and standard deviation. It is desired to determine the effect of initial sample size on a two-sided 90% prediction interval to contain the delivery time for a single future engine, using as criteria (a) the expected relative length and (b) the upper 95% prediction bound on the relative length of the two-sided prediction level.

From Table 10.1, the expected value of the relative length $E(L)$, for a two-sided 90% prediction interval to contain a single future observation, is 1.086 for $n = 15$ and 1.041 for $n = 30$. Harmonic interpolation for $n = 25$ yields $E(L) = 1.05$ (i.e., linear interpolation between $\frac{1}{30}$ and $\frac{1}{15}$ yields $\frac{1}{n} = \frac{1}{25}$). Similarly, harmonic interpolation in Table 10.2 for the 95% upper prediction bound on the relative length of a 90% prediction interval to contain a single future observation yields $n \cong 78$ for $\tilde{L}_U = 1.15$. Thus

- To obtain, for a single future observation, a two-sided 90% prediction interval whose expected length is 5% larger than that of the smallest achievable interval would require an initial sample size $n = 25$. Note that precision improves only slightly beyond $n = 20$. This is because beyond this point, the major part of the variability is not in the uncertainty in the initial data but in that for the single future observation.

- To obtain, for a single future observation, a two-sided 90% prediction interval whose length we can expect with 95% confidence to be no more than 15% larger than the length of the smallest achievable interval length would require an initial sample size $n = 78$.

10.2.4   One-Sided Prediction Bounds

The procedure for evaluating initial sample size for a one-sided prediction bound is similar to that for a two-sided prediction interval. In this case, however, relative length deals with that part of the interval (below or above the mean) of interest, and one uses $t_{(1-\alpha; n-1)}$ and $z_{(1-\alpha)}$ in place of $t_{(1-\alpha/2; n-1)}$ and $z_{(1-\alpha/2)}$ in Equations (10.2) and (10.3). As a result, in Figures 10.1 to 10.4 and Tables 10.1 to 10.6, the confidence levels 0.5, 0.8, 0.9, 0.95, and 0.99 are replaced by confidence levels of 0.75, 0.90, 0.95, 0.975, and 0.995, respectively.

10.2.5 Application to Other Prediction Intervals

The concepts presented in the previous sections can be readily applied to determine the initial sample size for other types of prediction intervals. Some specific cases are discussed below.

10.2.5.1 Simultaneous Prediction Intervals to Contain All of m Future Observations from a Normal Population

Using results from Section 4.8.1, the limiting length of a simultaneous two-sided prediction interval to contain all m future observations from a normal distribution, as the initial sample size n becomes large (i.e., known μ and σ), is

$$\mu \pm z_{(\delta)}\sigma,$$

where $\delta = (1 - \alpha/2)^{1/m}$. Thus, the ratio of the prediction interval length to the limiting interval length is

$$\frac{r_{(1-\alpha; m, n)}s_n}{z_{(\delta)}\sigma},$$

where $r_{(1-\alpha; m, n)}$ is the factor defined in Section 4.8.1 for obtaining a simultaneous two-sided prediction interval to contain all of m future observations from a normal population. As before, the expected relative length and upper prediction bound on the relative length can be obtained, respectively, by replacing s_n/σ by its expected value and by its appropriate upper prediction bound. The resulting expressions can then be used to assess the effect of the sample size on the relative length of the desired two-sided prediction interval, and to guide sample size determination. A one-sided prediction bound is handled similarly, but now one uses $\delta = (1 - \alpha)^{1/m}$.

10.2.5.2 Prediction Interval to Contain the Standard Deviation of a Future Sample

Using results from Section 4.9.1, the limiting two-sided probability interval to contain the standard deviation of a future sample as the previous sample size n becomes large (i.e., known σ) is

$$\left\{ \sigma \left[\frac{\chi^2_{(\alpha/2; m-1)}}{m-1} \right]^{1/2}, \sigma \left[\frac{\chi^2_{(1-\alpha/2; m-1)}}{m-1} \right]^{1/2} \right\}.$$

Thus, the ratio of the length of the prediction interval to its limiting length is

$$
\frac{s_n\left[\left(F_{(1-\alpha/2;\,m-1,\,n-1)}\right)^{1/2} - \left(F_{(\alpha/2;\,m-1,\,n-1)}\right)^{1/2}\right]}{\sigma\left[\left(\chi^2_{(1-\alpha/2;\,m-1)}\right)^{1/2} - \left(\chi^2_{(\alpha/2;\,m-1)}\right)^{1/2}\right]/(m-1)^{1/2}},
$$

where $F_{(\gamma;\,r_1,\,r_2)}$ is the 100γ percentile of the F distribution with r_1 numerator and r_2 denominator degrees of freedom. Expressions for the expected relative length and for an upper prediction bound on this length can be obtained by replacing s_n/σ by its expectation and by its upper prediction bound, respectively, as above. These results can then be used for sample size determination.

10.2.5.3   Other Cases

The preceding results can be easily generalized to determine sample size requirements for a prediction interval using a regression model (Section 11.4). Other straightforward extensions include determining sample size requirements for constructing the following:

- A prediction interval to contain the mean of a future sample from an exponential population (Section 11.3.4.2).
- A prediction interval to contain the difference between the means of two future samples (Section 11.5.2).
- A prediction interval to contain the ratio of two future sample standard deviations for a normal population, or the ratio of two future sample means from an exponential population (Section 11.5.2).

10.3   SAMPLE SIZE FOR DISTRIBUTION-FREE PREDICTION INTERVALS FOR k OF m FUTURE OBSERVATIONS

10.3.1   Tabular Method for Two-Sided Prediction Intervals

As indicated in Section 5.4.1, Tables A.20a through A.20c give the initial sample size n so that a two-sided prediction interval, that has as its endpoints the largest and the smallest observations of this initial sample, will enclose

- all m,
- at least $m - 1$, and
- at least $m - 2$

observations in a future sample of size m with $100(1 - \alpha)\%$ confidence for $(1 - \alpha) = 0.50, 0.75, 0.90, 0.95, 0.98, 0.99, 0.999$ and $m = 1(1)25(5)50(10)100$. These tables can be used directly to help choose the size of an initial sample that is to be used to set a prediction interval to contain all, or almost all, observations in a future sample from the same population.

We note that these sample size criteria are inherently different from those discussed in the preceding two sections in that they provide information about the *minimum* size sample that is needed to construct any such interval with the desired level of confidence. As noted in Chapter 5, if the initial sample is too small it is not possible to construct a distribution-free interval at the desired confidence level. Moreover, because the interval uses the endpoints of the sample, it may be unsatisfactorily large. In contrast, in our earlier discussion, we were concerned with taking a sufficiently large sample to satisfy specified requirements on precision relative to that under ideal conditions (i.e., an infinite sample size—providing exact knowledge of the distribution parameters).

10.3.2   Example for a Two-Sided Prediction Interval

Based on the measured values of the sample of $n = 100$ units given in Table 5.1, the manufacturer wants to find a distribution-free prediction interval to contain all of the measured values of a future sample of $m = 5$ units from the same population, without making any assumptions about the form of the distribution. From Table A.20a, we note for $m = 5$ that, even if one uses the extreme observations of the past sample, a minimum sample of size 193 would be required to obtain a 95% prediction interval. Thus, the past sample of size 100 would be inadequate. However, Table A.20a indicates that a 90% prediction interval can be obtained. In that case a sample of 93 observations would suffice.

10.3.3   Tabular Method for One-Sided Prediction Bounds

As indicated in Section 5.4.3, Tables A.21a through A.21c give the initial sample size n so that a one-sided lower (upper) prediction bound defined by the smallest (largest) observation of this initial sample will be exceeded by (will exceed)

- all m,
- at least $m - 1$, and
- at least $m - 2$

observations in a future sample of size m with $100(1 - \alpha)\%$ confidence for $1 - \alpha = 0.50, 0.75, 0.90, 0.95, 0.98, 0.99, 0.999$ and $m = 1(1)25(5)50(10)100$. These tables can be used directly to help choose the size of an initial sample for setting a one-sided upper (lower) distribution-free prediction bound to exceed (to be exceeded by) all, or almost all, observations in a future sample from the same population.

10.3.4 Example for One-Sided Prediction Bounds

A satellite will contain 12 rechargeable batteries of which 10 must survive for a time that is to be determined. The manufacturer needs a lower prediction bound that will, with 99% confidence, be exceeded by at least 10 of 12 failure times for the batteries that will be installed in a future single satellite. Because the batteries have at least two causes of failure and little is known about the life distribution, a distribution-free bound is to be used. Also, the batteries are very expensive, thus, only a limited number can be procured for the life test. The time of the first failure will be used for the lower prediction bound. Then the manufacturer needs to know how many randomly selected batteries to test to be 99% confident that at least 10 of the batteries in the future shipment of 12 will not fail prior to the time of the first failure in the initial sample. From Table A.21c, we obtain $n = 40$; thus, the first failure in a sample of $n = 40$ batteries will provide the desired prediction bound. We note that, in this example, testing can be terminated after the first failure. However, the results may be of limited value if the first failure occurs too early—other than telling us how little we know.

CHAPTER 11

A Review of Other Statistical Intervals

The previous chapters described statistical intervals for a variety of situations and areas of application involving inferences for a single population. This chapter overviews some other useful statistical intervals and provides further references. The discussion is far from exhaustive. Technical journals, such as the *IEEE Transactions on Reliability*, the *Journal of the American Statistical Association*, the *Journal of Quality Technology*, *Technometrics*, and many others, describe statistical intervals for numerous special applications. This chapter, therefore, provides information on only a few statistical intervals that are, or could be, widely used in practice, especially in quality control and reliability applications. Our coverage is somewhat more complete for tolerance and prediction intervals and confidence intervals on tail probabilities and percentiles than it is for other types of confidence intervals. For situations not covered in this book or in the references, consult an index of the previously mentioned journals or the *Current Index of Statistics*, a yearly keyword index published by the American Statistical Association and the Institute of Mathematical Statistics.

This chapter is organized as follows:

- Section 11.1 discusses simultaneous statistical intervals and reviews the relationships between various types of statistical intervals.
- Section 11.2 briefly discusses statistical intervals for censored data.
- Section 11.3 provides references for confidence, tolerance, and prediction intervals for various distributions that were not considered previously, and for some other special cases.
- Section 11.4 summarizes various confidence, tolerance, and prediction intervals for regression analysis problems.
- Section 11.5 summarizes some useful statistical intervals for comparing two or more populations or processes.

We restrict ourselves to discussing the existence of methods for obtaining such intervals and to giving references that provide additional information on theory and applications. Some of the discussion in this chapter is more advanced than that elsewhere in the book. Also, in view of the comprehensive nature of our review, we do not provide numerical examples. Many of the references provide added details and examples.

11.1 SIMULTANEOUS STATISTICAL INTERVALS AND RELATIONSHIPS

11.1.1 Simultaneous Intervals

Some practical problems require that more than one statistical interval (often many and, in some cases, an infinite number) be computed from the same data and be considered simultaneously. This was the case, for example, with the prediction intervals described in Section 4.8 to contain, with $100(1 - \alpha)\%$ confidence, all of m future observations from a normal distribution. We can view this as the construction of m intervals (one for each future observation), where we want to assert with $100(1 - \alpha)\%$ confidence that *all* of the intervals are correct. Thus, such intervals are called *simultaneous* prediction intervals. Similarly, one might want to compute simultaneous confidence intervals to contain each of the $\binom{k}{2} = k!/[2!(k - 2)!]$ possible differences between all pairs of means from k different populations. Miller (1981) treats, in detail, the theory and methods for making simultaneous inferences, including general methods for computing such intervals, and easy-to-use approximate methods, like the Bonferroni approximation, and their application. We review some of the basic practical methods; Miller (1981) provides further technical discussion and applications.

Assume that we want to combine k statistical intervals (which, for simplicity, we will refer to as "confidence intervals") that, individually, have confidence levels $1 - \alpha_1, \ldots, 1 - \alpha_k$. We want to know the confidence level, $1 - \alpha_J$, of the joint confidence statement (or how to compute the individual intervals so that they have the desired joint confidence level). The following three situations are of interest.

1. The intervals are *functionally dependent* on one another in the sense that the correctness of any one implies the correctness of all of the others (and conversely with regard to incorrectness). This would be the case, for example, if a set of three $100(1 - \alpha)\%$ confidence intervals for mean time to failure were obtained from the same set of data, using the same method, but expressed, respectively, in units of hours, days, and weeks. Less trivially, this is also the case when, with a single parameter distribution, one statistical interval or bound is computed from another statistical interval or bound. For

example, binomial confidence intervals for probabilities and binomial tolerance bounds are both calculated directly from the confidence intervals for the binomial parameters, as described in Sections 6.3 and 6.4, respectively. In such cases, the joint confidence $(1 - \alpha_J)$ for all of the intervals is just $(1 - \alpha_J) = (1 - \alpha_1) = \cdots = (1 - \alpha_k)$.

2. The intervals are statistically independent (i.e., the probability that any one is correct is equal to the conditional probability that it is correct, given the correctness of any combination of the other intervals). The assumption of independence is reasonable if each interval in the set is computed separately from independent sets of data (e.g., if various studies had been conducted independently and the data from each analyzed separately). It is, in general, *not* reasonable to assume independence if some or all of the intervals involve computations from the same data set (e.g., in the analysis of variance when confidence intervals are calculated for several group means, using the *same* pooled variance estimate, calculated over all of the groups). If the intervals *are* independent, then the joint confidence level is

$$(1 - \alpha_J) = (1 - \alpha_1)(1 - \alpha_2) \cdots (1 - \alpha_k).$$

For example, a set of three independent 95% confidence intervals would have a joint confidence level of $(0.95)^3 = 0.8574$ or 85.74%.

3. The intervals are neither functionally dependent nor statistically independent. In this case, the actual joint confidence level may be either less than or greater than the joint confidence level for the case of independence. For many special cases, it is possible to compute the exact confidence level, but, for others, the task is often analytically difficult or intractable or computationally intensive. Elementary probability theory, however, provides a simple, conservative lower bound on the actual confidence level for a joint confidence statement. In particular, the joint confidence level

$$(1 - \alpha_J) \geq 1 - \alpha_1 - \cdots - \alpha_k.$$

This is known as the Bonferroni bound, which was discussed briefly in Section 4.8.1. It provides a useful way for combining confidence statements to give a *conservative* bound for the actual joint confidence level. For example, in combining three 95% confidence intervals, the joint confidence level, as calculated from the Bonferroni bound, is at least $(1 - 0.05 - 0.05 - 0.05) = 0.85$ or 85.0%. Similarly, when combining two 99% intervals, we have a joint confidence level of at least $(1 - 0.01 - 0.01) = 0.98$ or 98%. In the latter case, the Bonferroni lower bound is close to the joint confidence level that one has under the assumption of independence, i.e., $(0.99)^2 = 0.9801$. The Bonferroni inequality is conservative in the sense that it provides confidence intervals that have confidence levels that are larger than the actual levels. The Bonferroni inequality works especially well (i.e., usually gives a close approximation to the nominal confidence level) when k is small (i.e., relatively few groups) and the $(1 - \alpha)$'s are close to 1 [see Hahn (1969)].

11.1.2 Some Relationships

We review here some useful relationships among different kinds of statistical intervals; some of these were introduced in earlier chapters.

1. A one-sided lower $100(1 - \alpha)\%$ confidence bound on the $100p$th percentile of a distribution is equivalent to a one-sided lower tolerance bound that one can claim with $100(1 - \alpha)\%$ confidence is exceeded by at least $100(1 - p)\%$ of the population. Similarly, a one-sided upper $100(1 - \alpha)\%$ confidence bound on the $100p$th percentile of a distribution is equivalent to a one-sided upper tolerance bound that one can claim with $100(1 - \alpha)\%$ confidence exceeds at least $100p\%$ of the population.

2. One-sided tolerance bounds can also be used to obtain *approximate* two-sided tolerance intervals. For example, suppose that $\underset{\sim}{T}_{p_L}$ is a lower tolerance bound that one can claim with $100(1 - \alpha_L)\%$ confidence is exceeded by at least $100p_L\%$ of the population, and suppose that $\tilde{T}_{p_U}$ is an upper tolerance bound that one can claim with $100(1 - \alpha_U)\%$ confidence exceeds at least $100p_U\%$ of the population. Then $[\underset{\sim}{T}_{p_L}, \tilde{T}_{p_U}]$ is an approximate two-sided tolerance interval that one can claim with $100(1 - \alpha_L - \alpha_U)\%$ confidence encloses at least $100(p_L + p_U - 1)\%$ of the sampled population. The actual confidence level for this two-sided tolerance interval is greater than $100(1 - \alpha_L - \alpha_U)\%$, and thus the interval is conservative (i.e., longer than necessary). This is an application of the Bonferroni approximation [as noted by Guenther (1972)]. When an exact two-sided tolerance interval is available (as it is, for example, for the normal and exponential distributions), then such an exact interval is more precise (i.e., shorter) than one from the preceding approximation.

3. One will find references to "γ-content tolerance intervals" and "β-expectation tolerance intervals" in the statistical literature, e.g., Guttman (1970). A "γ-content tolerance interval" [also sometimes referred to as a $(\gamma, 1 - \alpha)$ tolerance interval] is what we call a "tolerance interval," i.e., an interval that one can claim, with $100(1 - \alpha)\%$ confidence, encloses at least $100\gamma\%$ of a population. A "β-expectation tolerance interval" is what we call a "$100\beta\%$ prediction interval for a single future observation." Our terminology is consistent with much of the applications-oriented literature.

11.2 STATISTICAL INTERVALS FOR CENSORED DATA—AN OVERVIEW

Censored data arise when all that is known about some of the observations is that they exceed or are less than a known value. This is so, for example, for measurements that are below or above the endpoint of the scale of the measuring instrument.

"Right-censored" data are frequently encountered in product life data analysis. In that case, unfailed units at the time of analysis have right-censored lifetimes, because all that is known about their failure times is that they exceed (lie to the *right* of) the observed censoring time. Failure censoring (also known as Type II censoring) arises when it is decided initially that testing will be terminated after a prespecified number of failures has occurred. Time censoring (also known as Type I censoring) is more common in practice; this occurs when a life test is terminated after a prespecified elapsed time. Other kinds of censoring (e.g., left censoring and interval censoring) are also possible. For example, an observation is left censored if all that is known is that a failure had taken place at, or prior to, some initial inspection time. With interval-censored data, all that is known is that the observation is above some minimum value and below some maximum value. Chapter 1 of Nelson (1982) and Chapter 2 of Lawless (1982) contain examples and further discussion.

Special methods are needed for computing statistical intervals for censored data. We discuss some of these in Section 12.1. Exact confidence intervals for parameters and distribution percentiles are available (at least in theory) for the more popular statistical models for representing life data (e.g., the exponential, Weibull, and lognormal distributions), if there is only failure censoring. However, even when it is theoretically possible to obtain exact intervals, they may require factors that have been tabulated for only a few combinations of the needed inputs (i.e., distribution, sample size, number of failures, confidence level, etc.). Generally, complicated numerical methods or Monte Carlo simulations are required to obtain these factors. Alternately, when the needed factors are not available or when there is no exact theory, one can use large-sample approximate methods to compute confidence limits; see Section 12.1 for the main ideas and some other applications. General theory and applications for censored data and life data analysis, including methods for computing statistical intervals, are given by Mann, Schafer, and Singpurwalla (1974), Bain (1978), Lawless (1982), Nelson (1982), and Nelson (1990).

11.3 SOME FURTHER STATISTICAL INTERVALS
FOR A SINGLE POPULATION

11.3.1 Some General Results

This section gives references for confidence, tolerance, and prediction intervals for many of the commonly used statistical distributions that were not covered in the earlier chapters. Johnson and Kotz (1969, 1970a, 1970b) and

Patel, Kapadia, and Owen (1976) give theoretical properties for these and many other statistical distributions.

Chapter 7 of Patel, Kapadia, and Owen (1976) gives formulas and references for *confidence intervals* for a wide range of standard and nonstandard distributions. Most mathematical statistics texts [e.g., Mood, Graybill, and Boes (1974) and Kendall and Stuart (1973)] provide general theory and examples of using sample data to compute confidence intervals for distribution parameters and other quantities of interest for standard problems [see also the annotated categorized bibliography by Hahn and Meeker (1984) for other texts on basic statistical methods and mathematical statistics]. In Section 12.1.3 we discuss and give references to more advanced and general methods for computing confidence intervals.

General theory for *tolerance intervals* or regions is given by Guttman (1970). Jilek (1981) gives a bibliography containing 270 references on statistical tolerance intervals and regions. Jilek and Ackerman (1989) provide an update of this bibliography. Patel (1986) provides an extensive review of distribution-free and parametric tolerance intervals for a large number of distributions. In most cases, he provides computing formulas, descriptions of the methods, and references to needed tables. Also see the review by Guenther (1972). Many of the references in this section, dealing with tolerance intervals, were taken from these reviews. In Section 12.1.4 we discuss and give references to some of the more advanced and general methods for computing tolerance intervals.

Nelson (1968) outlines general theory and methods for computing *prediction intervals*. Hahn and Nelson (1973) review and describe many specific methods for computing prediction intervals, and Patel (1989) provides an updated review of the literature on prediction intervals. Many of the references in this section, dealing with prediction intervals, were taken from these review papers. In Section 12.1.5 we discuss and give references to more advanced and general methods for prediction intervals.

11.3.2 Further Statistical Intervals for the Normal Distribution

11.3.2.1 Data with Measurement Errors

Mee (1984b) shows how to compute confidence intervals for normal distribution tail probabilities and tolerance intervals when the observations are subject to measurement error. Also, see Hahn (1982) and Jaech (1984) for methods for computing tolerance intervals when there is measurement error and when one is interested in the true, and not the measured, values.

11.3.2.2 Some Special Normal Distribution Tolerance Intervals

Tietjen and Johnson (1979) show how to use sample data from a population that can be modeled by a normal distribution to compute upper tolerance bounds for the distribution of sample variances (or standard deviations) for random samples of a specified size from the same population. These bounds exceed, with stated confidence, a specified proportion of the sample standard deviations of a large number of future random samples from a normal population. The authors describe applications to setting control chart limits.

11.3.2.3 Some Special Normal Distribution Prediction Intervals

Hall and Prairie (1973) and Fertig and Mann (1977) give factors for computing one-sided prediction intervals to contain k out of m future observations from a normal population. Odeh (1990) gives similar factors for two-sided prediction intervals. Owen, Li, and Chou (1981) describe normal distribution prediction intervals to be used for screening nonconforming product, using a measured correlated variate. Hahn (1972a) treats simultaneous prediction intervals to contain the sample variances of all of m future samples from a previously sampled normal population. Nelson and Schmee (1981) give tables for computing prediction intervals for the longest-lived observation in a life test assuming a normal (or lognormal) time to failure distribution, based on early failure times from the same sample. This is useful in estimating the duration of an ongoing life test. Finally, Hahn (1972a) gives simultaneous prediction intervals to contain the ranges of all m future samples from a normal distribution.

11.3.3 Lognormal Distribution and Transformations

The probability density function for a lognormal distribution is

$$f(x;\mu,\sigma) = \frac{1}{\sqrt{2\pi}\,\sigma x}\exp\left[-\frac{1}{2}\left(\frac{\log_e(x)-\mu}{\sigma}\right)^2\right],$$

$$x > 0, \quad -\infty < \mu < \infty, \quad \sigma > 0.$$

Figure 11.1 shows lognormal probability density functions with different values of the shape parameter σ. Aitchison and Brown (1957) and Crow and Shimizu (1988) discuss this distribution in detail. The lognormal distribution is a popular model for various phenomena such as the size of bank accounts and of particles in the air or in water, and for situations where variables tend to increase relative to their current size [see Aitchison and Brown (1957) and Hahn and Shapiro (1967) for further discussion]. It is frequently used in environmental assessments and also to represent the time to failure of a product. The logarithm of a variable with a lognormal distribution has a normal distribution. Thus, methods for computing statistical intervals for the normal distribution can be used for the lognormal distribution. Tolerance limits, confidence limits for distribution percentiles, and prediction limits are calculated on the logarithms of the data, and are then converted back to the scale of the original data.

If lognormal data are censored, special statistical methods are needed. Schmee, Gladstein, and Nelson (1985) give tables for selected sample sizes

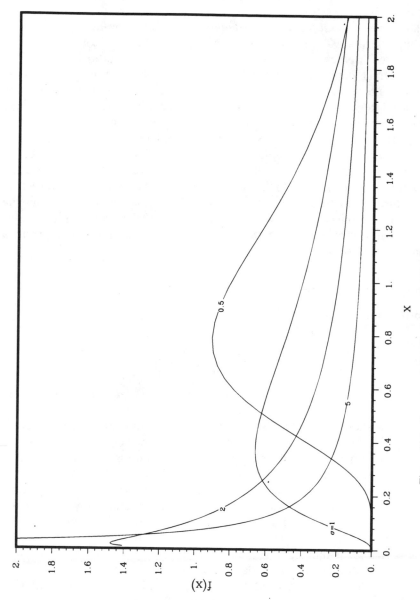

Figure 11.1 Lognormal distribution probability density functions with $\mu = 0$.

up to 100 for computing exact confidence intervals for lognormal distribution parameters from failure censored data. See Chapter 4 of Lawless (1982), Chapters 6, 7, and 8 of Nelson (1982), and Schneider (1986) for other tables, methods, and examples.

More generally, approximate normality is achieved by a transformation that generalizes taking logarithms of the data. A general family of transformations has been proposed by Box and Cox (1964) and is described in various texts [e.g., Box, Hunter, and Hunter (1978)]. Methods for constructing tolerance intervals, prediction intervals, and confidence intervals for probabilities and percentiles using such transformations are similar to those for the lognormal distribution. See Section 4.12 for a numerical example and further discussion.

11.3.4 Exponential Distribution

The one-parameter exponential distribution (see Figure 11.2) has the probability density function:

$$f(x;\theta) = \frac{1}{\theta}\exp\left(-\frac{x}{\theta}\right), \qquad x > 0, \quad \theta > 0.$$

where the scale parameter θ is also the distribution mean. This is an appropriate model for the lifetime of units that experience neither infant mortality nor wearout. That is, the probability of failure during a forthcoming period of time for an unfailed unit is independent of the age of the unit (e.g., the time to breakage of a glass in a restaurant). It may also be an appropriate model for the times between failures of a system that is made up of a large number of repairable components and that has reached stability [Drenick, (1960)]. The exponential is also the distribution of the times between occurrences in a homogeneous Poisson process. Section 11.3.14 gives a brief description of, and some useful references for, the homogeneous Poisson process. See Hahn and Shapiro (1967, page 105) for a further discussion of applications of the exponential distribution.

11.3.4.1 *Exponential Distribution Confidence and Tolerance Intervals*
Methods for setting confidence intervals for θ, the mean of the one-parameter exponential distribution, are treated, for example, in Chapter 3 of Bain (1978), Chapters 6, 7, and 8 of Nelson (1982), Nelson (1983), and Chapter 3 of Lawless (1982). Exact results are available for failure censoring or when there is no censoring. Approximate results must be used for other types of censoring. Because exponential distribution percentiles and tail probabilities are monotone functions of θ, confidence intervals on these can be obtained by substituting the endpoints of the confidence interval for θ into the expression for these quantities. In particular, the expression for the $100p$th

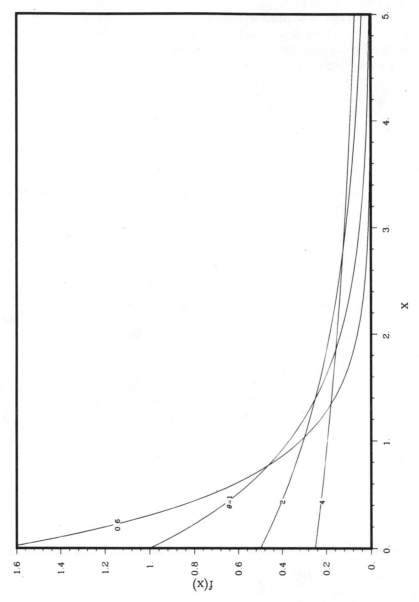

Figure 11.2 Exponential distribution probability density functions.

percentile of the exponential distribution is

$$Y_p = -\theta \log_e(1 - p).$$

Thus, a lower 95% confidence bound on the 90th percentile is

$$\underset{\sim}{Y}_{0.90} = -\underset{\sim}{\theta} \log_e(1 - 0.90)$$

where $\underset{\sim}{\theta}$ is the lower 95% confidence bound for θ.

Goodman and Madansky (1962) give methods for constructing one-sided tolerance bounds and two-sided tolerance intervals for the one-parameter exponential distribution. These methods are exact for complete data and for failure censored data.

11.3.4.2 *Exponential Distribution Prediction Intervals*

Hahn (1975) shows how to use sample data from an exponential population to obtain simultaneous prediction intervals to contain the means of all of m future samples from the same exponential population. These methods can also provide a prediction interval for the smallest or largest observation from a future sample of size m. The methods also apply for failure censored data.

Lawless (1971) gives a one-sided lower prediction bound that will, with specified confidence, be exceeded by at least k of m future observations from a previously sampled exponential distribution. Hall and Prairie (1973) also consider such limits, and Kaminsky (1977) compares different methods for setting such limits. Chapter 6 of Nelson (1982) describes other prediction intervals for the exponential distribution.

11.3.4.3 *Two-Parameter Exponential Distribution*

The probability density function for the two-parameter exponential distribution (see Figure 11.3) is

$$f(x; \gamma, \theta) = \frac{1}{\theta} \exp\left[-\left(\frac{x - \gamma}{\theta}\right)\right], \qquad x > \gamma, \quad \gamma > 0, \quad \theta > 0.$$

This distribution has the additional shift or threshold parameter γ, as well as the scale parameter θ. When the two-parameter exponential distribution is used to represent time to failure, failures cannot occur prior to time γ.

Chapter 3 of Lawless (1982), Chapter 3 of Bain (1978), and Cohen and Whitten (1988) provide methods for obtaining confidence intervals for the two-parameter exponential distribution. Guenther, Patil, and Uppuluri (1976), Dunsmore (1978), and Englehardt and Bain (1978a) discuss one-sided tolerance bounds for this distribution. Lawless (1977) gives methods for computing prediction intervals for the sample mean or for a specified ordered

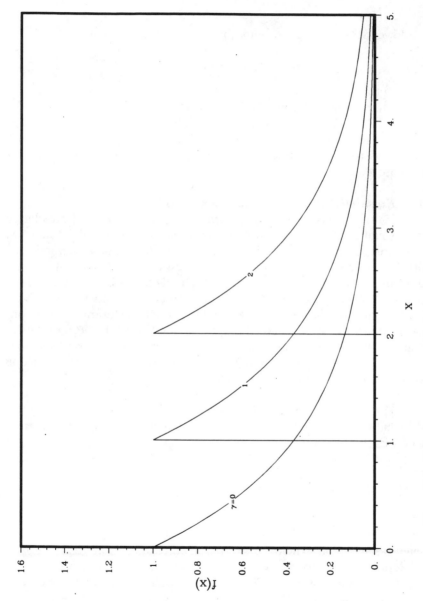

Figure 11.3 Two-parameter exponential distribution probability density functions with $\theta = 1$.

observation in a future sample from a two-parameter exponential distribution. Likes (1974) also gives a prediction interval for a specified ordered observation in a future sample from such a distribution.

11.3.5   Weibull and Other Extreme Value Distributions

The probability density function for the two-parameter Weibull distribution is

$$f(x; \alpha, \beta) = \frac{\alpha}{\beta} \left(\frac{x}{\alpha} \right)^{\beta-1} \exp\left[-\left(\frac{x}{\alpha} \right)^{\beta} \right], \qquad x > 0, \quad \alpha > 0, \quad \beta > 0.$$

Figure 11.4 shows Weibull probability density functions with scale parameter $\alpha = 1$ and with different values of the shape parameter β. This model is especially popular for representing time to failure [see, e.g., Hahn and Shapiro (1967), Lawless (1982), and Nelson (1982, 1983)]. However, the Weibull distribution is also used in other applications, for example, as the distributions of wind speed, the strength of a material, and, more generally, the distribution of the smallest observation in a large sample from certain distributions [including the smallest of any number of observations from a Weibull distribution; see Mann, Schafer, and Singpurwalla (1974, page 106)]. Hahn and Shapiro (1967, page 109) describe further applications of the Weibull distribution.

11.3.5.1   Relationship to Other Distributions
The Weibull distribution is closely related to several other distributions. In particular:

1. **The natural logarithm of a Weibull distributed variable** has a smallest extreme value distribution. Thus, statistical intervals for the smallest extreme value distribution can be transformed for use with the Weibull distribution and vice versa. The relationship is similar to that between the lognormal and the normal distributions; see Chapters 6, 7, and 8 of Nelson (1982) for illustrations. Also, Gumbel (1958), Galambos (1978), and Mann, Schafer, and Singpurwalla (1974) give further information on extreme value distributions.

2. **The largest extreme value distribution** is sometimes used to model the largest observation in a large sample or to predict the maximum of a large number of future observations. The observations from a largest extreme value distribution, when multiplied by -1, follow a smallest extreme value distribution. Thus, methods for the smallest extreme value distribution (and, therefore, for the Weibull distribution) can also be used for the largest extreme value distribution. Again, Gumbel

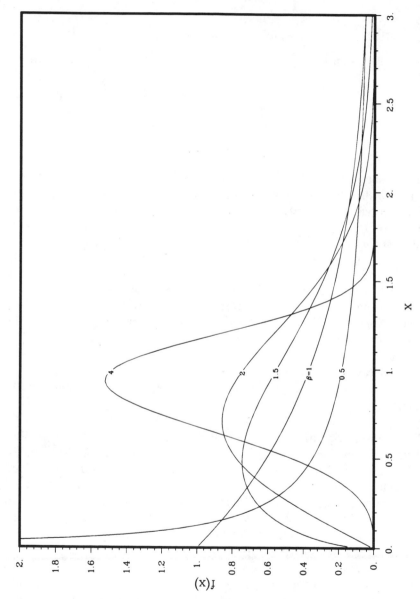

Figure 11.4 Weibull distribution probability density functions with $\alpha = 1$.

(1958), Galambos (1978), and Mann, Schafer, and Singpurwalla (1974) give further information.

3. A Weibull distribution with the shape parameter β equal to 1 is an exponential distribution.

4. A Weibull distribution with a *known* value of the shape parameter β is closely related to the exponential distribution. Specifically, if $x_1, x_2, \ldots, x_n$ follow a Weibull distribution, and if $x_i^* = x_i^\beta$, then $x_1^*, x_2^*, \ldots, x_n^*$ follow an exponential distribution.

Thus, the relatively simple procedures for obtaining statistical intervals for an exponential distribution can be readily extended to obtain statistical intervals for a Weibull distribution with known shape parameter. This is done by applying methods for the exponential distribution to the transformed values $x_i^* = x_i^\beta$. For example, to obtain a lower 95% confidence bound for the $100p$th percentile of such a distribution from a given set of observations $x_1, x_2, \ldots, x_n$, one proceeds as follows. First transform the data to obtain $x_1^* = x_1^\beta, x_2^*, \ldots, x_n^*$; then find $\underline{Y}_p^*$, the lower 95% confidence bound for the $100p$th percentile for an exponentially distributed variable, based upon the x_i^*; and finally obtain $\underline{Y}_p = (\underline{Y}_p^*)^{1/\beta}$, a lower 95% confidence bound for the $100p$th percentile of the original Weibull population.

Similar methods can be used to compute tolerance and prediction intervals and bounds for the Weibull population with a known (or assumed) value of β; see W. Nelson (1985) for details and more examples. The assumption that a parameter of a distribution is known exactly is a strong one, and, if wrong, can result in seriously incorrect inferences. On the other hand, if the assumption is close to being correct, simpler and more precise inferences are possible than if one had to use data to estimate β. In practice, one might compute statistical intervals for different reasonable assumed values of β and, thereby, assess the sensitivity of the resulting intervals to different possible values of β.

11.3.5.2 Weibull Distribution Confidence and Tolerance Intervals

Much of the literature on the Weibull distribution deals with failure censored data, because of the mathematical tractability of the analysis for this situation. For example, Bain (1978) provides tables of factors for one-sided tolerance bounds for Weibull and extreme value distributions. For complete data, the factors are given for sample sizes ranging from 5 to 120 (and for ∞) with many different proportions to be enclosed and several different levels of confidence. Less extensive tables are provided for failure censored data. Mann, Schafer, and Singpurwalla (1974) and Nelson (1982) provide similar tabulations. Bain and Englehardt (1981) provide alternative methods. Also see Johns and Lieberman (1966) for methods of computing confidence limits for Weibull model reliabilities and Mann and Fertig (1973, 1977) and Lawless (1975) for other methods and tables for computing tolerance intervals with the Weibull model.

Chapters 6, 7 and 8 of Nelson (1982) and Chapter 4 of Lawless (1982) give methods for computing confidence intervals for Weibull distribution parameters, distribution percentiles, and probabilities, with time or interval censoring, or for when needed tabulations are not available. These methods generally require sophisticated computer programs like PROC LIFEREG in SAS®, SYSTAT/SURVIVAL, LIMDEP, or CENSOR. Nelson (1990, page 237) describes and compares these and other programs for life data analysis.

11.3.5.3   Weibull Distribution Prediction Intervals

Mann and Saunders (1969) give one-sided lower prediction bounds for the smallest observation in a future sample from a Weibull distribution. They describe applications for setting warranty times for products. Mann (1970) gives extended tabulations for such a prediction bound, based on a linear combination of three particular order statistics from the initial sample. Mann and Fertig (1977) and Antle and Rademaker (1972) provide similar prediction intervals, based on maximum likelihood estimators. Also, see Fertig, Meyer, and Mann (1980), Englehardt and Bain (1979), Lawless (1973), Englehardt and Bain (1982), and Kushary and Mee (1990) for other methods and tabulations for computing Weibull distribution prediction intervals. Gumbel (1958, page 234), suggests an approximate prediction interval to contain an extreme observation from the largest extreme value distribution.

11.3.5.4   Three-Parameter Weibull Distribution

The probability density function for the three-parameter Weibull distribution is

$$f(x; \gamma, \alpha, \beta) = \frac{\alpha}{\beta} \left(\frac{x - \gamma}{\alpha} \right)^{\beta - 1} \exp\left[-\left(\frac{x - \gamma}{\alpha} \right)^{\beta} \right],$$

$$x > \gamma, \quad \gamma > 0, \quad \alpha > 0, \quad \beta > 0.$$

This distribution, like the two-parameter exponential, includes an additional "threshold" parameter γ (prior to which failures cannot occur in product life applications). Cohen and Whitten (1988) describe methods that can be used to compute confidence intervals for the parameters of this distribution. Jones and Scholtz (1983) give a method for obtaining approximate lower tolerance bounds for a three-parameter Weibull distribution. Also see Chapter 5 of Lawless (1982) for a simple method for making inferences with this model and for other references.

11.3.6 Logistic and Log-Logistic Distributions

The probability density function of the logistic distribution is

$$f(x;\mu,\sigma) = \frac{\exp\left(\dfrac{x-\mu}{\sigma}\right)}{\sigma\left[1+\exp\left(\dfrac{x-\mu}{\sigma}\right)\right]^2},$$

$$-\infty < x < -\infty, \quad -\infty < \mu < \infty, \quad \sigma > 0,$$

where μ and σ are location and scale parameters, respectively. Figure 11.5 shows the logistic probability density function with $\mu = 0$ and several values of σ. As can be seen from this figure, the logistic distribution is shaped much like the normal distribution, but has slightly longer tails.

Hall (1975) gives one-sided tolerance intervals for a logistic distribution. The intervals are based on best linear unbiased estimators. Balakrishnan and Fung (1991) develop one-sided tolerance bounds and two-sided tolerance intervals for the logistic distribution, also based on best linear unbiased estimators. They also discuss applications to acceptance sampling plans. Bain (1978) gives tolerance bounds based on maximum likelihood estimators and failure censoring. These methods are exact for complete data and for failure censored data. Bain, Balakrishnan, Eastman, Engelhardt, and Antle (1991) give methods, based on maximum likelihood estimators, for computing confidence intervals for the logistic distribution parameters and the logistic survival function. They also give methods for computing upper and lower tolerance bounds for the logistic distribution. See Chapter 5 of Lawless (1982) for approximate methods for computing confidence intervals with other kinds of censoring. The logarithm of a variable with a log-logistic distribution has a logistic distribution. Thus, statistical intervals for the logistic distribution can be used for the log-logistic distribution and vice versa. The relationship is like that between the lognormal and the normal distributions. See Balakrishnan (1991) for related material and further information about the logistic distribution.

11.3.7 Gamma Distribution

The probability density function of the gamma distribution is

$$f(x;\theta,\kappa) = \frac{x^{\kappa-1}}{\theta^\kappa \Gamma(\kappa)} \exp\left[-\left(\frac{x}{\theta}\right)\right], \quad x > 0, \quad \theta > 0, \quad \kappa > 0,$$

where θ and κ are scale and shape parameters respectively. Figure 11.6 shows the gamma probability density function for $\theta = 1$ and several values of the shape parameter κ. Like the Weibull distribution, the gamma distribution

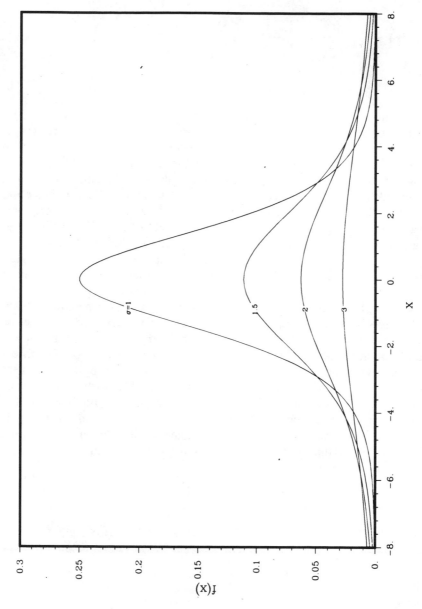

Figure 11.5 Logistic distribution probability density functions with $\mu = 0$.

218

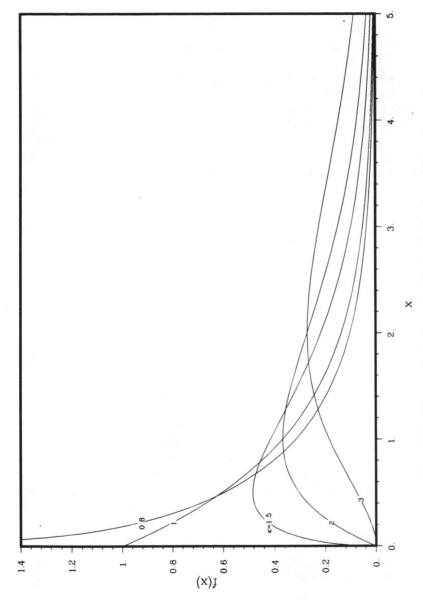

Figure 11.6 Gamma distribution probability density functions with $\theta = 1$.

219

is used for representing time to failure (but is, perhaps, not quite as popular for this use). See Hahn and Shapiro (1967, page 83) for further applications of the gamma distribution.

Guenther (1972) shows how to compute one-sided tolerance bounds and two-sided tolerance intervals and prediction intervals for a single future observation (referred to as β-expectation tolerance intervals) for the gamma distribution.

Large-sample approximate methods are generally used to compute statistical intervals for both complete and censored data from a gamma distribution. Section 5.1 of Lawless (1982) and Chapter 5 of Bain (1978) give approximate methods for computing confidence intervals for the parameters and percentiles of the gamma distribution with censored data. Shiue and Bain (1986) give approximate prediction intervals for a single future observation and for the average of m future observations from a gamma distribution.

11.3.8   Generalized Gamma Distribution

The probability density function of the generalized gamma distribution is

$$f(x;\theta,\beta,\kappa) = \frac{\beta x^{\beta\kappa-1}}{\theta^{\beta\kappa}\Gamma(\kappa)} \exp\left[-\left(\frac{x}{\theta}\right)^{\beta}\right], \quad x > 0, \quad \theta > 0, \quad \beta > 0, \quad \kappa > 0,$$

where θ is a scale parameter and β and κ are shape parameters. When $\beta = 1$, the generalized gamma reduces to the standard gamma distribution. When $\kappa = 1$, one obtains the Weibull distribution, and as $\kappa \to \infty$, the distribution approaches the lognormal. Figure 11.7 shows the generalized gamma density function for $\theta = 1$ and for several combinations of values of κ and β. This distribution is also sometimes used as a model for life data. Farewell and Prentice (1977) give an alternative parameterization that is especially useful for computing maximum likelihood estimates. See Lawless (1982) for a more complete description of the properties of this distribution, for an outline of some estimation methods, and for more references.

Lawless (1982) also provides methods for computing confidence intervals for the generalized gamma distribution. Bain and Weeks (1965) give one-sided tolerance bounds for this distribution and for the various special case distributions that arise if any two of the model parameters are known.

11.3.9   The Inverse Gaussian Distribution

The probability density function of the inverse Gaussian distribution is

$$f(x;\mu,\theta) = \left(\frac{\theta}{2\pi x^3}\right)^{1/2} \exp\left[-\frac{\theta(x-\mu)^2}{2\mu^2 x}\right], \quad x > 0, \quad \mu > 0, \quad \theta > 0,$$

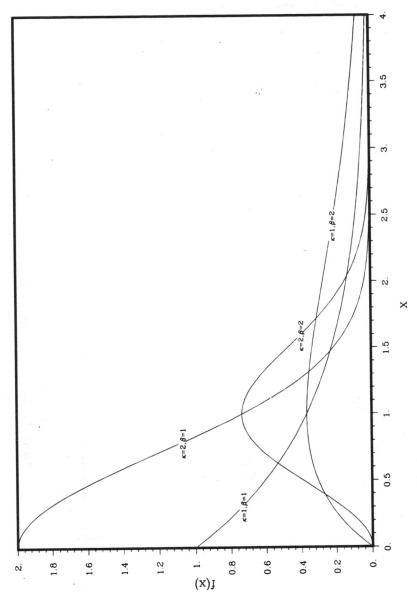

Figure 11.7 Generalized gamma distribution probability density functions with $\theta = 1$.

$\kappa=2, \beta=1$

$\kappa=2, \beta=2$

$\kappa=1, \beta=2$

$\kappa=1, \beta=1$

X

f(x)

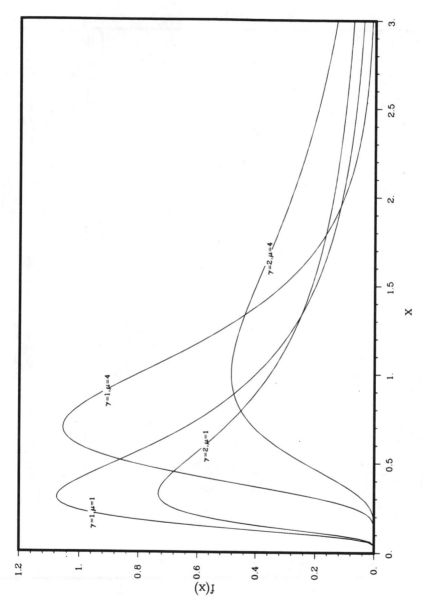

Figure 11.8 Inverse Gaussian distribution probability density functions.

where μ and θ are parameters of the distribution. Figure 11.8 shows the inverse Gaussian probability density function for different values of μ and σ. The inverse Gaussian is the distribution of the time that it takes a Brownian motion process with constant positive drift and variance to first cross a fixed positive boundary.

Folks and Chhikara (1978) and Chhikara and Folks (1989) review the properties and applications of the inverse Gaussian distribution. They also give exact methods for setting confidence intervals for the parameters of this distribution and for other quantities of interest. Chhikara and Guttman (1982) give a prediction interval for a future observation, and Padgett (1982) gives a prediction interval for the mean of several future observations from an inverse Gaussian distribution. Also, Padgett and Tsoi (1986) give prediction intervals for future observations from an inverse Gaussian distribution. With censored data, large-sample approximate methods (see Section 12.1.5) are needed. See Chhikara and Folks (1989) for other information and references for this distribution.

11.3.10 Life Distributions with Increasing Hazard Functions

Hanson and Koopmans (1964) provide a tolerance interval for a time to failure distribution that is known to have an increasing hazard function (sometimes called its "failure rate"). Patel (1980a) extends this method to provide a tolerance interval for a distribution with an increasing hazard function which, in addition, has an unknown threshold parameter (no units may fail at a time less than the value of the threshold parameter). Saunders (1968) gives a conservative lower prediction bound for the smallest observation in a future sample from any life distribution with an increasing hazard function. The procedure is conservative in the sense that the associated level of confidence is at least the stated value. Patel (1980b) and Ng (1984) give some similar procedures.

11.3.11 Uniform Distribution

The probability density function of the uniform distribution is

$$f(x;\theta) = \frac{1}{\theta}, \qquad 0 < x < \theta, \quad \theta > 0,$$

where θ is a parameter of the distribution. Figure 11.9 shows the uniform probability density function.

The uniform distribution can be used to represent the time of occurrence of an event for which all occurrence times are equally likely over some time interval, or a random occurrence over a line where all points on the line are equally likely to be occurrence points. Some examples of variables that one would expect to be, at least approximately, uniformly distributed are the

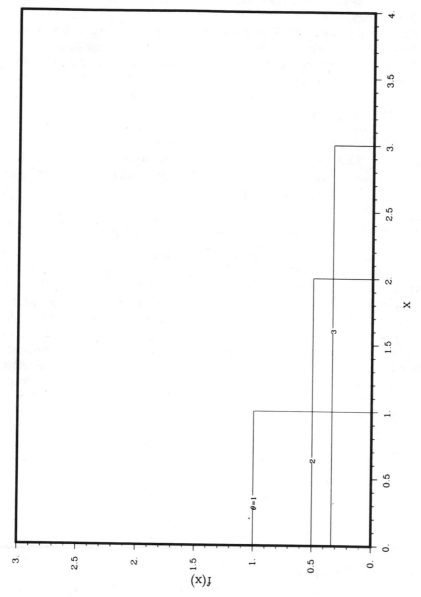

Figure 11.9 Uniform distribution probability density functions.

specific instant during a (365-day) calendar year that a random event (such as natural birth) occurs, and the point in a coil of wire where the highest thickness is measured.

Kendall and Stuart (1973, page 138) give a confidence interval for the parameter θ. Faulkenberry and Weeks (1968) give tolerance intervals for a uniform distribution.

11.3.12 Double Exponential Distribution

The probability density function of the double exponential distribution (also known as the Laplace distribution) is

$$f(x;\mu,\sigma) = \frac{1}{2\sigma} \exp\left(-\frac{|x-\mu|}{\sigma}\right),$$

$$-\infty < x < -\infty, \quad -\infty < \mu < \infty, \quad \sigma > 0,$$

where μ and σ are location and scale parameters, respectively. Figure 11.10 shows the double exponential probability density function for $\mu = 0$ and several values of σ.

Bain and Englehardt (1973) show how to compute confidence intervals for the parameters of the double exponential distribution. Kapperman (1977) gives methods for obtaining tolerance intervals for the double exponential distribution.

11.3.13 System Availability

For a repairable system, operating times between failures and times to repair are random quantities. The availability of such a system is the fraction of the time that the system is operating. Nelson (1970) gives methods for computing one-sided prediction bounds and two-sided prediction intervals for the availability of such a system over a specified future time period. The bounds and intervals are based on the assumptions that times to failure and times to repair are independently and exponentially distributed.

11.3.14 Nonhomogeneous Poisson and Weibull Processes

Chapter 7 gives methods for obtaining statistical intervals for situations in which the Poisson distribution can be used as a model for the number of occurrences over some fixed amount of time or space. The closely related homogeneous Poisson process model is used to describe the distribution of events that occur randomly and with a constant intensity over some interval of time or space. For a precise definition, see Cox and Lewis (1966, Chapter 2) or Kempthorne and Folks (1971, page 196). If the other assumptions of the model hold, but the intensity changes (e.g., as a function of time), the model

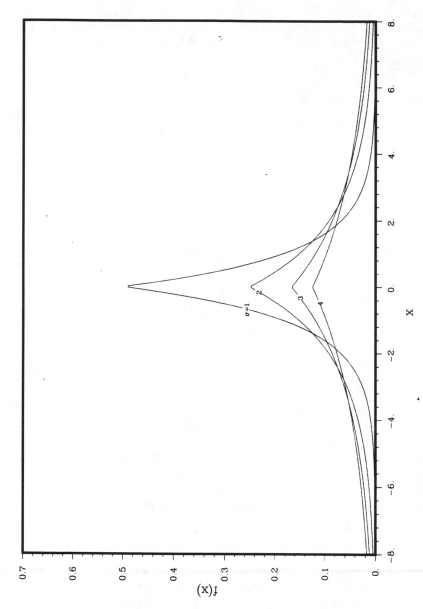

Figure 11.10 Double exponential (Laplace) distribution probability density functions with $\mu = 0$.

226

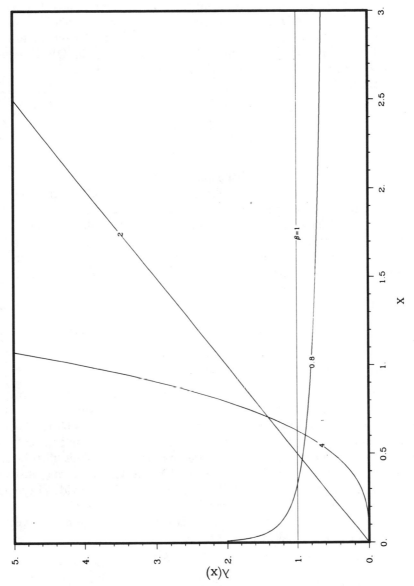

Figure 11.11 Weibull (power law) process intensity functions with $\alpha = 1$.

227

is known as a nonhomogeneous Poisson process. Cox and Lewis (1966) present inferential methods for nonhomogeneous Poisson processes and other similar point process models.

The Weibull process (not to be confused with the Weibull distribution), also known as the "power law" process, is a special nonhomogeneous Poisson process. It is frequently used, for example, to model times between failures of some repairable systems. Ascher and Feingold (1984) review statistical models and methods for such repairable systems.

The intensity function of the Weibull process is

$$\lambda(t; \alpha, \beta) = (\beta/\alpha)(t/\alpha)^{\beta-1}, \qquad t > 0, \quad \alpha > 0, \quad \beta > 0,$$

where α and β are scale and shape parameters, respectively. Figure 11.11 shows the Weibull process intensity function for $\alpha = 1$ and several values of the shape parameter β. Section 4.5 of Bain outlines some of the basic theoretical properties of the Weibull process model. Finkelstein (1976) gives confidence intervals for the parameters of this model. Also, Lee and Lee (1978) give other confidence intervals and a prediction interval for test completion time. Crow (1982) discusses applications to reliability growth and confidence bounds for the mean time between failures from such a process. Englehardt and Bain (1978b) give one-sided prediction bounds for future observations from a Weibull process.

11.3.15 Multivariate Normal Distribution

The multivariate normal distribution is a frequently used model to represent the joint distribution of multiple characteristics (or vectors) on the same observational unit. This distribution is discussed in numerous books devoted to multivariate analysis [e.g., Johnson and Wichern (1988)], and these books also describe confidence intervals for the parameters of this distribution. Theory for computing tolerance regions for the multivariate normal distribution is given by Guttman (1970). Hall and Sheldon (1979) provide alternative methods for constructing tolerance regions. Chew (1966) provides prediction intervals to contain a future observation vector from a past sample of vectors from a multivariate normal distribution.

11.3.16 Negative Binomial Distribution

The negative binomial distribution is the model for the number of independent (success/failure) trials until the occurrence of k "successes" when the success probability is p for each trial. The negative binomial probability function is

$$\Pr(x = x'; k, p) = \binom{x' - 1}{k - 1} p^k (1 - p)^{x' - k},$$

$$x' = k, k + 1, \ldots, \quad k \geq 1, \quad 0 \leq p \leq 1.$$

For more background and information on this distribution, see, for example, Guttman, Wilks, and Hunter (1982, page 36). Scheaffer (1976) provides four different methods for computing approximate confidence limits for p. He also discusses prediction intervals for this distribution.

11.3.17 Confidence Intervals for the Number of Nonconforming Units in a Lot

Assume a sample of n units is selected at random without replacement from a lot of N units, D of which are nonconforming, where D is unknown. Based on d, the observed number of nonconforming units in the sample, it is desired to find a $100(1 - \alpha)\%$ confidence interval for D. This problem is addressed by Tomsky, Nakano, and Iwashika (1979) who provide tables for N from 2 to 100, for various values of d and n. Odeh and Owen (1983) provide tables for $N = 400(200)2000$. These intervals are based upon the hypergeometric distribution. Also, Buonaccorsi (1987) investigates and compares several methods of computing confidence intervals for a population proportion, based on a sample from a finite population. As the lot size N increases, the methods for proportions, based on the binomial distribution, described in Chapter 6 apply; see also the discussion on infinite population assumptions in Section 1.10.

11.4 STATISTICAL INTERVALS FOR LINEAR REGRESSION ANALYSIS

Regression analysis is a widely used statistical technique to relate the mean of a response variable (also sometimes known as the "dependent variable") to one or more explanatory variables (sometimes called "independent variables" or "covariates"). Given sample data and an assumed relationship between the response and the explanatory variables, the method of least squares is often used to estimate unknown parameters in the relationship. There are many applications for regression analysis and the resulting statistical intervals. Most texts on statistical methods devote one or more chapters to regression analysis, and there are a number of books that deal exclusively with this subject. See, for example, Neter, Wasserman, and Kutner (1990) for a detailed elementary exposition, and Draper and Smith (1981) or Seber

(1977) for a more advanced discussion. Also, Hahn and Meeker (1984) provide an annotated bibliography of books on regression analysis.

As in other situations, the validity of statistical intervals generally depends on the correctness of the assumed model. The commonly used linear regression model assumes that the observed response variable y follows a normal distribution with mean

$$\mu = \beta_0 + \beta_1 x_1 + \cdots + \beta_p x_p;$$

here the x_j's are known values of the explanatory variables and the β_j's are unknown coefficients that are to be estimated from the data using the method of least squares. The standard model also assumes that the observed y values are statistically independent with a constant variance σ^2 for a given set of x_j's. In some applications, these x_j's might be known functions of some explanatory variables [e.g., $1/\text{temperature}$, log(voltage), $(\text{time})^2$, etc].

Users of regression analysis should watch for departures from the model assumptions. Methods for doing this can involve graphical analysis of the residuals (i.e., the differences between the observed and the predicted response values) from the fitted model, as described in texts on applied regression analysis. Some ways of handling possible departures from the assumed model follow:

1. If the mean of y cannot be expressed as a linear function of the parameters, special (usually iterative) nonlinear least squares methods for estimating the parameters may be required; see, for example, Gallant (1987), Bates and Watts (1988), or Seber and Wild (1989).

2. If σ^2 is not the same for all observations, a transformation [e.g., a Box–Cox (1964) transformation, see Section 4.12.1] of the response variable might be appropriate. Sometimes the method of weighted least squares is used. See Carroll and Ruppert (1988) or Sections 2.11 and 5.2 and 5.3 of Draper and Smith (1981). In other cases, it might be desirable to model both the mean and the standard deviation as separate functions of the explanatory variable [see Nelson (1984)].

3. If the observed response values (y's) are not statistically independent, either generalized least squares [see Chapter 5 of Graybill (1976)] or time series analysis methods [see Box and Jenkins (1976), Fuller (1976), or Pankratz (1983)] might be appropriate.

4. If some of the values of the response variable are censored, or if they do not follow a normal distribution, the method of maximum likelihood, rather than the method of least squares, may be appropriate for estimating the parameters. See Chapter 6 of Lawless (1982) or Nelson (1990) for details.

5. If the observed values of the explanatory variables contain significant measurement error, the methods given by Fuller (1987) might be appropriate.

The rest of this section provides references to methods for computing statistical intervals for the standard linear regression model, when all of the assumptions hold. For other situations, some of the references given above provide similar methods. Some of the procedures require factors that are too numerous to tabulate and thus require specialized computer software [e.g., Eberhardt, Mee, and Reeve (1989)]. In Chapter 12 we will describe some other more general methods that can be applied to regression analysis, but these also generally require special computer software.

11.4.1 Confidence Intervals for Linear Regression Analysis

Many books on statistical methods and specialized texts on regression analysis give details on the construction of confidence intervals for

- The parameters $(\beta_0, \beta_1, \ldots, \beta_p)$ of the regression model.
- The expected value or mean $\mu = \beta_0 + \beta_1 x_1 + \cdots + \beta_p x_p$ of the response variable for a specified set of conditions for the explanatory variable(s).
- The variance σ^2 (or standard deviation σ) of the observations for a given set of x_j's.

See, for example, Chapter 3 and 7 of Neter, Wasserman, and Kutner (1990) and Chapter 2 of Draper and Smith (1981). Also, Thomas and Thomas (1986) give confidence bounds for the percentiles of the assumed normal distribution of the response variable for a specified set of conditions for the explanatory variable(s).

11.4.2 Tolerance Intervals for Linear Regression Analysis

Tolerance intervals for the response variable for one or more conditions of the explanatory variable are *not* provided in most of the standard textbook chapters and books on regression analysis. [One exception is Graybill (1976).] This may be because special factors (based on the noncentral t distribution) are required to compute these intervals. Lieberman and Miller (1963) give tabulations for, and examples of, simultaneous tolerance intervals for the distribution of a response variable for a range of conditions for the explanatory variable(s). Chapter 3 of Miller (1981), Wallis (1951), Bowden (1968), Turner and Bowden (1977, 1979), and Limam and Thomas (1988a) describe other methods of computing simultaneous tolerance bands for a regression model. Mee, Eberhardt, and Reeve (1989) describe a procedure for computing simultaneous tolerance intervals for a regression model and describe applications to calibration problems.

11.4.3   Prediction Intervals for Regression Analysis

Many books on statistical methods and specialized texts on regression analysis show how to compute a prediction interval for a single future observation on a response variable for a specified set of conditions for the explanatory variable(s). In fact, in introductory textbooks, regression analysis is the only situation where prediction intervals are generally discussed. The methods for a single future observation easily extend to a prediction interval for the mean of m future response variable observations [see, e.g., Chapter 2 of Draper and Smith (1981), and Chapters 3 and 7 of Neter, Wasserman, and Kutner (1990)]. Approximate simultaneous prediction intervals to contain all of m future observations for a regression model are described in Lieberman (1961). Exact intervals and improved approximations are given by Hahn (1972b), and a related application is discussed by Nelson (1972a). Chapter 3 of Miller (1981) provides a comprehensive discussion of these and related methods.

11.5   STATISTICAL INTERVALS FOR COMPARING POPULATIONS AND PROCESSES

Designed experiments are often used to compare two or more competing products, designs, treatments, packaging methods, etc. Statistical intervals are useful for presenting the results of such experiments. The simplest comparison—namely, that where independent random samples are taken from each of two populations or processes that can be described by normal distributions—is emphasized in this section. References dealing with the use of statistical intervals to compare more than two normal distributions are also provided. The statistical methods for (fixed effects) analysis of variance (ANOVA) and analysis of covariance models frequently used in the analysis of such data are special cases of the more general regression analysis methods that were outlined in the previous section. Thus, these methods can usually also be applied to the problems of this section; see Mendenhall (1968) or Neter, Wasserman and Kutner (1990) for details. Also, Miller (1981) provides details about many more advanced (especially simultaneous) statistical intervals for comparing two or more populations or processes.

Chapters 3, 4, 5, and 6 of Lawless (1982) and Chapters 10, 11, and 12 of Nelson (1982) discuss methods for comparing various nonnormal distributions and for making comparisons when the data are censored.

11.5.1   Confidence Intervals for Comparing Populations

Most texts on elementary statistical methods show how to compute confidence intervals for the difference between the means of two randomly sampled populations or processes, assuming normal distributions and using various assumptions concerning the variances of the two distributions, based upon either paired or unpaired observations. Simultaneous confidence intervals for the differences among pairs of means for more than two normal distributions are also discussed in various texts; see, for example, Mendenhall (1968). These methods for comparing means are also approximately correct for many situations where the underlying distributions are not normal, especially when the deviations from normality are not too severe and/or the sample sizes are not very small. Many texts also show how to compute confidence intervals for the ratio of the variances (or standard deviations) of two normal distributions. See, for example, Guttman, Wilks, and Hunter (1982), Ostle and Mensing (1975), or Bowker and Lieberman (1972).

When a normal distribution cannot be assumed, distribution-free procedures may be used. For example, Gibbons (1975) shows how to obtain a distribution-free confidence interval for the difference between the location parameters of two populations and on the ratio of their scale parameters.

11.5.2 Prediction Intervals for the Comparison of Future Samples from Two Populations

Hahn (1977) shows how to obtain a prediction interval on the difference between future sample means from two previously sampled populations, assuming normal distributions can be used to describe the populations. Meeker and Hahn (1980) present similar prediction intervals for the ratio of exponential distribution means and normal distribution standard deviations. Both papers describe applications where one desires to make a statement about a future comparison of two populations, based upon the results of past samples from the same populations. Such an interval might, for example, be desired by a manufacturer who wishes to predict the results of a future comparison, to be conducted by a regulating agency, of two previously tested products.

11.5.3 Tolerance Intervals for a Particular Subpopulation or a Mixture of Several Normal Subpopulations

Tolerance intervals for a subpopulation corresponding to any particular group in a "fixed effects" analysis of variance model can be computed by using the methods referenced in Section 11.4, because, as previously indicated, fixed effects analysis of variance models are special cases of linear regression models. Mee (1989) gives a method for computing a tolerance interval for a population, based on a stratified sampling scheme (i.e., when samples are taken from each subpopulation).

In "random effects" analysis of variance models [see Chapter 12 of Mendenhall (1968)], one is dealing with two or more sources of random variability from a single population. These might, for example, consist of a random sample of batches, as well as a random sample of units within batches. Lemon (1977) gives an approximate procedure for constructing one-sided tolerance bounds for such situations, assuming the batch means follow a normal distribution, and Mee and Owen (1983) provide improved factors. The Mee and Owen (1983) tolerance bounds are exact if the ratio of the within batch variance to the between batch variance is known; otherwise, their factors provide approximate bounds. Mee (1984a) gives similar factors for two-sided tolerance intervals (referred to as a β-content tolerance interval) and for two-sided prediction intervals for a single observation from the entire population (referred to as a β-expectation tolerance interval). Limam and Thomas (1988b) provide methods for constructing similar tolerance intervals for a one-way random effects model with explanatory variables.

CHAPTER 12

Other Methods for Setting Statistical Intervals

Up to this point, most of the methods that we have presented for computing statistical intervals have been based on:

1. The exact sampling distribution of the sample data or resulting computed statistics (or simple approximations for these).
2. The "classical" or "frequentist" approach to making inferences [for which intervals are computed such that in repeated sampling from the same population, $100(1 - \alpha)\%$ of the intervals will, in the long run, result in correct statements].

This chapter outlines some useful alternative methods for computing statistical intervals. Specifically:

- Section 12.1 describes "large-sample" methods for computing approximate statistical intervals. These methods are quite general and are especially useful when "exact" methods are not available. Large sample methods are widely used in various applications, particularly for nonlinear regression, time series analysis, and in the analysis of censored data. For the most part, they require the use of sophisticated computer programs.
- Section 12.2 briefly describes some of the basic concepts of Bayesian statistical intervals. This approach allows one to incorporate prior information (which in many cases is subjective), along with sample information, to construct statistical intervals.

The discussion in this chapter, like that in Chapter 11, is somewhat more advanced than that in most other chapters.

12.1 LARGE-SAMPLE APPROXIMATE STATISTICAL INTERVALS

There are a number of different methods for computing statistical intervals. The choice among these depends on the question to be answered, on the nature of the data (e.g., whether or not there are censored observations), and on the assumed statistical model for the process being studied. Most of the methods presented in this book have been derived from exact statistical sampling theory; these are exact in the sense that if the assumed model is correct, the resulting interval will have exactly the nominal confidence level. In some cases (e.g., discrete models), exact sampling theory leads to approximate conservative intervals in that the actual confidence level is at least as large as the nominal value. However, many important statistical problems do not lend themselves to exact sampling theory solutions, or such solutions are too difficult (either theoretically or computationally) to be useful. In such cases, we often rely on approximate methods, most of which are based on relatively simple "large-sample" theory. The statistical intervals derived from such theory are approximate because the actual level of confidence associated with a statistical interval is only approximately equal to the nominal value. The approximation generally improves as the sample size increases.

The sample size required for an interval based on large-sample theory to provide adequate results depends on the specific situation. Sometimes, samples as small as 20 or 30 are sufficiently large to provide approximations that are good enough for practical purposes. Often, however, 50 or more observations are needed to obtain reasonable results. For problems with censored data, however, the effective sample size should be measured mainly by the number of uncensored observations, rather than the total sample size; see Schmee, Gladstein, and Nelson (1985), Billman, Antle, and Bain (1972), and Ostrouchov and Meeker (1988). In general, the adequacy of a large-sample approximation can be assessed by a Monte Carlo simulation. Such simulations can often be performed in the programming language of one of the more versatile statistical computing packages (e.g., SAS® or S).

One should always keep in mind that, for large-sample methods, just as for those based on exact sampling theory, the results are also often approximate because the assumed statistical model provides only an approximation to the true model (see earlier discussions, especially in Chapter 1). Thus, it is quite possible that the (usually unknown) bias introduced by an inadequate model creates more inaccuracy than that introduced by using large-sample theory.

12.1.1 Approximate Statistical Intervals in Statistical Computing Packages

Various widely used statistical computing packages (e.g., BMDP®, MINITAB®, and SAS®) generally use exact statistical methods when these are available. In more complicated situations, approximate methods are used.

The following are some typical procedures involving approximate methods that are available in advanced statistical packages.

1. Procedures for nonlinear regression analysis include BMDP® programs 3PR and PAR and SAS® procedure NLIN. See, for example, Gallant (1987), Bates and Watts (1988), or Seber and Wild (1989) for general theory and applications of nonlinear regression.

2. Procedures for fitting autoregressive-moving average (i.e., Box–Jenkins) time series models include BMDP® program 2T, the MINITAB® ARIMA command, and the SAS® ARIMA procedure. See Fuller (1976) and Box and Jenkins (1976) for general theory and applications.

3. Procedures for fitting models (including regression models with other than normally distributed errors) to censored data include BMDP® program 2L, SAS® procedure LIFEREG, SYSTAT/SURVIVAL, and special computer packages like CENSOR [see Meeker and Duke (1981)]. See Nelson (1982, 1990) and Lawless (1982) for general theory and applications.

Most statistical packages like SAS® and BMDP® provide point estimates and standard errors for different quantities of interest (e.g., model parameters or distribution percentiles) in these cases, and often do not provide confidence intervals. One can, however, use the formulas in Section 12.1.3.1, which require only simple calculations, using the point estimates and the standard errors of the quantities of interest, to obtain approximate intervals and bounds. Other more sophisticated approximate methods of computing statistical intervals or bounds (e.g., those described in Sections 12.1.3.2 and 12.1.3.3) may require special computer software or can be implemented in one of the statistical programming languages like SAS/IML® or S [see Becker, Chambers, and Wilks (1988)].

12.1.2 Preliminaries

In large-sample theory evaluations, one usually assumes that, as under exact theory, the data were generated by some random process (often simple random sampling from a particular population or process). To draw conclusions or make statements or predictions about unknown properties of this process (e.g., its mean), based on the data, one may, as in earlier chapters, make some very general assumptions (e.g., the observations are randomly selected from a continuous distribution) or assume a specific model (e.g., the data come from a normal distribution whose mean is related to some known explanatory variables through an assumed regression relationship of a specified form, but with unknown parameters). Generally, the statistical model contains unknown parameters for which one may desire point or interval estimates from the available data. Also, we are frequently interested in obtaining point and interval estimates of some functions of these parameters, such as:

- Distribution percentiles.
- Probabilities of specified events.
- Predictions for future observations.

In regression problems, these estimates depend on the values of the explanatory variables.

Methods of estimation, like least squares (commonly used in linear and nonlinear regression analysis and in time series analysis) or maximum likelihood (commonly used in the analysis of censored data or when fitting models with other than normally distributed random variation, and, less commonly, in time series analysis), provide estimates of model parameters and an estimate of the covariance matrix of the estimated parameters (i.e., a matrix that shows the variances of all parameter estimators, and the covariances among all pairs of such estimators). See Chapter 9 of Cox and Hinkley (1974), Appendices C and E of Lawless (1982), or Chapters 8 and 12 of Nelson (1982) for a description of general maximum likelihood theory. For a discussion of the general theory of nonlinear least squares, see, for example, Gallant (1987), Bates and Watts (1988), or Seber and Wild (1989).

12.1.3   Methods of Computing Confidence Intervals

There are several methods for computing approximate confidence intervals. These differ with respect to their statistical accuracy and precision, difficulty of computation, and their availability in popular statistical computing packages.

12.1.3.1   *Approximate Confidence Intervals Based on the Large-Sample Normal Distribution Theory*

The classical approach uses a (multivariate) normal distribution approximation for the sampling distribution(s) of the estimator of the unknown model parameter(s). In particular, let θ denote an unknown quantity of interest. This may be a distribution parameter or a function of one or more parameters, such as a percentile, or a distribution tail probability. Often, it can be assumed that if the sample size is large, an appropriate estimator $\hat{\theta}$ of θ approximately follows a normal distribution, with the approximation improving as the size of the sample increases. Then an approximate two-sided $100(1 - \alpha)\%$ confidence interval for θ is

$$\left[\underset{\sim}{\theta}, \tilde{\theta}\right] = \hat{\theta} \pm z_{(1-\alpha/2)}s_{\hat{\theta}} \qquad (12.1)$$

where $z_{(\gamma)}$ is the 100γth percentile of the standard normal distribution and $s_{\hat{\theta}}$, the "standard error" of $\hat{\theta}$, is an estimate of the standard deviation of (the sampling distribution of) $\hat{\theta}$. The value $s_{\hat{\theta}}$ is calculated from the known relationship between θ and the parameters of the distribution. This generally involves the variances, and, sometimes, the covariances among the estimators of the unknown distribution parameters; these are usually estimated from the data [see page 517 of Lawless (1982), Chapter 8 of Nelson (1982), or Chapter 5 of Nelson (1990)]. One-sided $100(1 - \alpha)\%$ confidence bounds are obtained by substituting α for $\alpha/2$, and using the appropriate lower or upper endpoint of the two-sided interval. This is the simplest and most commonly used method for using large-sample theory to compute approximate confidence intervals for parameters or other quantities of interest.

If θ is a quantity that must be positive (e.g., a percentile of a distribution, such as the Weibull, for time to failure), it is common to assume that $\log(\hat{\theta})$, rather than $\hat{\theta}$, approximately follows a normal distribution. Then an approximate two-sided $100(1 - \alpha)\%$ confidence interval for θ is

$$\left[\underline{\theta}, \bar{\theta} \right] = \left[\hat{\theta}/q, \hat{\theta}q \right]$$

where $q = \exp(z_{(1-\alpha/2)}s_{\hat{\theta}}/\hat{\theta})$. Again, one-sided $100(1 - \alpha)\%$ confidence bounds are computed by substituting α for $\alpha/2$ in this expression, and choosing the appropriate endpoint of the two-sided interval. When θ must be positive, this approach often gives better results than using $\hat{\theta}$ and $s_{\hat{\theta}}$ directly in Equation (12.1). This is because the sampling distribution of $\log(\hat{\theta})$ is unrestricted, and thus can be expected to be more symmetric than that of $\hat{\theta}$, thus, likely providing an improved approximation. Also this approach, unlike Equation (12.1), always gives positive endpoints for the confidence interval.

Similarly, when θ is restricted to be between 0 and 1 (e.g., θ is the proportion of units in the population whose value exceeds a specified threshold), a confidence interval based on $\log[\hat{\theta}/(1 - \hat{\theta})]$ will generally be better than one based on $\hat{\theta}$. A two-sided $100(1 - \alpha)\%$ confidence interval based on this transformation is

$$\left[\underline{\theta}, \bar{\theta} \right] = \left[\frac{\hat{\theta}}{\hat{\theta} + (1 - \hat{\theta})q}, \frac{\hat{\theta}}{\hat{\theta} + (1 - \hat{\theta})/q} \right]$$

where $q = \exp\{(z_{(1-\alpha/2)}s_{\hat{\theta}})/[\hat{\theta}(1 - \hat{\theta})]\}$, and a one-sided confidence bound is obtained similarly. See Nelson (1982, page 383) and Sections 3.2 and 8.1 of Lawless (1982) for further discussion and examples.

The same concept can be used to compute confidence intervals for a parameter or other quantity θ which is restricted to some other specified range $a \le \theta \le b$ by basing the interval on the assumption that

$$\hat{L} = \log\left(\frac{\hat{\theta} - a}{b - \hat{\theta}}\right)$$

approximately follows a normal distribution. The resulting two-sided $100(1 - \alpha)\%$ confidence interval for θ is computed as

$$\left[\underset{\sim}{\theta}, \tilde{\theta}\right] = \left[\frac{\exp(\underset{\sim}{L})}{1 + \exp(\underset{\sim}{L})}, \frac{\exp(\tilde{L})}{1 + \exp(\tilde{L})}\right],$$

where

$$\left[\underset{\sim}{L}, \tilde{L}\right] = \left[\hat{L} - q, \hat{L} + q\right]$$

and

$$q = \frac{z_{(1-\alpha/2)}s_{\hat{\theta}}(b - a)}{(\hat{\theta} - a)(b - \hat{\theta})}.$$

One-sided confidence bounds are again obtained in a similar manner. The use of the preceding transformations eliminates the possibility that the endpoint of a confidence interval will lie outside of the range of possible values of θ (e.g., a probability larger than 1 or a negative bound on a variance, or time to failure). Some computer programs automatically use such transformations to compute confidence intervals [e.g., CENSOR, described by Meeker and Duke (1981)].

12.1.3.2 Approximate Confidence Intervals Based on Inverting Likelihood Ratio Tests

Although large-sample normal-theory confidence intervals are easier to compute, more accurate approximate confidence intervals can often be obtained, especially for relatively small-sample situations, by a method that involves inverting a likelihood ratio test for the quantity of interest. In particular, a likelihood ratio based $100(1 - \alpha)\%$ confidence interval for an unknown quantity θ is the set of all values of θ_0 for which the hypothesis H_0: $\theta = \theta_0$ cannot be rejected with a likelihood ratio test, with a significance level (i.e., probability of rejecting H_0 when it is true) equal to α. General theory for likelihood ratio tests is given in various texts on mathematical statistics, such as Hogg and Craig (1978) and Cox and Hinkley (1974).

The superiority of this approximate method over the approximate normal-theory approach has been shown by Monte Carlo simulation for a number of different models and types of data; see Lawless (1982, page 178), Meeker (1987), Donaldson and Schnabel (1987), Ostrouchov and Meeker

(1988), and Vander Wiel and Meeker (1990). Also, confidence intervals based on inverting likelihood ratio tests do not require the transformations discussed in the last section [in fact, maximum likelihood estimates are invariant to such transformations; see page 339 of Cox and Hinkley (1974)], and the endpoints of the intervals will always be within the allowable range of the parameter. See Hall (1987) for theory comparing likelihood-ratio-based confidence intervals with other kinds of confidence intervals.

Inverting a likelihood ratio test is likely to require between 5 and 20 times the computer time needed to compute confidence limits using the large sample normal-theory methods, but this may not be a serious problem with modern computing resources. However, few general computer programs are, as yet, available to do the necessary computations. Venzon and Moolgavkar (1988) and Cook and Weisberg (1988) describe efficient methods for computing likelihood-ratio-based confidence intervals.

12.1.3.3 *Confidence Intervals Based on Inverting Goodness of Fit Tests*
One can compute confidence intervals or bounds by inverting goodness of fit tests. This approach was suggested by Kempthorne (1971). It is described by Kempthorne and Folks (1971) and, in detail, by Easterling (1976), who illustrates the procedure with a series of examples using parametric models. These procedures are useful for simultaneously assessing model adequacy and statistical uncertainty. When the goodness of fit test is approximate, the resulting confidence intervals or bounds (also called consonance intervals or bounds) are also approximate.

Hall and Wellner (1980) and Nair (1981) suggest the inversion of goodness of fit tests to obtain approximate nonparametric simultaneous confidence bands for a survival function based on right censored data. Nair (1984) compares several such methods. Weston and Meeker (1991) suggest and evaluate a refinement that can improve the large-sample approximation.

12.1.3.4 *Approximate Confidence Intervals Based on the Bootstrap*
Another alternative for computing approximate confidence or other statistical intervals is to use the "bootstrap" method. This is a simulation-based resampling procedure of the available data that uses a Monte Carlo estimate of the distribution of an appropriate estimating statistic to set confidence limits for an unknown quantity of interest. There are several parametric and nonparametric ways to do this. See, for example, Efron and Gong (1983), Efron and Tibshirani (1986), Efron (1987), and Buckland (1984) for discussions on how to use the bootstrap to set confidence intervals. Buckland (1985) provides a FORTRAN algorithm that computes several different kinds of bootstrap confidence intervals. Other applications are beginning to appear in the data analysis literature.

The accuracy (i.e., approximation to the nominal confidence level) of bootstrap methods is often close to or better than that of inverting a likelihood ratio test [see Efron (1985) and Efron (1987)]. However, this result should be evaluated for specific situations. Bootstrap methods can require even more computing than inverting a likelihood ratio test, and up to hundreds to thousands of times more computer time than using simple normal-theory methods. However, with modern computing resources, even this may not be a severe problem, and the improvement may often be worth the extra cost. Presently, special computer programs (or special procedures written in the language of versatile statistical packages like SAS® or S) are needed to compute bootstrap confidence intervals.

12.1.4   Approximate Large-Sample Tolerance Intervals

Because one-sided tolerance bounds for a distribution are equivalent to one-sided confidence bounds for distribution percentiles, the approximate large-sample methods just described can be used to compute approximate one-sided tolerance bounds. (The parameter θ in this case is the distribution percentile to be estimated.) To construct two-sided tolerance intervals, the Bonferroni approximation (see Section 11.1.1) can be applied to the combined one-sided lower and upper tolerance bounds.

12.1.5   Approximate Large-Sample Prediction Intervals

In general, the theory for statistical prediction intervals is complicated. For situations for which no exact procedure is available, Atwood (1984) gives a rather general method that uses maximum likelihood estimates and large-sample theory to compute prediction intervals that approximately account for uncertainty in the parameter estimates.

With large initial samples, approximate prediction intervals can sometimes be computed by assuming that the sample data estimate the model parameters without error. Then a prediction interval is simply based on a known probability distribution. In Chapter 1 we called such an interval a "probability interval." Such intervals tend to be unconservative, i.e., shorter than they should be, because they consider only the uncertainty in the future observations and ignore the variability in the estimates from the past data. This large-sample approximation tends to be justified when the uncertainty in the random quantity to be predicted dominates that in the estimates of the unknown model parameters (from the past data), as is often the case for large initial samples. In practice, the approximation tends to be crude when $n < 30$ and adequate for $n > 60$. The most common application of this approximation is in time series forecasting [see Box and Jenkins (1976) and Pankratz (1983)].

12.2   BAYESIAN STATISTICAL INTERVALS

The statistical intervals discussed up to this point are computed and evaluated from the point of view described in Section 1.15; that is, if one were to repeat the random sampling process many times, each time constructing a similar interval, in the long run, $100(1 - \alpha)\%$ of the intervals will be correct [i.e., actually contain the quantity (or quantities) they are claimed to contain], where the confidence level $[100(1 - \alpha)\%]$, is chosen (or at least known) by the analyst. This is the "frequentist," "classical," or "sampling theory" approach to statistical inference.

Bayesian methods provide a useful alternative approach that allows the analyst to incorporate additional information (i.e., beyond the available data) into an inferential or decision-making process. Uncertainty is expressed in terms of a probability distribution for the unknown parameter or parameters of an assumed statistical model. Bayes' theorem [see a standard text on probability theory, or Box and Tiao (1973), Press (1989), or Guttman, Wilks and Hunter (1982)] is used to combine sample information with prior information, which is expressed in terms of (an often subjective) "prior distribution." This leads to a "posterior distribution" that provides the updated information about the true values of the model parameters. The posterior distribution can be used to make the desired inferences, including the construction of Bayesian confidence, tolerance, and prediction intervals.

For certain problems and assumed prior distributions, Bayesian techniques give the same results as the sampling theory approach [e.g., a confidence interval to contain the mean of a normal distribution—see page 203 of Guttman, Wilks, and Hunter (1982)]. However, the interpretation given to these results is different in that one, for example, makes statements directly about the probability of a parameter or function of parameters being within the calculated Bayesian interval. Also, there are important differences of opinion on valid sources of prior information and on what should be done when there is no quantifiable prior information. For further discussion, see Section 7 of Guttman (1970), Chapter 1 of Box and Tiao (1973), Press (1989), Lindley (1965, 1970), or Martz and Waller (1982). Also, Hahn and Raghunathan (1988) show how to use Bayesian methods to combine information from several different sources and give applications to prediction problems and other industrial applications.

Another approach, known as "empirical Bayes," uses data from a previous sample or from a sequence of previous samples to obtain a prior distribution [see Maritz (1970) or Casella (1985)].

12.2.1 Bayesian Confidence Intervals

Bayesian confidence intervals (regions) for distribution or model parameters are obtained by finding an interval (region) of the posterior distribution to contain the parameter(s) or other quantity(ies) of interest with a specified probability. Basic theory, methods, and examples for computing Bayesian confidence intervals for normal distribution means and variances (or standard deviations) and for the binomial parameter p are given in Chapter 9A of Guttman, Wilks, and Hunter (1982). For theory and applications related to reliability problems and frequently used life distribution models (exponential, Weibull, gamma, etc.), see Martz and Waller (1982). Also, Tsokos and Shimi (1977a, 1977b) contain papers that show how to compute Bayesian confidence (and other) intervals for some models frequently used in reliability analysis. Raghunathan and Rubin (1988) use Bayesian methods and the SIR algorithm to compute a confidence bound on the proportion of a population below a specified limit and to assess the sensitivity to assumptions about the shape of the distribution. The SIR algorithm, described by Rubin (1988), is a Monte Carlo based method for computing Bayesian confidence intervals. Gelfand and Smith (1990) discuss this and other methods for computing Bayesian posterior densities for a variety of possible models.

12.2.2 Bayesian Tolerance Intervals

Section 9 of Guttman (1970) shows how to construct Bayesian tolerance intervals in general and how to do so for some specific distributions, including the normal distribution. This book also describes the relationship between Bayesian and sampling theory tolerance intervals. Also see Martz and Waller (1982) and Section 9 of Lindley (1970). Miller (1989) gives an empirical Bayes procedure for computing tolerance intervals for a normal population, using past data to estimate the population standard deviation.

12.2.3 Bayesian Prediction Intervals

Section 8 of Guttman (1970) shows how to compute Bayesian prediction intervals for a single future observation (which he calls β-expectation tolerance intervals). The discussion is both general and specific for some distributions, including the normal distribution. The relationship between Bayesian and sampling theory prediction intervals is also shown. Other applicable general results for Bayesian prediction intervals are given by Aitchison and Sculthorpe (1965) and in the books by Weiss (1961), Aitchison and Dunsmore (1975), and in Section 9 of Lindley (1970). Thatcher (1964) compares Bayesian and sampling theory prediction intervals for a future sample from a binomial population. Additional results concerning Bayesian prediction intervals for the binomial distribution are given by Bratcher, Schucany, and Hunt (1971). Evans, Jones, and Owen (1976) show how to use and interpret the Bayesian predictive distribution.

12.2.4 Bayesian Intervals for Linear Regression Models

Bayesian statistical intervals can also be computed for the situations described in Section 11.4. Zellner (1971) gives details for the application of Bayesian statistical techniques to the simple linear regression model, and the use of Bayesian techniques with more general econometric models. Also see Box and Tiao (1973), Broemeling (1985), and Press (1989).

CHAPTER 13

Case Studies

13.1 INTRODUCTION

This chapter presents a series of case studies that illustrate the methods in this book. They are a representative sample of frequently occurring problems that we have encountered recently. We present these problems as they were presented to us, rather than in a "clean" textbook style. Then we describe our proposed "solution." We stress the basic underlying assumptions and the practical aspects of using and interpreting statistical intervals.

We do not illustrate all of the topics covered in the earlier chapters—only some that we have encountered recently. Thus, there is some repetition of techniques, but each example has some new feature.

In some cases, the names of the product or the units of measurement were changed, or the available data were multiplied by a scale factor, to protect proprietary information. In other cases, it was necessary, for similar reasons, to replace real data with simulated data.

13.2 EXAMPLE 1: DEMONSTRATION THAT THE OPERATING TEMPERATURE OF MOST MANUFACTURED DEVICES WILL NOT EXCEED A SPECIFIED VALUE

13.2.1 Problem Statement

The designers of a solid-state electronic device wished to "demonstrate that the surface temperature of most devices will not exceed 180 °C in operation," based on measurements on a sample of such devices.

13.2.2 Some Basic Assumptions

To make the demonstration, it is necessary that the selected devices be a random sample from the production process. This assumption deserves careful scrutiny. If, for example, early prototype devices are used in the

245

demonstration test, inferences from the test might not apply to subsequent production. In this example, special care was taken to assure that the test units were randomly selected from those made in a pilot production process that closely simulated actual manufacturing conditions. For example, raw materials were obtained from the same sources as those used in production. Also, the operating conditions and performance measurements for the demonstration test must be the same as those to be encountered in operation. Thus, additional efforts were made to have the test condition simulate as closely as possible, the actual field environment, and to use comparable measuring instruments. We also assume that the production process is stable, now and in the future. One should keep in mind that, as indicated in Chapter 1, if these assumptions are not met, the resulting statistical intervals (in this analytic study) express only one part of the total uncertainty and are likely to be too short.

13.2.3 Statistical Problem

After discussion with the device designers, the following more statistically precise problem statement was agreed upon. It is desired to show, with 90% confidence, that p_C, the proportion of units produced by the process with surface temperatures less than $L = 180$ °C, is at least $p_c^s = 0.99$ (or some other high proportion); that is, show with high confidence that $p_C =$ Pr(Temp $< 180) > 0.99$.

There are two general methods of making such a demonstration:

1. Make no assumption about the form of the statistical distribution of the surface temperatures. In this case, one dichotomizes the data by classifying each sampled device as either conforming (temperature less than or equal to 180 °C) or as nonconforming (temperature greater than 180 °C). Then one can use the procedures for proportions (i.e., those based on the binomial distribution) given in Chapter 6 to compute a lower confidence bound on p_C.

2. Assume that the operating temperatures (or some transformation of the operating temperatures) follow a particular probability distribution (such as the normal distribution) and use the temperature readings to find a lower confidence bound for p_C. Thus, for a normal distribution, one would use the procedures given in Section 4.5.

In either case, if the lower 90% confidence bound for p_C exceeds 0.99, the needed demonstration is achieved.

The second method provides a more efficient use of the data—especially if all, or the great majority, of the values are well within bounds—if one can assume an appropriate distribution for the operating temperatures. Of course, when one can only observe that a unit is conforming or nonconforming, the

EXAMPLE 1   **247**

second approach is not applicable. This occurs, for example, with a photoflash which, when tested, either works or does not. We will use both approaches and compare the findings.

13.2.4 Results from a Preliminary Experiment

The available data were limited to a random sample of only six devices from the pilot production process. These yielded the following surface temperature readings (in °C):

$$170.5, 172.5, 169.5, 174.0, 176.0, 168.0.$$

These data are plotted on normal probability paper in Figure 13.1. There is no obvious deviation from normality. The sample, however, is clearly too small to draw any definitive conclusions about the underlying distribution. We observe that none of the six observations was above the 180 °C threshold. From the data, we calculate the sample mean and standard deviation to be

$$\bar{x} = \frac{170.5 + 172.5 + \cdots + 168.0}{6} = 171.75,$$

and

$$s = \left(\frac{(170.5 - 171.75)^2 + \cdots + (168.0 - 171.75)^2}{6 - 1} \right)^{1/2} = 2.98.$$

We will use distribution-free and normal-theory methods to obtain lower confidence bounds for p_C, the proportion of conforming devices (i.e., those with operating temperatures less than 180 °C) for the sampled process.

13.2.5 Distribution-Free Lower Confidence Bound on the Proportion Conforming

Because there were $x = 6$ conforming devices in a sample of size $n = 6$, the point estimate of the proportion conforming is $\hat{p}_C = x/n = 6/6 = 1.0$. Using the methods of Section 6.2, a lower $100(1 - \alpha)\%$ confidence bound for p_C is computed as

$$\underset{\sim}{p_C} = \left\{ 1 + \frac{(n - x + 1) F_{(1-\alpha;\, 2n-2x+2,\, 2x)}}{x} \right\}^{-1}$$

where $F_{(1-\alpha;\, r_1, r_2)}$ is the $100(1 - \alpha)$th percentile of the F distribution (see Tables A.9a to A.9f) with r_1 numerator and r_2 denominator degrees of freedom. With $x = 6$ and using $F_{(0.90;\, 2,12)} = 2.807$, a lower 90% confidence

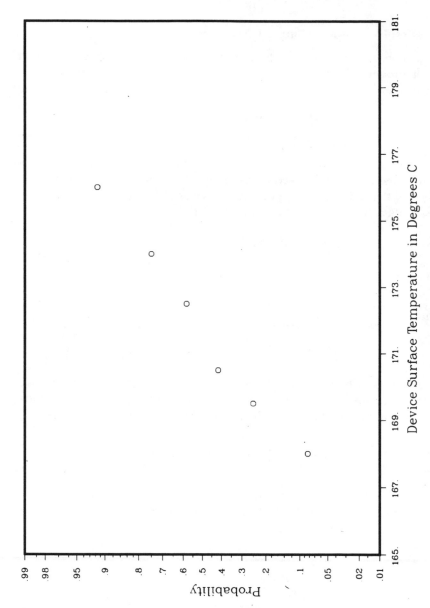

Figure 13.1 Normal probability plot of device surface temperature readings.

EXAMPLE 1   **249**

bound for p_C is

$$p_C = \left\{1 + \frac{(6 - 6 + 1)2.807}{6}\right\}^{-1} = 0.68.$$

(This value can also be obtained from Table A.23a or from Figure 6.1a.) Thus, all we can say with 90% confidence, using a distribution-free approach, is that at least 68% of the devices for the sampled process have temperatures below 180 °C. More generally, lower confidence bounds for various confidence levels are

Confidence Level	Lower Confidence Bound on p_C
75%	0.79
90%	0.68
95%	0.61
99%	0.46

Therefore, even though all of the test units were in conformance, we cannot claim that the true p_C exceeds the specified proportion $p_C^S = 0.99$ with 90% (or even with 75%) confidence. This, of course, is not surprising due to the small sample size (i.e., $n = 6$). Indeed, these results are the best that one can obtain with a sample of six devices, using distribution-free methods. This fact, however, was known before taking the sample. (See Sections 9.6 and 13.2.8.)

13.2.6 Lower Confidence Bound for the Proportion Conforming Assuming a Normal Distribution

We now assume that a normal distribution with an unknown mean and standard deviation adequately describes the distribution of the surface temperatures of the manufactured devices. We can use the sample estimates $\bar{x} = 171.75$ and $s = 2.98$ to compute a point estimate for $p_C = \Pr(y < 180)$. Substituting $\bar{x}$ for the mean and s for the standard deviation of the normal distribution, we get the point estimate $\hat{p}_c = \Phi[(L - \bar{x})/s] = \Phi[(180 - 171.75)/2.98] = \Phi(2.77) = 0.9972$. Also, under the normal distribution assumption, one can use the methods given in Section 4.5.1 and Table 7 of Odeh and Owen (1980) to compute a lower 90% confidence bound for the process proportion less than 180 °C. This would also be a lower 90% confidence bound on the probability that a single randomly selected device will have a surface temperature less than 180 °C. To proceed, we first compute

$$K = \frac{\bar{x} - L}{s} = \frac{171.75 - 180}{2.98} = -2.77.$$

Because the table gives lower bounds on $\Pr(y > L)$ and we are seeking a lower bound on $\Pr(y < L)$, we enter the table with $-K = 2.77$. In particular, using Table 7.3.1 with O & O's ETA $= 0.90$, $n = 6$, and interpolating between $(2.6, 0.91098)$ and $(2.8, 0.92904)$ for $-K = 2.77$, we obtain the lower 90% confidence bound $p_C = 0.93$. Thus we are 90% confident that the proportion of devices below 180 °C is at least 0.93. Lower confidence bounds for various confidence levels are

Confidence Level	Lower Confidence Bound on p_C
75%	0.98
90%	0.93
95%	0.87
99%	0.74

Even though these bounds are more favorable than those computed without making any distributional assumptions, they are still not good enough to achieve the desired demonstration. We need emphasize that the limited data do *not* contradict the claim that 99% of the devices from the process are in conformance, since, after all, our point estimate is 99.72%. Rather, in this example, in which the burden of proof was placed on the designers, the limited sample of size $n = 6$ was just not big enough to achieve the desired demonstration with 90% confidence. (Unlike the distribution-free case, this was not known prior to obtaining the data.) Moreover, under the normal distribution assumption the data *do* allow one to claim with 75% confidence that at least 98% of the devices in the sampled population have temperatures below 180 °C.

13.2.7 An Alternative Approach: Normal Distribution Tolerance Bound for Device Temperatures

It was not possible to demonstrate with 90% confidence that at least 99% of the devices from the sampled process meet the 180 °C requirement, based on the sample of six devices, even under normal distribution assumptions. Thus, the designers asked: What surface temperature value can be demonstrated with 90% confidence to be met by at least 99% of the devices? This new question calls for a one-sided upper tolerance bound. One now uses the methods described in Section 4.6.3, with $n = 6$, $\bar{x} = 171.75$, $s = 2.98$, $p = 0.99$, $1 - \alpha = 0.90$ and $g'_{(0.90; 0.99, 6)} = 4.243$ from Table A.12d. Then an upper 90% tolerance bound to exceed the surface temperatures for at least 99% of the devices from the sampled process is

$$\tilde{T}_{0.99} = \bar{x} + g'_{(1-\alpha; p, n)}s = 171.75 + 4.243(2.98) = 184.4.$$

Thus we can claim, with 90% confidence, that at least 99% of the devices

EXAMPLE 1   **251**

from the sampled process have surface temperatures that are less than 184.4 °C.

13.2.8 Sample Size Requirements

The analyses failed to demonstrate that 99% of the devices from the sampled process have surface temperatures below 180 °C. Thus it is clear that a larger sample will be required to achieve the desired demonstration. The designers now need to know how large a random sample from the process is required for this purpose.

Assuming that p_C is really greater than the specified value $p_C^S = 0.99$, we need a sample that is large enough to demonstrate this fact with some specified probability [i.e., to have $\Pr(\tilde{p}_C > p_C^S) > p_{DEM}$]. A very large sample is required if the actual (unknown) p_C is close to (but greater than) $p_C^S = 0.99$. On the other hand, the sample size could be smaller if p_C is very close to 1 (e.g., $p_C = 0.9999$).

For analysis method 1 (the distribution-free approach), Figure 9.2g shows p_{DEM}, the probability of a successful demonstration at the 90% confidence level that p_C is at least 0.99, as a function of p_C, the process proportion conforming, and the sample size n. For example, if the actual proportion of units below 180 °C in the sampled process is really $p_C = 0.996$, a sample size of $n = 1538$ has a probability $p_{DEM} = 0.95$ of resulting in a successful demonstration at the 90% confidence level. Moreover, for the demonstration to be successful, it is necessary that no more than $c = 10$ of the 1538 sampled devices have measured temperatures greater than 180 °C (i.e., one can show, using the binomial distribution, that this outcome would have a probability of approximately 0.95 if the true conformance probability were 0.996).

Graphs like Figure 9.2g are easy to construct with a computer program. We provide such graphs for 14 combinations of problem parameters (i.e., confidence level, and p_C^S, the specified proportion conforming) in Chapter 9. As described in Section 9.6.3, the needed sample size can also be found from Table A.30. In particular, we enter Table A.30 to determine the necessary sample size so as to demonstrate with confidence level $(1 - \alpha) = 0.90$ that the conformance probability is greater than $p' = p_C^S = 0.99$, subject to the requirement that $\delta = 1 - p_{DEM} \leq 0.05$ when the actual conformance probability is $p^* = 0.996$. The table gives the necessary sample size as 1538.

For analysis method 2 (i.e., assuming that the temperatures follow a normal distribution with unknown mean μ and standard deviation σ), the probability of a successful demonstration, p_{DEM}, again depends on the confidence level to be associated with the demonstration test, on the sample size n, and on p_C, the actual unknown proportion conforming. Figure 9.1g shows the probability of a successful demonstration at the 90% confidence level as a function of the actual proportion conforming, for various sample sizes. In this case, the demonstration would be successful if $\tilde{T}_{0.99}$ (computed as in Sections 4.6.3 or 13.2.7) is less than 180 °C or, equivalently, if $\underset{\sim}{p_C}$

(computed as in Sections 4.5.1 or 13.2.6) is greater than 0.99. For example, we can see from this graph that a sample of $n \cong 325$ units gives $p_{DEM} \cong 0.95$ when $p_C = 0.996$. The required sample size using method 2 is considerably smaller than that required for method 1 ($n = 1538$) which, however, did not require the assumption of a normal distribution. More generally, comparison of Figures 9.2g and 9.1g shows the potential gain from using the actual measurements and assuming a normal distribution when this assumption is warranted, as compared to a distribution-free approach.

As described in Section 9.3.3, the needed sample size under a normal distribution assumption can also be found from Table A.29. We enter Table A.29 with confidence level $(1 - \alpha) = 0.90$, to demonstrate that the conformance proportion is greater than $p' = p_C^S = 0.99$ so that $\delta = 1 - p_{DEM} \leq 0.05$ when the actual conformance proportion is $p^* = 0.996$. The table gives the necessary sample size as 329. The slight difference between this and the sample size 325 obtained from Figure 9.1g is due to our inability to interpolate much more than two significant digits from the graph.

We now compute the upper tolerance bound $\tilde{T}_{0.99}$ (Section 4.6.3) for use in the demonstration test. To obtain the needed factor, enter Table A.12d with $(1 - \alpha) = 0.90$, $p' = 0.99$, and interpolating for $1/n = 1/329$ between $(1/240, 2.497)$ and $(1/480, 2.444)$ gives $g'_{(0.90; 0.99, 329)} \approx 2.468$ (interpolation in $1/n$ in Table A.12d is generally more accurate than linear interpolation for $n > 60$). The demonstration will be successful if the upper 90% tolerance bound to exceed 99% of the sampled process, calculated from the data on the 329 randomly selected devices, i.e.,

$$\tilde{T}_{0.99} = \bar{x} + 2.468s,$$

is less than the specified limit of 180 °C.

13.3 EXAMPLE 2: FORECASTING FUTURE DEMAND FOR SPARE PARTS

13.3.1 Background and Available Data

A company manufactures replacement bearings for an electric motor. Demand has been stable over recent years. However, the company is planning to discontinue production of this product. Before so doing, they wish to make enough bearings so they can claim, with 95% confidence, that demand can be met for at least 7 years. The numbers of bearings sold in each of the past 5 years (in thousands of units) were

$$27.7, 37.1, 35.7, 30.8, 32.7.$$

These data are graphed against time in Figure 13.2.

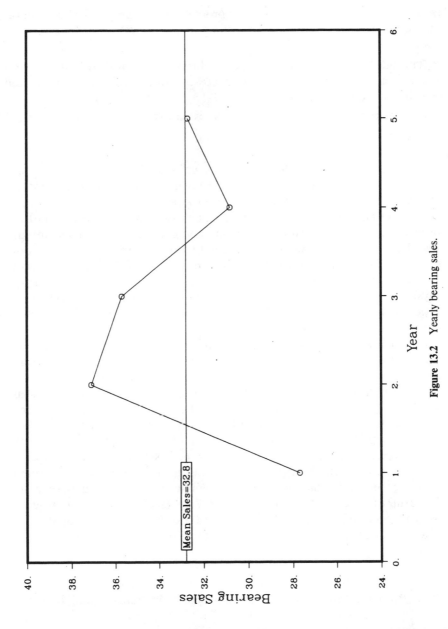

Figure 13.2 Yearly bearing sales.

253

13.3.2 Assumptions

Based on previous history with similar products and from physical considerations, it was deemed reasonable to assume that:

- The number of units sold per year follows a normal distribution with a mean and standard deviation that are constant from year to year, and will continue to be so.
- The number of units sold each year is statistically independent of the number sold in any other year.

Under these assumptions, the number of units sold in each of the past 5 years and each of the next 7 years can be regarded as independent observations from the same normally distributed process.

Often, a time series (i.e., a sequence of observations taken over time) will exhibit some trend or correlation among consecutive observations. In such cases, the preceding assumptions would not be true. One might then account for trend by using regression analysis (if extrapolation of the trend appears to be justified) or, more generally, by using special techniques for the statistical analysis of time series; see, for example, Box and Jenkins (1976). (However, one generally needs much more data than available here to obtain meaningful estimates of the nature of the trend or general correlation structure.) Although there are statistical tests and informal graphical means of checking for departures from the important stated assumptions [see, e.g., Chapter 4 in Neter, Wasserman, and Kutner (1990)], it is possible to detect only very extreme departures with just 5 observations. No such gross departures are evident in Figure 13.2. However, we need emphasize that these assumptions are critical for our analysis, and are often not satisfied in practice. Also, even if such assumptions were reasonable in the past, they might not hold in the future. For example, the fact that the producer is stopping production might itself impact sales. In the problem at hand, however, sales of the product containing the bearing had been without significant trend for many years and the bearings in motors that are less than 7 years old (which would be less heavily represented in the field in the future) rarely failed. Thus, demand for replacement bearings could be expected to remain fairly constant over the next 7 years.

13.3.3 Prediction and Upper Prediction Bound for Future Total Demand

From given data for the past 5 years, the sample mean is

$$\bar{x} = \frac{27.7 + 37.1 + 35.7 + 30.8 + 32.7}{5} = 32.80,$$

EXAMPLE 2   **255**

and the sample standard deviation is

$$s = \left(\frac{(27.7 - 32.80)^2 + \cdots + (32.7 - 32.80)^2}{5 - 1} \right)^{1/2} = 3.77.$$

Under the stated assumptions, $\bar{x}$ provides a prediction for the average yearly demand. Thus, $7(32.8) = 229.6$ provides a prediction for the total 7-year demand (in thousands of units) for the replacement bearings. However, because of statistical variability in both the past and the future yearly demands, the actual total demand would be expected to differ from this prediction. Using the method outlined in Section 4.7 and $t_{(0.95; 4)} = 2.132$, a one-sided upper 95% prediction bound for the mean of the yearly sales for the next $m = 7$ years is

$$\tilde{\bar{y}} = \bar{x} + t_{(1-\alpha; n-1)} \left(\frac{1}{m} + \frac{1}{n} \right)^{1/2} s = 32.80 + 2.132 \left(\frac{1}{7} + \frac{1}{5} \right)^{1/2} (3.77) = 37.5.$$

Thus, an upper 95% prediction bound for the total 7-year demand is 7(37.5), or 262,500 bearings. That is, although our prediction for the 7-year demand is 229,600 bearings, we can claim, with 95% confidence, that the total demand for the next 7 years will not exceed 262,500 bearings. At the same time, if the producer actually built 262,500 bearings, we would predict that the inventory would, most likely, last for 262.5/32.80, or approximately 8, years. In passing, we note that a one-sided *lower* 95% prediction bound for the total demand for the next 7 years is 196,700 bearings.

13.3.4   An Alternative Upper Prediction Bound Assuming a Poisson Distribution for Demand

An alternative upper prediction bound for the total demand in the next 7 years can be obtained by assuming that yearly demand can be modeled with a Poisson distribution with a constant rate (at least for the past 5 years) and that this will continue to be so (at least for the next 7 years). In this case, $\hat{\lambda} = \bar{x} = 32.8$ is an estimate for the yearly demand rate. Thus, a point prediction for the demand (in thousands of units) in the next 7 years is again $7(32.80) = 229.6$. Using the method given in Section 7.5.3, with $n = 5$, $m = 7$, and $z_{(0.95)} = 1.645$, an approximate upper 95% prediction bound for total demand for the next 7 years is

$$\tilde{y} = m\hat{\lambda} + z_{(1-\alpha)} m \left\{ \hat{\lambda} \left(\frac{1}{m} + \frac{1}{n} \right) \right\}^{1/2}$$

$$= 7(32.80) + (1.645)7 \left\{ (32.80) \left(\frac{1}{7} + \frac{1}{5} \right) \right\}^{1/2}$$

$$= 229.6 + 38.6 = 268.2.$$

These results agree reasonably well with those obtained under the assumption that yearly demand follows a normal distribution. One would, of course, not expect full agreement because somewhat different models have been assumed. As frequently happens, it is not clear from the underlying situation, or the limited data, which model is more appropriate. Therefore, in this case, as in many others, it is useful to compute intervals under various plausible models and compare the results. However, we should note that, for this problem, the Poisson distribution approach is subject to similar restrictive assumptions as those described for the normal distribution in Section 13.3.2.

13.4 EXAMPLE 3: ESTIMATING THE PROBABILITY OF PASSING AN ENVIRONMENTAL EMISSIONS TEST

13.4.1 Background

A manufacturer must submit three engines, randomly selected from production, for an environmental emissions test. To pass the test, the measurement for a particular pollutant on each of three engines to be tested must be less than 75 ppm. Based on the measurements in Table 13.1 from a previous test, involving 20 (presumably randomly selected) engines, the manufacturer wants to construct a lower (i.e., worst case) 95% confidence bound for the probability of passing the test.

13.4.2 Basic Assumptions

We make the important assumption that the 20 past engines and the 3 future engines are all randomly selected from the same "in statistical control" production process. The reasonableness of this assumption needs to be carefully evaluated based on an understanding of the problem and, possibly, a plot of the available data against manufacturing order. In this case, such a plot, shown in Figure 13.3, indicates no obvious trends or other nonrandom behavior.

Table 13.1 Engine Emissions Measurements*

73.2	67.8	68.5	73.8	69.3
70.9	65.4	71.2	72.4	69.6
67.1	69.2	66.5	72.9	75.4
74.2	69.1	64.0	68.9	70.2

*Measurements are in manufacturing sequence (reading across).

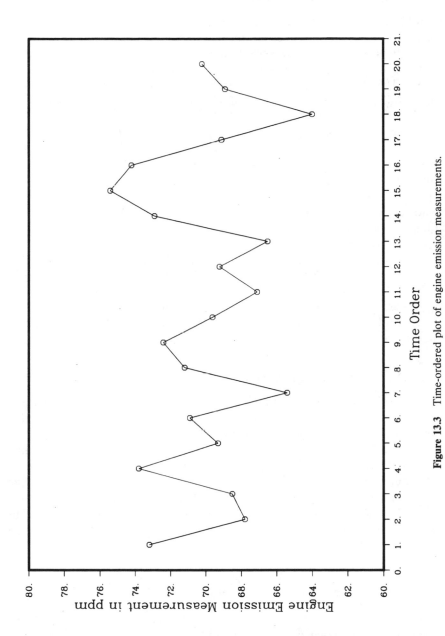

Figure 13.3 Time-ordered plot of engine emission measurements.

13.4.3 Distribution-Free Approach

The following approach requires no assumptions about the form of the underlying distribution of emissions measurements. Because the values for 19 out of 20 of the previously tested engines did not exceed the 75-ppm threshold, an estimate of the proportion of conforming engines is $\hat{p}_C = 19/20 = 0.95$. Thus, a point estimate of the probability that all three future engines will meet the specified limit of 75 ppm is $\hat{p}_{\text{DEM}} = (\hat{p}_C)^3 = (0.95)^3 = 0.86$.

A lower 95% confidence bound on p_C can be obtained by using the methods for proportions described in Section 6.2. For example, entering Table A.23b with $x = 19$ and $n = 20$ and confidence level $(1 - \alpha) = 0.95$, we find $\underset{\sim}{p}_C = 0.784$. Thus, under the stated assumptions, we can claim with 95% confidence that the proportion of conforming engines is at least 0.784. The lower 95% confidence bound on the probability that *all three* future engines will conform is then

$$\underset{\sim}{p}_{\text{DEM}} = \left(\underset{\sim}{p}_C\right)^3 = (0.784)^3 = 0.48 .$$

Thus, based on the limited available data, we are 95% confident that the probability of passing the test is at least 0.48—unfortunately, not a very high number.

13.4.4 Normal Distribution Approach

Figure 13.4, is a normal probability plot of the data in Table 13.1. This plot suggests that a normal distribution adequately models such emission measurements. From the data, $\bar{x} = 69.98$ and $s = 3.04$. A point estimate for $p_C = \Pr(x < 75)$, assuming that the emission measurements follow a normal distribution, is $\hat{p}_C = \Phi[(L - \bar{x})/s] = \Phi[(75 - 69.98)/3.04] = \Phi(1.65) \approx 0.95$. Under the same normal distribution assumption, one can use the methods of Section 4.5.1 and Table 7 of Odeh and Owen (1980) to compute a lower 95% confidence bound for the proportion of conforming units. This would also be a lower 95% confidence bound on the probability that a single randomly selected unit conforms to the specified limit. First compute

$$K = \frac{\bar{x} - L}{s} = \frac{69.98 - 75}{3.04} = -1.65.$$

Because we want a lower confidence bound on $\Pr(x < 75)$, we need to enter Table 7.4.4 of Odeh and Owen (1980) with $-K = 1.65$. Interpolating for $-K = 1.65$ with $n = 20$ with O & O's ETA $= 1 - \alpha = 0.95$, we obtain $\underset{\sim}{p}_C = 0.856$ as the lower 95% confidence bound on the probability that a single engine will meet the specified limit. Then a lower 95% confidence bound on

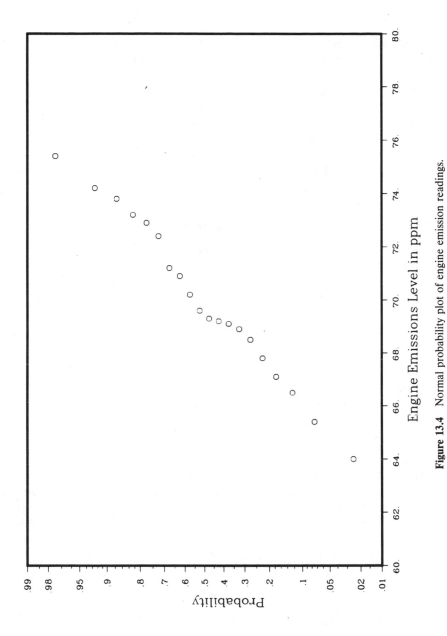

Figure 13.4 Normal probability plot of engine emission readings.

259

the probability of all three units meeting specifications is

$$p_{\text{DEM}} = (p_C)^3 = (0.856)^3 \cong 0.63.$$

Thus, we can say, with 95% confidence, that the probability that all three engines will conform to the specification is at least 0.63—still not a very high value.

This method gives a somewhat more precise bound on the desired probability than does the distribution-free method. The price paid for this gain is the need to assume that the measurements follow a normal distribution. Needless to say, neither confidence bound was very comforting to the experimenters. Greater assurance might be gained by increasing the sample size, or possibly, by decreasing the measurement error associated with the pollution readings.

13.4.5   Finding an Upper Prediction Bound on the Emission Level

Because it was not clear that all three submitted engines would pass the emissions test at a threshold of 75 ppm, it was desired to determine a new threshold value for which we could be highly confident of passing the test for all three future engines. This calls for finding an upper prediction bound that will, with a specified degree of confidence, not be exceeded by the future measurement on each of the three engines.

For the distribution-free approach given in Section 5.4.3, one can use Equation (5.7) to compute the confidence level associated with using the largest of the $n = 20$ previous observations as an upper prediction bound to exceed the $m = 3$ future observations. The confidence level is

$$1 - \alpha = \frac{n}{n + m} = \frac{20}{20 + 3} = 0.87.$$

Thus we can be 87% confident that the emission levels of all three future observations will be less than $x_{(20)} = 75.4$ ppm.

We note that the preceding distribution-free approach does not allow us to construct an upper 95% prediction bound to exceed the next three observations, unless we take a larger sample. However, if we can assume that the emission readings can be adequately represented by a normal distribution, the methods outlined in Section 4.8.2 can be used to compute such a bound. For this example, using

$$n = 20, \quad m = 3, \quad 1 - \alpha = 0.95, \quad \text{and} \quad r'_{(0.95;3,20)} = 2.331$$

in Table A.14, we obtain

$$\tilde{y}_3 = \bar{x} + r'_{0.95;3,20)}s = 69.98 + (2.331)(3.04) = 77.1.$$

That is, we are 95% confident that all three future engines will have emissions less than 77.1 ppm.

EXAMPLE 4 **261**

13.5 EXAMPLE 4: PLANNING A DEMONSTRATION TEST TO VERIFY THAT A RADAR SYSTEM HAS A SATISFACTORY DETECTION PROBABILITY

13.5.1 Background and Assumptions

The manufacturer of an airplane radar system needs to demonstrate with a high degree of confidence that the system can, under specified conditions, detect a target at a distance of 35 miles with a detection probability p_D that exceeds the specified value $p_D^s = 0.95$. The verification test consists of a sequence of passes under the specified conditions, between the radar-equipped airplane and an approaching target plane, both traveling approximately the same path for each pass. The pass is deemed to be a success if detection occurs before the planes are 35 miles apart, and a failure otherwise. The demonstration involves calculating a lower confidence bound ($\underset{\sim}{p_D}$) for the probability p_D of detection at a distance exceeding 35 miles, and will be deemed successful if $\underset{\sim}{p_D} > p_D^s = 0.95$. The problem is to determine the required number of passes, i.e., the sample size. It is assumed that each pass will give an independent observation of miles to detection from the defined process consisting of all similar passes, i.e., same direction, weather, system configuration, etc. (As always, properly defining this process to assure that it really represents the situation of interest, and planning the passes accordingly is critical.) After the passes have taken place, one might consider two possible methods for computing $\underset{\sim}{p_D}$:

1. Make no assumption about the form of the distribution of miles to detection and use the observed proportion of successes, i.e., the proportion of passes exceeding 35 miles; in this case, one uses the binomial distribution to compute a lower confidence bound on p_D, the detection probability, using the procedure outlined in Section 6.2.
2. Assume that miles to detection (or some transformation, such as log miles) follows a normal distribution, and use the method described in Section 4.5 to compute $\underset{\sim}{p_D}$.

It was decided to base the sample size on the first method because (a) there was little prior information about the form of the distribution of miles to detection and (b) it leads to a more conservative procedure (i.e., fewer assumptions, and a larger sample size). If the resulting data supported the assumption of a normal distribution, then the second method might be used in the subsequent data analysis (thus giving more precise assessments).

13.5.2 Choosing the Sample Size

Assume that it is desired to demonstrate, with 90% confidence, that the detection probability p_D is greater than $p_D^s = 0.95$. One would clearly wish successful demonstration if p_D "appreciably exceeds" 0.95, where the term "appreciably exceeds" still requires elaboration.

Using the methods outlined in Section 9.6, based on the binomial distribution model, the probability of a successful demonstration is graphed in Figure 9.2e as a function of the true detection probability for various number of passes (n). Thus we see, for example, if $n = 77$ passes are used, the demonstration $p_D^S > 0.95$ will be successful if there are no more than $c = 1$ passes that *fail* to detect the target at a distance of 35 miles or more. Further, if the actual detection probability is $p_D = 0.98$ (and, thus, the probability of an unsuccessful pass is 0.02), then the probability of a successful demonstration (i.e., $c = 0$ or 1) for $n = 77$ is

$$p_{\text{DEM}} = \text{Pr}(\underset{\sim}{p}_D > 0.95) = \text{Pr}(x \le 1) = B(1; 77, 0.02) = 0.543,$$

where x is the number of unsuccessful passes. A probability of 0.543 of successful demonstration would generally not be regarded as adequate. However, we similarly determine from Figure 9.2e that if we felt that the actual probability of detection is really as high as 0.995, the probability of a successful demonstration with $n = 77$ is 0.94, which would likely be regarded as satisfactory. Thus, the required sample size to achieve a high probability of demonstration when p_D exceeds 0.95 depends heavily on the specification of the expected, but obviously unknown, true probability of detection.

We also note from Figure 9.2e, that for the demonstration probability to be about 0.90 when the true detection probability is 0.98, $n = 234$ passes are required. In this case, the demonstration will require 7 or fewer unsuccessful passes. The actual demonstration probability when $p_D = 0.98$ (i.e., the probability of obtaining seven or fewer passes without a detection in the 234 passes) is

$$p_{\text{DEM}} = \text{Pr}(\underset{\sim}{p}_D > 0.95) \doteq \text{Pr}(x \le 7) = B(7; 234, 0.02) = 0.89998.$$

Alternately, we can use Table A.30 to determine the required sample size. Entering this table, (1) to demonstrate with confidence level $(1 - \alpha) = 0.90$ that the detection probability p_D is greater than $p' = p_D^S = 0.95$, while (2) assuring that the probability of failing to achieve demonstration is $\delta = 1 - p_{\text{DEM}} < 0.10$ when the actual detection probability is $p^* = 0.98$, gives the required sample size as $n = 258$. This is greater than the sample size indicated by Figure 9.2e because $n = 234$ gives only $p_{\text{DEM}} = 0.89998$ (slightly less than the desired p_{DEM} of 0.90), while $n = 258$ is the smallest sample size to give $p_{\text{DEM}} \ge 0.90$, and gives an actual p_{DEM} of

$$p_{\text{DEM}} = \text{Pr}(\underset{\sim}{p}_D > 0.95) = \text{Pr}(x \le 8) = B(8; 258, 0.02) = 0.923.$$

Also, a sample of $n = 258$ will permit demonstration as long as there are eight or fewer, rather than seven or fewer, unsuccessful passes.

EXAMPLE 5   263

13.6   EXAMPLE 5: ESTIMATING THE PROBABILITY OF EXCEEDING A REGULATORY LIMIT

13.6.1   Background and Assumptions

A company has taken readings on the concentration level of a chemical compound at a particular point in a river. One reading (actually an average of five measurements) was taken during the first week of the quarter during each of the past 27 quarters. The data are given in Table 13.2 (which also shows the day of the week when the measurement was taken) and are plotted against time in Figure 13.5. The company was asked to use the past data to estimate the probability that a future reading will exceed the regulatory limit of 300 ppm, even though all of the 27 past readings have been appreciably

Table 13.2   Chemical Concentration Readings

Observation Number	Quarter	Day in Week of Measurement (during first week of quarter)	Concentration Level Reading
1	2	1	48
2	3	3	94
3	4	1	112
4	1	5	44
5	2	3	93
6	3	4	198
7	4	2	43
8	1	1	52
9	2	3	35
10	3	5	170
11	4	1	25
12	1	3	22
13	2	2	44
14	3	4	16
15	4	5	139
16	1	2	92
17	2	5	26
18	3	1	116
19	4	4	91
20	1	4	113
21	2	5	14
22	3	1	50
23	4	3	75
24	1	5	66
25	2	1	43
26	3	2	10
27	4	5	83

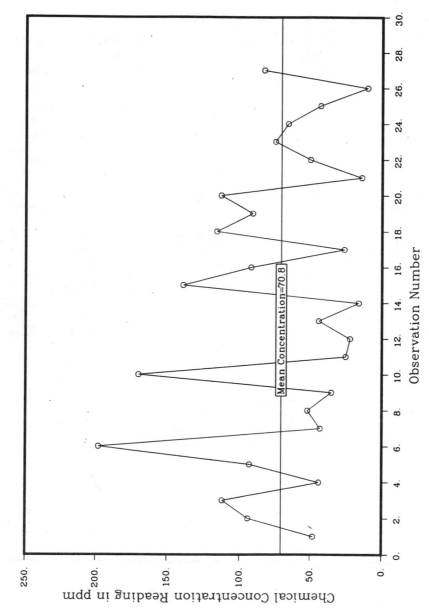

Figure 13.5 Plot of quarterly readings of chemical concentration levels.

EXAMPLE 5   265

below this value. In order to respond, it will be necessary to use a statistical model to describe the relationship between the past and future readings. The simplest such model assumes that all past and future readings are random observations from the same process. However, because we are dealing with a time series, generated by a process that might change over time (due, e.g., to changes in production level, pollutant processing methods, etc.), this model may not be appropriate. The physical process must be reviewed and the data carefully checked to assess the existence of a trend, a cyclical or a seasonal pattern, or other correlations, among the observations. A trend might be present if the mean of the process is changing with time, due, for example, to changes in production level or pollution abatement measures. Seasonal effects might occur because of differences in concentration due to seasonal variations in production or the impact of changes in weather conditions. Differences might arise due to varying levels of production on different days of the week. Fortunately, there were no physical reasons to expect the simple model not to apply in this example. However, it is hoped that there might be a reduction in pollution levels in the future. In this case, the results obtained under the assumed model would tend to be overly conservative in the sense that they are likely to overpredict future levels of pollution. In any case, an empirical assessment of the validity of the assumptions, based on the past data, is in order.

13.6.2 Preliminary Graphical Analysis

Figure 13.5 shows appreciable variability in the readings, but it does not give any clear indication of a trend or cyclical pattern. Figures 13.6 and 13.7 provide a histogram and a normal probability plot, respectively, of the data. Because the histogram is not symmetric and because the points on the normal probability plot deviate significantly from a straight line, there is some evidence that such readings are not normally distributed. In particular, the data are appreciably skewed to the right and there are some large extreme observations. (Checks did not suggest any errors in the data.) Figures 13.8, 13.9, and 13.10 are displays of the natural logarithms of the data that

```
READING

MIDDLE OF    NUMBER OF
INTERVAL     OBSERVATIONS
   0.         0
  20.         6      ******
  40.         6      ******
  60.         3      ***
  80.         2      **
 100.         4      ****
 120.         3      ***
 140.         1      *
 160.         0
 180.         1      *
 200.         1      *
```

Figure 13.6 Histogram of chemical concentration readings.

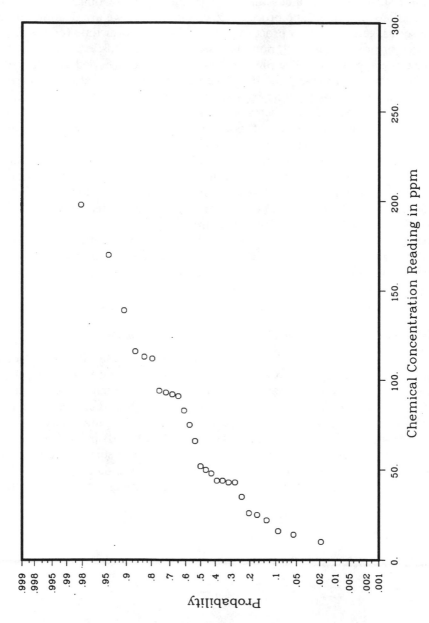

Figure 13.7 Normal probability plot of chemical concentration readings.

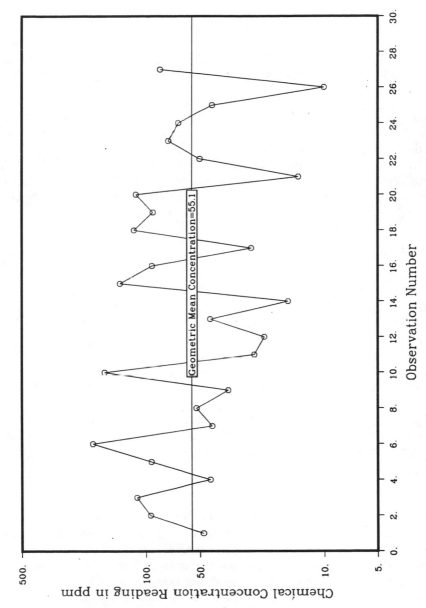

Figure 13.8 Log plot of quarterly readings of chemical concentration levels.

READING

MIDDLE OF INTERVAL	NUMBER OF OBSERVATIONS	
2.4	1	*
2.8	2	**
3.2	3	***
3.6	5	*****
4.0	4	****
4.4	6	******
4.8	4	****
5.2	2	**

Figure 13.9 Histogram of log chemical concentration readings.

correspond to Figures 13.5, 13.6, and 13.7, respectively. We note from these plots that the histogram of the logs is more symmetric and the corresponding probability plot is much closer to a straight line than the original data, indicating that the readings may be better approximated by a *lognormal*, than by a normal, distribution. (A lognormal distribution is frequently found to be an appropriate model in pollution assessment problems.) Thus, henceforth, we will consider the (natural) logarithms of the data.

13.6.3 Formal Test for Trend

Formal statistical procedures can be used to supplement the graphical analyses. For example, examination of Figure 13.8 might suggest to some that there is a linear (downward) trend over time in the plotted values. One might check for such a time trend in the data by fitting a simple linear regression of the readings (in this case, the logarithms of the readings) versus time. If a trend is present and can be assumed to continue into the future, the regression model [or some further generalization, such as those discussed in books or time series analysis, e.g., Box and Jenkins (1976) and Pankrantz (1983)] might be used for forecasting.

Figure 13.11 gives a summary of the simple regression analysis for the logarithm of concentration versus time, computed with the MINITAB computer program. The assumed model for this analysis is

$$\log(\text{concentration}) = \beta_0 + \beta_1(\text{time}) + \varepsilon$$

where time $= 1, 2, \ldots, 27$ (i.e., the 27 quarters for which data were available) and ε is a random noise (or error) term which is assumed to follow a normal distribution with a mean of 0 and a standard deviation that is constant over time. We also assume that the ε's are independent of each other. (This assumption was subsequently checked; see Section 13.6.4.) A two-sided 95% confidence interval for β_1, the slope of the regression line is

$$b_1 \pm t_{(1-\alpha/2;\, n-2)} s_{b_1} = -0.021 \pm 2.060(0.019) = [-0.060, 0.018]$$

where b_1 is the least squares estimate of β_1, s_{b_1} the standard error of this

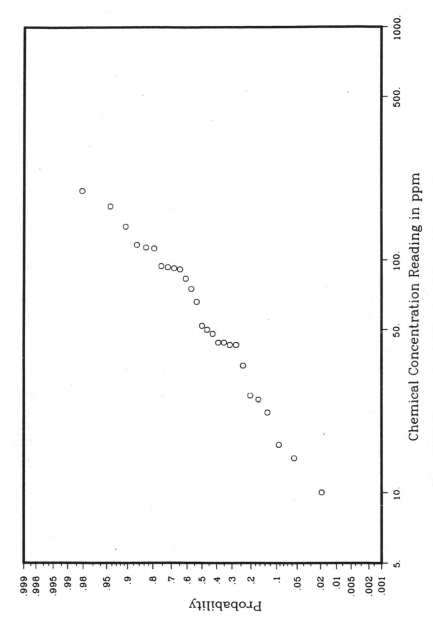

Figure 13.10 Lognormal probability plot of chemical concentration readings.

269

```
THE REGRESSION EQUATION IS
Y =    4.31 - 0.0214 X

                                      ST. DEV.   T-RATIO =
           COLUMN    COEFFICIENT      OF COEF.   COEF/S.D.
           --          4.3087          0.3042      14.16
      X1   TIME       -0.02137         0.01899     -1.13

THE ST. DEV. OF Y ABOUT REGRESSION LINE IS
S = 0.7686
WITH (27-2) = 25 DEGREES OF FREEDOM

R-SQUARED =  4.8 PERCENT
R-SQUARED =  1.0 PERCENT, ADJUSTED FOR D.F.

ANALYSIS OF VARIANCE

      DUE TO     DF         SS       MS=SS/DF
      REGRESSION  1       0.7482      0.7482
      ERROR      25      14.7680      0.5907
      TOTAL      26      15.5162
```

Figure 13.11 Summary of regression analysis of log chemical concentration readings versus time.

estimate, and $t_{(1-\alpha/2;\,n-2)}$ the $100(1 - \alpha/2)$th percentile of a Student t distribution with $n - 2$ degrees of freedom, i.e., $t_{(0.975;\,25)} = 2.060$ from Table A.7; see the brief discussion in Section 11.4.1 and the references given there. Because this confidence interval on the slope coefficient contains the value 0, there is no statistical evidence of a trend in the data.

13.6.4 Formal Tests for Periodicity and Autocorrelation

Checks for periodicity were first performed by using an analysis of variance to test for differences among the four quarters of the year and among the different days of the week. No statistically significant differences were found.

Figure 13.12 gives a table and graph of sample autocorrelations $r_1, r_2, \ldots, r_{15}$. For a time series with a constant mean and standard deviation

```
           -1.0 -0.8 -0.6 -0.4 -0.2  0.0  0.2  0.4  0.6  0.8  1.0
           +----+----+----+----+----+----+----+----+----+----+
     1 -0.126                 [      XXXX          ]
     2 -0.227                 [    XXXXXXX         ]
     3 -0.004                 [       X            ]
     4  0.208                 [      XXXXXX        ]
     5  0.120                 [      XXXX          ]
     6 -0.353               [ XXXXXXXXXX           ]
     7 -0.144                 [     XXXXX          ]
     8 -0.044                [      XX             ]
     9  0.116                [      XXXX           ]
    10  0.139                [      XXXX           ]
    11 -0.317                [  XXXXXXXXX          ]
    12  0.164                [      XXXXX          ]
    13  0.200                [      XXXXXX         ]
    14  0.115               [       XXXX           ]
    15 -0.106               [      XXXX            ]
```

Figure 13.12 Sample autoacceleration function of the log chemical concentration readings.

EXAMPLE 5   **271**

(often referred to as "stationary"), r_k is an estimate of the correlation between observations that are k time periods apart. The set of values $r_k, k = 1, 2, 3, \ldots$ is known as the sample autocorrelation function (ACF) and is an important tool for modeling time series data. The ACF can be computed easily with many of the popular data analysis computer programs (e.g., SAS® and MINITAB®). Although exact sampling theory for these statistics is complicated, in large samples of independent observations (say, $n > 60$), the r_k can be assumed to be approximately normally distributed with a standard error approximated by "Bartlett's formula" [see, e.g., page 35 of Box and Jenkins (1976) or page 68 of Pankrantz (1983)]. Although crude for the current sample of size 27, this approximation can be used to roughly assess the statistical significance of the correlations or to construct approximate confidence intervals to contain the true correlations. This provides an approximate formal check for the assumption that the readings are uncorrelated. The brackets shown in Figure 13.12 indicate approximate limits outside of which a sample autocorrelation would be statistically significant at a 5% significance level. We note that the estimated autocorrelations are all contained within their bracketed bounds (equivalently, confidence intervals for the true correlations all enclose 0). Thus there is no evidence of autocorrelation within the quarterly data.

If any of the above tests had given positive results, it would be an indication that more sophisticated time series analyses might be required. In this case, all tests came out negative, as expected from physical considerations, and so it seems satisfactory to proceed using simple methods. At the same time, we need to note that

- The power of a statistical test to establish significance is highly dependent upon sample size ($n = 27$ is quite modest).
- The fact that certain patterns were not exhibited in the past does not guarantee that they will not happen in the future.

13.6.5 Distribution-Free Binomial Model

Without making any assumptions about the form of the distribution of the readings, it is possible to estimate p_G, the probability of exceeding the regulatory limit of 300 ppm. We assume only that both the past and the future readings are independently and randomly chosen from the same stationary process. Because none of the 27 past quarterly readings exceeded 300 ppm, an estimate of this probability is $\hat{p}_G = 0/27 = 0$. The methods given in Section 6.2, based on the binomial distribution, can be used to compute an upper confidence bound on this probability. In particular, using the calculational method outlined in Section 6.2.2, an upper 95% confidence bound on the probability of exceeding the regulatory limit on a randomly selected day (at least in the first week of a forthcoming quarter) is

$$\tilde{p}_G = \left\{ 1 + \frac{n - x}{(x + 1) F_{(1-\alpha; 2x+2, 2n-2x)}} \right\}^{-1}$$

$$= \left\{ 1 + \frac{27 - 0}{(0 + 1) F_{(0.95; 2, 54)}} \right\}^{-1}$$

$$= 0.105 ,$$

using $F_{(0.95; 2, 54)} = 3.168$. Thus, we are 95% confident that p_G is less than $\tilde{p}_G = 0.105$. Alternately, we could have obtained this value from Figure 6.1b.

The result might seem a little disappointing, in light of the data. Even though there were no readings above, or even close to, 300 ppm for the sampled day in each of the past 27 quarters, the data have not established with 95% confidence that the probability is satisfactorily small, even if we assume that there will be no change in the process. Of course, this analysis ignores the actual values of the readings (other than whether or not they exceed 300 ppm). An alternative analysis (see below) that assumes a distributional model would be expected to be more informative (i.e., provide a tighter bound for p_G).

13.6.6 Lognormal Distribution Model

Our graphical analyses of the 27 readings indicated that the logarithms of the readings might be modeled adequately by a normal distribution. The sample mean and standard deviation of the log readings are $\bar{x} = 4.01$ and $s = 0.773$, respectively. A point estimate for the proportion of days that the limit will be exceeded, assuming that the logarithm of the chemical concentration readings follows a normal distribution, is $\hat{p}_G = \widehat{\Pr}[x \geq \log(300)] = 1 - \Phi[(\log(L) - \bar{x})/s] = 1 - \Phi[(5.70 - 4.01)/0.773] = 1 - \Phi(2.19) = 0.0143$. Under the same normal distribution assumption, one can use the methods given in Section 4.5.1 and Table 7 of Odeh and Owen (1980) to compute an upper 95% confidence bound for the proportion of days that the limit will be exceeded. This would also be an upper 95% confidence bound on the probability that the limit will be exceeded on a single randomly selected day. Specifically, first compute

$$K = \frac{\bar{x} - \log(L)}{s} = \frac{4.01 - 5.70}{0.773} = -2.19 .$$

Because we want an upper confidence bound on the probability of exceeding a specified value, we enter Table 7.4.4 of Odeh and Owen (1980) with O & O's ETA $= 1 - \alpha = 0.95$, and $n = 27$. From the table we read 0.94482 and 0.92436, corresponding to $-K = 2.2$ and $-K = 2.0$, respectively. Linear

EXAMPLE 6   273

interpolation for $-K = 2.19$ gives 0.94380, and the upper 95% confidence bound for p_G is $\tilde{p}_G = 1 - 0.94380 = 0.05620$. Thus, we are 95% confident that the probability of a reading exceeding 300 ppm is less than 0.056. This value is smaller, and, thus more satisfactory than the upper 95% confidence bound of 0.105, which was obtained with the distribution-free binomial distribution model. This improvement was obtained in return for the (not unreasonable) assumption that the readings follow a lognormal distribution.

In passing, we note that if we had incorrectly assumed a normal (rather than a lognormal) distribution for the readings, we would have obtained the upper 95% confidence bound for p_G to be $\tilde{p}_G = 1 - 0.99985 = 0.00015$ or 0.015%. The contrast between 0.00015 and 0.05620 (a ratio of about 375) illustrates our statement in Section 4.10, that confidence bounds on probabilities in the tail of a distribution are not robust to an incorrect distributional assumption.

13.7 EXAMPLE 6: ESTIMATING THE RELIABILITY OF A CIRCUIT BOARD

13.7.1 Background and Assumptions

A company manufactures a circuit board that contains 110 similar integrated circuit "chips" that must operate in a field environment for 30,000 hours. Successful operation requires that all chips in a board operate without failure in service. It is reasonable to assume that chips on the same board fail independently of one another. This assumption would be incorrect if, for example, failures are caused by shocks that affect more than one chip, or if failure of one chip increases the stress on the others. However, the assumption appears reasonable because failure from internal defects in the chips is the dominant failure mode. Thus, one can assess life in service from tests on chips, rather than requiring tests on boards. In fact, an accelerated test on individual chips has been developed for this purpose. This test simulates the 30,000 hours of operation in a normal service environment by a 1000-hour dynamic high temperature–high humidity exposure at 85 °C and 85% relative humidity.

Due to a (hopefully) one-time manufacturing problem, some of the chips in a special shipment of 50,000 chips may be prone to failure during field service (such chips will, henceforth, be referred to as being "defective"). To estimate p_C, the proportion of defective chips in the shipment, a random sample of 1000 chips was selected from the inventory of 50,000 chips, and these chips were subjected to the 1000 hour accelerated test. Two chips failed the test.

We note that in this example, unlike most of the others in this chapter, the sample is from a well-defined population concerning which we wish to draw inferences—namely the 50,000 chips in inventory from this shipment. Thus, in this sense, using the terminology of Chapter 1, this is an enumerative study. (This would not be the case if we had wished to draw inferences about future chips from the process from which this shipment came.) However, because (1) the chips are being tested in an accelerated test environment that is meant to simulate operational conditions, and (2) testing on chips is to be used to draw conclusions about results on boards under the assumption of independent failures, one might argue that, in totality, this is an analytic, rather than an enumerative, study. We will not get hung up in this discussion on terminology but simply emphasize the importance of these fundamental assumptions to drawing conclusions from this evaluation.

We also note that we are dealing here with a finite population of 50,000 chips. However, our sample of 1000 chips is a small percentage of the population (appreciably less than 10%), and, therefore, the finiteness of the population can be ignored for practical purposes, as indicated in Chapter 1. If this had not been the case—say, for example, there had been only 5000 chips in the shipment—then the methods referenced in Section 11.3.17 could be applied to draw inferences about the unsampled chips, instead of those to be described below. [The "finite population correction factor" method mentioned in Section 1.10, if it were applied with Equation (6.2), would not provide an adequate approximate confidence interval, in this case. This is because Equation (6.2) is based upon a normal distribution approximation to the sampling distribution of $\hat{p}$, and the approximate confidence interval method in Equation (6.2) is inappropriate for a situation with only two nonconforming units; see Section 6.2.4.]

The following information is desired for the in-service operating conditions, together with appropriate statistical bounds:

a. An estimate of the proportion of defective chips in the shipment.
b. An estimate of the proportion of boards (each containing 110 chips from the shipment) that will contain one or more defective chips.
c. An estimate of the probability that at least 9 of 10 boards that use chips from the shipment, and are to be installed in a system, will operate successfully in service.

13.7.2 Estimate of Proportion of Defective Chips

An estimate of the true proportion (p_c) of defective chips for the shipment is $\hat{p}_c = 2/1000 = 0.002$, or 0.2%. Using the methods given in Section 6.2, a two-sided 90% confidence interval for p_C is

$$\left[\underline{p}_C, \tilde{p}_C \right] = [0.0003555, 0.006283].$$

Thus, we are 90% confident that the percentage of defective chips in inventory is between 0.036 and 0.63%.

EXAMPLE 6   **275**

13.7.3 Estimating the Probability That an Assembled Circuit Board Will Be Defective

A board is defective if it contains one or more defective chips. The number of defective chips on a board follows a binomial distribution with parameters p_C and $n = 110$. Because the circuit board contains 110 chips, the probability that a board is not defective is the probability that none of the 110 chips is defective, assuming that the only reason for board failure is the independent failure of the chips. Under the previously stated assumptions of independence, this probability is

$$(1 - p_C)^{110}.$$

Thus, the probability that the board is defective (i.e., has one or more defective chips) is

$$p_B = 1 - (1 - p_C)^{110}.$$

An estimate of this probability is, therefore,

$$\hat{p}_B = 1 - (1 - \hat{p}_c)^{110} = 1 - (1 - 0.002)^{110} = 0.20.$$

Because p_B is a monotonic increasing function of p_C (see Section 6.3), confidence bounds for p_B can be obtained directly from those for p_C. In particular, the endpoints of a 90% confidence interval for the probability that a circuit board will be defective are

$$\underset{\sim}{p}_B = 1 - \left(1 - \underset{\sim}{p}_C\right)^{110} = 1 - (1 - 0.0003555)^{110} = 0.038357,$$

$$\tilde{p}_B = 1 - (1 - \tilde{p}_C)^{110} = 1 - (1 - 0.006283)^{110} = 0.50008.$$

Thus, we are 90% confident that the percentage of defective boards constructed from chips in inventory is between 3.8 and 50.0%. This long interval is not unreasonable when we recognize that the available data are on the equivalent of only about nine boards.

13.7.4 Estimating System Reliability

A system that contains 10 circuit boards requires at least 9 such boards to operate successfully. We assume that the number of boards that fail in service follows a binomial distribution with parameters $n = 10$ and p_B. This model would apply under assumptions similar to those stated in Section 13.7.1.

Following the approach outlined in Section 6.3, the probability of successful system operation is

$$p_D = \Pr(0 \text{ boards fail}) + \Pr(1 \text{ board fails})$$

$$= (1 - p_B)^{10} + 10(p_B)^1(1 - p_B)^9. \tag{13.1}$$

Because p_D is a monotonically decreasing function of p_B, a lower (upper) confidence bound on p_D is obtained by substituting the corresponding upper (lower) confidence bound for p_B into Equation (13.1). That is,

$$\underset{\sim}{p}_D = (1 - \tilde{p}_B)^{10} + 10\tilde{p}_B(1 - \tilde{p}_B)^9$$

$$= (1 - 0.50008)^{10} + 10(0.50008)(1 - 0.50008)^9 = 0.011,$$

$$\tilde{p}_D = (1 - \underset{\sim}{p}_B)^{10} + 10\underset{\sim}{p}_B(1 - \underset{\sim}{p}_B)^9$$

$$= (1 - 0.038357)^{10} + 10(0.038357)(1 - 0.038357)^9 = 0.946.$$

Thus, based on the fact that the random sample of 1000 chips contained 2 defective units, we can say, with 90% confidence, that the probability of successful system operation is between 0.011 and 0.946! Thus, our evaluation has been highly uninformative. This is not very surprising because the available data on 1000 chips are used to draw conclusions, with a high degree of confidence, about a system involving 1100 chips.

13.8   EXAMPLE 7: USING SAMPLE RESULTS TO ESTIMATE THE PROBABILITY THAT A DEMONSTRATION TEST WILL BE SUCCESSFUL

Audio quality performance scores have been obtained on a random sample from production of 16 high fidelity speakers. The data are shown in Table 13.3. A future demonstration test will be successful if the score for *each* of 32 additional randomly selected speakers exceeds 450. We need to estimate p_{DEM}, the probability of passing the demonstration test. The process has been established to be stable, i.e., in statistical control, and our analysis is based on the important assumption that this will continue to be the case.

Table 13.3   Audio Quality Performance Measurements

552	586	702	722
742	790	800	838
838	921	960	981
994	1035	1110	1405

EXAMPLE 7

277

13.8.1 Distribution-Free Binomial Model Approach

Because all of the initial 16 units had performance scores above 450, an estimate of p_G, the proportion of units from production with scores above 450, is $\hat{p}_G = 16/16 = 1.0$. Using the methods discussed in Section 6.2, a 90% lower confidence bound for p_G is

$$
\begin{aligned}
\underset{\sim}{p}_G &= \left\{ 1 + \frac{(n - x + 1)F_{(1-\alpha; 2n-2x+2, 2x)}}{x} \right\}^{-1} \\
&= \left\{ 1 + \frac{(16 - 16 + 1)F_{(0.90; 2, 32)}}{16} \right\}^{-1} \\
&= \left\{ 1 + \frac{(1)2.477}{16} \right\}^{-1} \\
&= 0.866.
\end{aligned}
$$

That is, we are 90% confident that at least 86.6% of the units from production would, if measured, score above 450.

The probability that all 32 future units will score higher than 450 is $p_{DEM} = (p_G)^{32}$, and, because $\hat{p}_G = 1.0$, the resulting point estimate is also $\hat{p}_{DEM} = 1.0$. Because p_{DEM} is an increasing function of p_G, a lower 90% confidence bound for p_{DEM} can be obtained by using $\underset{\sim}{p}_G$ in this formula in place of p_G. That is,

$$
\underset{\sim}{p}_{DEM} = (\underset{\sim}{p}_G)^{32} = (0.866)^{32} = 0.010.
$$

Thus, we are 90% confident that the probability of passing the demonstration test is at least 1%! This result is not very useful except to tell us that we cannot be assured of a successful demonstration about 32 future units from successful go/no-go data on only 16 past units.

13.8.2 Normal Theory Approach

Figure 13.13 is a normal probability plot of the 16 scores for the initially tested units. Except for the largest observation (1405), the data seem to be well represented by a normal distribution. Thus, this distribution may provide a reasonable model, at least for the lower tail of the distribution of audio performance scores—and it is this lower tail that is of concern to us. Sample statistics for these data are $\bar{x} = 873.5$ and $s = 211.5$. (Note, however, that both of these values may be inflated estimates if the largest observation is erroneous—a topic that we will investigate further in Section 13.8.3.) A point estimate for the probability that the performance of a single randomly selected unit will exceed 450, assuming that the performance scores follows a normal distribution, is $\hat{p}_G = \widehat{\Pr}(x \geq 450) = 1 - \Phi[(450 - \bar{x})/s] = 1 - \Phi[(450 - 873.5)/211.5] = 1 - \Phi(-2.002) \approx 0.9773$. Under the normal

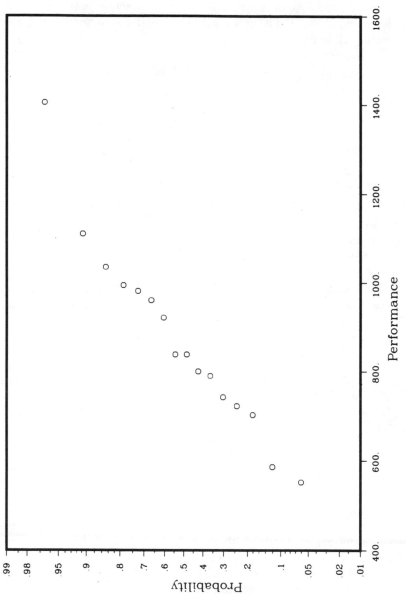

Figure 13.13 Normal probability plot of audio quality performance measurements.

EXAMPLE 7   **279**

distribution assumption, one can use the methods given in Section 4.5.1 and Table 7 of Odeh and Owen (1980) to compute a lower 90% confidence bound for $p_G = \Pr(x \geq 450)$. In particular, entering Table 7.3.3 of Odeh and Owen (1980) with $n = 16$, O & O's ETA $= 1 - \alpha = 0.90$, and

$$K = \frac{\bar{x} - L}{s} = \frac{873.5 - 450}{211.5} = 2.002,$$

one finds $p_G = 0.9206$. That is, assuming that the data are a random sample from a normal distribution, we can be 90% confident that the proportion of units with scores above $L = 450$ is at least 0.9206.

Following Section 13.8.1, we compute a lower 90% confidence bound for p_{DEM} as

$$\underline{p}_{DEM} = (\underline{p}_G)^{32} = (0.9206)^{32} = 0.07.$$

We can similarly calculate the upper 90% confidence bound on p_{DEM} to be $\tilde{p}_{DEM} = 0.84$, resulting in a two-sided 80% confidence interval of 0.07 to 0.84. This tells us that the available data are insufficient for our purposes, and suggests that additional sampling is needed to be able to draw any definitive conclusions about passing the demonstration test. (This would not have been the case if $\tilde{p}_{DEM}$ had turned out to be a sufficiently small value, or $\underline{p}_{DEM}$ a sufficiently large value.)

13.8.3   Testing Sensitivity of the Conclusions to Changes in the Extreme Observation(s)

Because the largest observation (i.e., 1405) was suspiciously large and because it tended to contradict the assumption of a normal distribution for the performance scores, it is worthwhile to investigate the sensitivity of our conclusions to this observation. Thus, the following alternative analyses were performed.

1. The largest observation was ignored altogether and the analysis was repeated. This would be a reasonable approach if the extreme observation were totally incorrect (e.g., due to a data recording mistake that is completely independent of the true value) and if the correct observation could not be recovered. It would not be correct otherwise.

2. The largest observation was treated as a "censored" observation with a value equal to or larger than the second largest observation (a value of 1110). In this case, we assume that we do not know the exact value of the largest observation, but believe that its value was no less than that of the second largest observation. This (like the next method) may be a reasonable approach if we do not want one, or a few large, extreme observations in the upper tail of the distribution to unduly affect the inferences concerning the lower tail of the distribution (where the normal distribution seems to provide a good model).

**Table 13.4 Maximum Likelihood Estimates and Approximate Two-Sided
90% Confidence Intervals [in brackets] for μ, σ, p_G and p_{DEM}**

Estimates of	μ	σ	p_G	p_{DEM}
Original data	874	205	0.9806	0.54
16 observations	[808, 939]	[163, 257]	[0.925, 0.9952]	[0.084, 0.858]
Largest observation	838	157	0.9932	0.81
Ignored	[786, 890]	[124, 198]	[0.958, 0.9990]	[0.251, 0.968]
Censoring	860	175	0.9905	0.74
at 1110	[804, 916]	[138, 222]	[0.950, 0.9983]	[0.193, 0.946]
(1 censored				
observation)				
Censoring	857	171	0.9912	0.75
at 1000	[801, 914]	[132, 223]	[0.948, 0.9986]	[0.178, 0.956]
(3 censored				
observations)				
Censoring	861	178	0.9895	0.71
at 900	[795, 927]	[127, 249]	[0.931, 0.9985]	[0.100, 0.952]
(7 censored				
observations)				
Censoring	826	144	0.9955	0.87
at 800	[765, 887]	[98, 211]	[0.946, 0.9996]	[0.171, 0.9885]
(9 censored				
observations)				

3. The largest observations were assumed to have been (right) censored at
   several other scores: 800, 900, 1000, and 1100. Such censoring might
   provide reasonable estimates in the "lower part" of the distribution (as
   desired here) if one felt that the normal distribution provides a good
   representation of that part of the distribution, but not of the distribution as a whole. See Oppenlander, Schmee, and Hahn (1988) for
   further exposition of this approach.

   The analysis for simply dropping the suspect observation is the same as in
Section 13.8.2. However, as indicated in Chapters 11 and 12, special statistical methods for analyzing censored data are required for the other two
approaches. The analyses for methods 2 and 3 were conducted using
CENSOR, a computer program that performs maximum likelihood estimation with censored data [see Meeker and Duke (1981) for a description of
CENSOR and Lawless (1982) or Nelson (1982) for a description of the theory
and methods for analyzing censored data].
   The results of these alternative analyses are compared with those of the
original analyses in Table 13.4. This tabulation, in addition to the estimates
and approximate 90% confidence intervals for p_G and p_{DEM}, also provides
maximum likelihood estimates and 90% confidence intervals for the normal

EXAMPLE 8   **281**

distribution parameters μ and σ. (The estimates for p_G and p_{DEM} for the original data do not agree exactly with those in the previous section. This is because the method of maximum likelihood was used in their calculation, so as to make the results directly comparable with those for the other analyses, which require this method.)

The results, at first glance, suggest a fairly large difference among the estimates. However, this is overshadowed by the continued large statistical uncertainty, reflected by the lengths of the confidence intervals.

Some of the censored data analyses provide confidence intervals that are somewhat shorter than those from the original data. This is because the outlier in the original data inflated the estimate of σ and the lengths of the confidence intervals are approximately proportional to this estimate. One disadvantage of the censoring procedure is that some subjectivity is needed to choose a censoring point. A good rule of thumb for doing this, when interest centers on the lower (upper) tail of the distribution, is to let a normal probability plot serve as a guide, and to censor those observations that cause departure from linearity in the upper (lower) tail in the plot. Of course, wherever possible, physical considerations should enter into the choice of the censoring point.

In practice, we frequently do not know which specific model is appropriate and which analysis is best. Thus, performing a variety of analyses, as we have done here, provides useful insights into the "robustness" of the results under varying assumptions. Examination of the maximum likelihood estimates and 90% confidence intervals for p_{DEM} in Table 13.4 all lead us to the same conclusion irrespective of the approach used; i.e., the available data are insufficient to allow us to draw any definitive conclusions about the outcome of the demonstration test.

13.9 EXAMPLE 8: ESTIMATING THE PROPORTION WITHIN SPECIFICATIONS FOR A TWO-VARIABLE PROBLEM

13.9.1 Problem Description

Specifications for an electronic device require:

- Forward voltage must exceed 0.50 volts,
- Breakdown voltage must exceed 95 volts,
- A device may not have both forward voltage below 0.55 volts *and* reverse breakdown voltage below 100 volts.

The available data consist of a sample of ten devices randomly selected from the process that builds the device. This has resulted in the measurements given in Table 13.5. Based upon this information, it is desired to obtain an upper 90% confidence bound on the proportion of devices outside of specifi-

Table 13.5   Electronic Device Forward and Breakdown Voltage Measurements

Device Number	Forward Voltage	Breakdown Voltage
1	0.52	101
2	0.65	110
3	0.57	97
4	0.53	98
5	0.59	105
6	0.64	107
7	0.60	100
8	0.48	93
9	0.60	105
10	0.54	102

cations (i.e., the proportion nonconforming) from the sampled process. Assumptions, concerning the "representativeness" of the sample, similar to those described for the previous examples, apply here also.

13.9.2   Solution

If one can represent the measurements by a bivariate normal distribution, one can then use tabulations for this distribution to get a point estimate of the process proportion nonconforming (using the sample estimates in place of the unknown distribution parameters). However, obtaining a confidence interval or bound on this proportion, based on bivariate normal distribution assumptions, is a complex problem, especially since the forward voltage and breakdown voltage measurements are clearly correlated. Thus, we propose, instead, a much simpler, first cut solution that, though less efficient statistically, also does not require any assumptions about the form of the statistical distribution.

We note that two of the ten sampled devices are nonconforming (in particular device number 4 fails to meet the third requirement, and device number 8 fails on all three requirements). Thus, the observed proportion of nonconforming units is 0.2. From this information, we use Figure 6.1a or Table A.23a to obtain the desired upper 90% confidence bound on the process proportion noncomforming to be 0.45.

As in previous examples, the information lost by ignoring the actual measurements, and using only the information of whether or not a device is nonconforming, depends on the specific situation. Thus, if all devices had been well within the acceptance region, this first cut simple approach would have resulted in a greater loss of information than was the case with the data at hand.

EXAMPLE 9   **283**

13.10   EXAMPLE 9: DETERMINING THE MINIMUM SAMPLE SIZE FOR A DEMONSTRATION TEST

13.10.1   Problem Description

A manufacturer feels that a production process provides essentially 0% nonconforming units, with regard to a long list of specifications, some of which require a destructive test to evaluate. An unconvinced customer, however, before accepting the product, requires the manufacturer to demonstrate "with 90% confidence" that the process results in no more than 5% nonconforming units.

Thus, each unit in a random sample from the process is to be evaluated and classified as conforming or nonconforming. Because this involves an expensive series of tests, the manufacturer wishes to minimize the required random sample size to achieve the desired demonstration.

13.10.2   Solution

The manufacturer continues to be convinced that the process yields essentially no nonconforming units. This means that a random sample would also be free of such units. One would then use the methods given in Chapter 6 to obtain an upper confidence bound on the process proportion nonconforming, based upon the selected sample size. Thus, one can use the results of Chapter 6 in reverse to find how large a sample is needed.

In particular, from Figure 6.1a one notes that an upper 90% confidence bound of 0.05 is achieved with a sample of about size 50—if that sample, indeed, has no nonconforming units. Thus, the minimum required sample size is approximately 50. Actually, more precise methods lead to a required sample size of 45 units (as can also be seen from Table A.18).

13.10.3   Further Comments

The desired demonstration will be achieved only if the sample really results in zero nonconforming units. For this to be likely, the true process proportion noncomforming must, indeed, be quite small. In fact, from the lowest curve in Figure 9.2e (which is expressed in terms of the process proportion conforming), we note that even if the process proportion conforming is as large as 0.99, there is only a 64% chance of successful demonstration. In fact, we see from this curve that a process conformance rate of close to 0.998 is required for there to be a 90% chance that the demonstration test will be successful. This is why in discussing sample size requirements in Section 9.6, we did not use a "minimum sample size" approach. Instead, we required specification of not only the process proportion conforming that is to be demonstrated, but also of the proportion conforming for which we desire a high probability that the demonstration test be successful.

Epilogue

We have now, reluctantly, come to the end of our story. We say "reluctantly," not because we lament the end of this project (on which we have been working, though not always very actively, for 12 years), but because there is much more to be said. For example, we have limited consideration to mainly simple univariate (single variable) situations. Statistical intervals, in fact, illustrate the proverbial "bottomless pit" with seemingly never-ending opportunities for expanding upon and adding to the exposition. However, the fundamental intervals that we have presented can, by and large, be extended to more complex situations. Thus, we will conclude by reemphasizing some of our major ideas, and adding a few anecdotes about:

- The importance of calculating the "right" statistical interval.
- The role of statistical intervals versus other forms of inference.
- The limitations of statistical inference.

STATISTICAL INTERVALS: *VIVE LA DIFFÉRENCE*!

We have presented a wide variety of statistical intervals, and tried to explain the situations for which each is appropriate, how each is calculated, and the underlying assumptions. Some of these—such as confidence intervals for the population or process mean or standard deviation (assuming a normal distribution) or for a population or process proportion (assuming a binomial distribution)—are well known to users of statistical methods. Others, such as a confidence interval for the proportion below or above a threshold, or a prediction interval to contain one or more future observations, are surprisingly unfamiliar, even to many professional statisticians. We say "surprisingly" because, as evidenced by the case studies in Chapter 13, these intervals are frequently needed in applications. We have suggested that the reasons for this unfamiliarity include tradition, the relatively advanced nature of the underlying mathematics (generally *not* needed to *use* the intervals), and the complexity of the calculations (mostly irrelevant, given the tables, figures, and computer routines provided here and elsewhere).

In this regard, we recount our experience in interviewing, for positions in industry, promising recent Ph.D.'s in statistics. As part of the screening process, we often ask the following two questions, typical of those we are asked by our clients:

- An appliance, built in large quantities, is required to have a noise level of less than 50 decibels. A sample of eight units has resulted in the following readings:

  46.4, 46.7, 46.9, 47.0, 47.0, 47.2, 47.6, and 48.1 decibels.

  What can one conclude with a "high degree of assuredness" about the percentage of units manufactured during the year that fail to meet the 50-decibel threshold? What important assumptions, that are implicit in our inferences, do we need emphasize?
- Consider again the preceding noise measurements. However, now assume that a single added appliance is to be selected. What can one say, with a high degree of confidence, about the maximum noise that one may reasonably expect from this ninth appliance, and what added assumptions does this require?

We find that the great majority of interviewees either are unable to tell us how they would go about constructing the desired intervals or, worse still, answer incorrectly (by, e.g., proposing a confidence interval on the mean in response to the second question). In contrast, diligent readers of this book will have no trouble passing our test!

THE ROLE OF STATISTICAL INTERVALS

We feel strongly that, before reporting or using any form of statistical inference, an analyst should carefully examine and plot the data. Numerous methods and software for exploratory data analysis are available for this purpose and should be applied before one proceeds to more sophisticated evaluations, should these seem necessary.

However, when we do try to draw formal conclusions about a population or process from an appropriately selected random sample, statistical intervals play a central role in quantifying uncertainty, and provide an important supplement to point estimates. In our experience, such intervals are much more useful than significance or hypothesis tests. As previously indicated, we believe that this is because few statistical hypotheses hold exactly. Moreover, one can reject almost any statistical hypotheses by taking a sufficiently large sample—and avoid disproving a hypothesis by having a small enough sample, or even no sample at all.

THE LIMITATIONS OF STATISTICAL INFERENCE

Starting with Chapter 1, and throughout this book, we have stressed the basic assumptions underlying statistical inferences about a sampled population or process, especially in analytic studies. We conclude with a recent example.

In a mail survey, the 730 member families of a religious congregation were asked the following question:

> "On an overall basis, do you feel the minister is doing a good job (answer yes or no)?"

Among the 105 respondents, 58 answered "yes," and 47 said "no." Because the results were from "sample data," one of us was asked to make a statement that incorporated the "statistical uncertainty" about the proportion of families in the congregation that favored the minister. In this example, there was a well-defined population (the 730 families) which also comprised the sample frame. In fact, if all families had responded, one would have data from the entire population and there would be no statistical uncertainty, at least with regard to the stated viewpoints at the time of the survey. Moreover, *if* the respondents could be considered as randomly selected from the congregation, one could apply the methods presented in Chapter 6 for drawing inferences about the population proportion (perhaps including an appropriate finite population adjustment for the fact that the sample constituted an appreciable part of the population).

In reality, however, the sample was far from random; the respondents, in fact, were self-selected. Thus, those that felt most strongly about the issue, and, perhaps, those most active in the congregation (and in organizing the survey) were the ones who were most likely to respond, or urge their similarly viewed friends to do so. Without further study, little can be said about how representative this nonrandom sample really is of the population as a whole. Therefore, we felt that it would be misleading to calculate a statistical interval to contain the proportion of all congregants favoring the minister. Instead, we proposed that the results be presented as they stand, with appropriate comments concerning their possible inadequacy and/or provide encouragement for recanvassing nonrespondents in a follow-up study.

Where does this leave us? Despite our enthusiasm for statistical intervals, we feel that there are numerous situations where the practitioner is better served by not calculating such intervals and by emphasizing instead the limitations of the available information, perhaps suggesting how improved data can be obtained. Moreover, when statistical intervals are calculated in such situations, one need stress that they provide only a lower bound on the total uncertainty.

Learning is an iterative process. Sometimes, the available data are sufficient to draw meaningful conclusions; statistical intervals may then play an important role in quantifying uncertainty. Often, however, current information provides only a stepping stone to further study. In such cases, statistical intervals may be useful in describing what is known (or, indeed, unknown) at present. They can then guide the practitioner in deciding on the next step in the investigation.

APPENDIX A

Tables

Table A.1a Factors for Calculating Two-Sided 95% Statistical Intervals for a Normal Distribution

Number of Given Observations	Factors for Confidence Intervals for the Mean μ	Factors for Tolerance Intervals to Contain at Least $100p\%$ of the Distribution			Factors for Simultaneous Prediction Intervals to Contain all m Future Observations						Factors for Prediction Intervals to Contain the Mean of $n = m$ Future Observations
		p			m						
n		0.90	0.95	0.99	1	2	5	10	20	n	
4	1.59	5.37	6.34	8.22	3.56	4.41	5.56	6.41	7.21	5.29	2.25
5	1.24	4.29	5.08	6.60	3.04	3.70	4.58	5.23	5.85	4.58	1.76
6	1.05	3.73	4.42	5.76	2.78	3.33	4.08	4.63	5.16	4.22	1.48
7	0.92	3.39	4.02	5.24	2.62	3.11	3.77	4.26	4.74	4.01	1.31
8	0.84	3.16	3.75	4.89	2.51	2.97	3.57	4.02	4.46	3.88	1.18
9	0.77	2.99	3.55	4.63	2.43	2.86	3.43	3.85	4.25	3.78	1.09
10	0.72	2.86	3.39	4.44	2.37	2.79	3.32	3.72	4.10	3.72	1.01
12	0.64	2.67	3.17	4.16	2.29	2.68	3.17	3.53	3.89	3.63	0.90
15	0.55	2.49	2.96	3.89	2.22	2.57	3.03	3.36	3.69	3.56	0.78
20	0.47	2.32	2.76	3.62	2.14	2.48	2.90	3.21	3.50	3.50	0.66
25	0.41	2.22	2.64	3.46	2.10	2.43	2.83	3.12	3.40	3.49	0.58
30	0.37	2.15	2.55	3.35	2.08	2.39	2.78	3.06	3.33	3.48	0.53
40	0.32	2.06	2.45	3.22	2.05	2.35	2.73	2.99	3.25	3.49	0.45
60	0.26	1.96	2.34	3.07	2.02	2.31	2.67	2.93	3.17	3.53	0.37
∞	0.00	1.64	1.96	2.58	1.96	2.24	2.57	2.80	3.02	∞	0.00

The two-sided 95% statistical interval is $\bar{x} \pm c_{(0.95;\,n)}s$ where $c_{(0.95;\,n)}$ is the appropriate tabulated value and $\bar{x}$ and s are, respectively, the sample mean and standard deviation of a sample of size n. A similar table first appeared in Hahn (1970b). Adapted with permission of the American Society for Quality Control.

Table A.1b Factors for Calculating Two-Sided 99% Statistical Intervals for a Normal Distribution

Number of Given Observations	Factors for Confidence Intervals for the Mean μ	Factors for Tolerance Intervals to Contain at Least $100p\%$ of the Distribution			Factors for Simultaneous Prediction Intervals to Contain all m Future Observations						Factors for Prediction Intervals to Contain the Mean of $n = m$ Future Observations
		p			m						
n		0.90	0.95	0.99	1	2	5	10	20	n	
4	2.92	9.42	11.12	14.41	6.53	7.94	9.88	11.32	12.70	9.41	4.13
5	2.06	6.65	7.87	10.22	5.04	5.97	7.25	8.22	9.15	7.25	2.91
6	1.65	5.38	6.37	8.29	4.36	5.07	6.06	6.80	7.53	6.25	2.33
7	1.40	4.66	5.52	7.19	3.96	4.56	5.38	6.01	6.62	5.69	1.98
8	1.24	4.19	4.97	6.48	3.71	4.24	4.95	5.50	6.04	5.32	1.75
9	1.12	3.86	4.58	5.98	3.54	4.01	4.66	5.15	5.63	5.07	1.58
10	1.03	3.62	4.29	5.61	3.41	3.85	4.44	4.89	5.34	4.89	1.45
12	0.90	3.28	3.90	5.10	3.23	3.63	4.15	4.54	4.94	4.65	1.27
15	0.77	2.97	3.53	4.62	3.07	3.43	3.89	4.23	4.58	4.44	1.09
20	0.64	2.68	3.18	4.17	2.93	3.25	3.65	3.96	4.26	4.26	0.90
25	0.56	2.51	2.98	3.91	2.85	3.15	3.53	3.81	4.08	4.17	0.79
30	0.50	2.39	2.85	3.74	2.80	3.09	3.45	3.71	3.97	4.12	0.71
40	0.43	2.25	2.68	3.52	2.74	3.01	3.35	3.60	3.84	4.07	0.61
60	0.34	2.11	2.51	3.30	2.68	2.94	3.26	3.49	3.71	4.05	0.49
∞	0.00	1.64	1.96	2.58	2.58	2.81	3.09	3.29	3.48	∞	0.00

The two-sided 99% statistical interval is $\bar{x} \pm c_{(0.99;n)}s$ where $c_{(0.99;n)}$ is the appropriate tabulated value and $\bar{x}$ and s are, respectively, the sample mean and standard deviation of a sample of size n. A similar table first appeared in Hahn (1970b). Adapted with permission of the American Society for Quality Control.

Table A.2a Factors for Calculating Two-Sided 95% Statistical Intervals for a Normal Distribution

Number of Given Observations	Factors for Confidence Intervals for the Standard Deviation σ		Factors for Simultaneous Prediction Intervals to Contain all $m = n$ Future Observations	
	Factor for Calculating		Factor for Calculating	
n	Lower Limit	Upper Limit	Lower Limit	Upper Limit
4	0.57	3.73	0.25	3.93
5	0.60	2.87	0.32	3.10
6	0.62	2.45	0.37	2.67
7	0.64	2.20	0.41	2.41
8	0.66	2.04	0.45	2.23
9	0.68	1.92	0.47	2.11
10	0.69	1.83	0.50	2.01
12	0.71	1.70	0.54	1.86
15	0.73	1.58	0.58	1.73
20	0.76	1.46	0.63	1.59
25	0.78	1.39	0.66	1.51
30	0.80	1.34	0.69	1.45
40	0.82	1.28	0.73	1.38
60	0.85	1.22	0.77	1.29
∞	1.00	1.00	1.00	1.00

The two-sided 95% statistical interval is $[c_{L(0.95;\,n)}s, c_{U(0.95;\,n)}s]$ where $c_{L(0.95;\,n)}$ and $c_{U(0.95;\,n)}$ are the appropriate tabulated factors and s is the sample standard deviation of a sample of size n. A similar table first appeared in Hahn (1970b). Adapted with permission of the American Society for Quality Control.

Table A.2b Factors for Calculating Two-Sided 99% Statistical Intervals for a Normal Distribution

Number of Given Observations	Factors for Confidence Intervals for the Standard Deviation σ		Factors for Simultaneous Prediction Intervals to Contain all $m = n$ Future Observations	
	Factor for Calculating		Factor for Calculating	
n	Lower Limit	Upper Limit	Lower Limit	Upper Limit
4	0.48	6.47	0.15	6.89
5	0.52	4.40	0.21	4.81
6	0.55	3.48	0.26	3.87
7	0.57	2.98	0.30	3.33
8	0.59	2.66	0.34	2.98
9	0.60	2.44	0.37	2.74
10	0.62	2.28	0.39	2.56
12	0.64	2.06	0.43	2.31
15	0.67	1.85	0.48	2.07
20	0.70	1.67	0.54	1.85
25	0.73	1.56	0.58	1.72
30	0.74	1.49	0.61	1.64
40	0.77	1.40	0.66	1.52
60	0.81	1.30	0.71	1.40
∞	1.00	1.00	1.00	1.00

The two-sided 99% statistical interval is $[c_{L(0.99;\,n)}s, c_{U(0.99;\,n)}s]$ where $c_{L(0.99;\,n)}$ and $c_{U(0.99;\,n)}$ are the appropriate tabulated factors and s is the sample standard deviation of a sample of size n. A similar table first appeared in Hahn (1970b). Adapted with permission of the American Society for Quality Control.

Table A.3a Factors for Calculating One-Sided 95% Statistical Bounds for a Normal Distribution

Number of Given Observations	Factors for Confidence Bounds for the Mean μ	Factors for Tolerance Bounds to exceed (be exceeded by) at Least 100p% of the Distribution			Factors for Simultaneous Prediction Bounds to exceed (be exceeded by) all m Future Observations						Factors for Prediction Bounds to exceed (be exceeded by) the Mean of $n = m$ Future Observations
			p					m			
n		0.90	0.95	0.99	1	2	5	10	20	n	
4	1.18	4.16	5.14	7.04	2.63	3.40	4.47	5.28	6.06	4.21	1.66
5	0.95	3.41	4.20	5.74	2.34	2.95	3.79	4.42	5.03	3.79	1.35
6	0.82	3.01	3.71	5.06	2.18	2.72	3.43	3.97	4.49	3.58	1.16
7	0.73	2.76	3.40	4.64	2.08	2.57	3.22	3.70	4.17	3.45	1.04
8	0.67	2.58	3.19	4.35	2.01	2.47	3.07	3.52	3.95	3.37	0.95
9	0.62	2.45	3.03	4.14	1.96	2.40	2.97	3.38	3.79	3.32	0.88
10	0.58	2.35	2.91	3.98	1.92	2.35	2.89	3.28	3.67	3.28	0.82
12	0.52	2.21	2.74	3.75	1.87	2.27	2.78	3.14	3.50	3.24	0.73
15	0.45	2.07	2.57	3.52	1.82	2.20	2.67	3.01	3.34	3.21	0.64
20	0.39	1.93	2.40	3.30	1.77	2.13	2.57	2.89	3.19	3.19	0.55
25	0.34	1.84	2.29	3.16	1.74	2.09	2.52	2.82	3.11	3.20	0.48
30	0.31	1.78	2.22	3.06	1.73	2.07	2.48	2.78	3.05	3.21	0.44
40	0.27	1.70	2.13	2.94	1.71	2.04	2.44	2.72	2.99	3.24	0.38
60	0.22	1.61	2.02	2.81	1.68	2.01	2.40	2.67	2.92	3.30	0.31
∞	0.00	1.28	1.64	2.33	1.64	1.95	2.32	2.57	2.80	∞	0.00

The one-sided 95% statistical bound is $\bar{x} + c'_{(0.95;\,n)}s$ or $\bar{x} - c'_{(0.95;\,n)}s$ where $c'_{(0.95;\,n)}$ is the appropriate tabulated value and $\bar{x}$ and s are, respectively, the sample mean and standard deviation of a sample of size n. A similar table first appeared in Hahn (1970b). Adapted with permission of the American Society for Quality Control.

Table A.3b Factors for Calculating One-Sided 99% Statistical Bounds for a Normal Distribution

Number of Given Observations	Factors for Confidence Bounds for the Mean μ	Factors for Tolerance Bounds to exceed (be exceeded by) at least $100p\%$ of the Distribution			Factors for Simultaneous Prediction Bounds to exceed (be exceeded by) all m Future Observations						Factors for Prediction Bounds to exceed (be exceeded by) the Mean of $n = m$ Future Observations
		p			m						
n		0.90	0.95	0.99	1	2	5	10	20	n	
4	2.27	7.38	9.08	12.39	5.08	6.31	8.07	9.43	10.76	7.63	3.21
5	1.68	5.36	6.58	8.94	4.10	4.94	6.13	7.04	7.95	6.13	2.37
6	1.37	4.41	5.41	7.33	3.63	4.30	5.22	5.94	6.64	5.41	1.94
7	1.19	3.86	4.73	6.41	3.36	3.93	4.70	5.30	5.90	5.00	1.68
8	1.06	3.50	4.29	5.81	3.18	3.69	4.37	4.90	5.43	4.73	1.50
9	0.97	3.24	3.97	5.39	3.05	3.52	4.14	4.62	5.09	4.55	1.37
10	0.89	3.05	3.74	5.07	2.96	3.39	3.97	4.41	4.85	4.41	1.26
12	0.78	2.78	3.41	4.63	2.83	3.22	3.74	4.13	4.52	4.23	1.11
15	0.68	2.52	3.10	4.22	2.71	3.07	3.53	3.88	4.22	4.08	0.96
20	0.57	2.28	2.81	3.83	2.60	2.93	3.34	3.65	3.95	3.95	0.80
25	0.50	2.13	2.63	3.60	2.54	2.85	3.24	3.52	3.80	3.89	0.70
30	0.45	2.03	2.52	3.45	2.50	2.80	3.17	3.44	3.71	3.86	0.64
40	0.38	1.90	2.36	3.25	2.46	2.74	3.09	3.35	3.59	3.83	0.54
60	0.31	1.76	2.20	3.04	2.41	2.68	3.02	3.26	3.49	3.84	0.44
∞	0.00	1.28	1.64	2.33	2.33	2.57	2.88	3.09	3.29	∞	0.00

The one-sided 99% statistical bound is $\bar{x} + c'_{(0.99;\,n)}s$ or $\bar{x} - c'_{(0.99;\,n)}s$ where $c'_{(0.99;\,n)}$ is the appropriate tabulated value and $\bar{x}$ and s are, respectively, the sample mean and standard deviation of a sample of size n. A similar table first appeared in Hahn (1970b). Adapted with permission of the American Society for Quality Control.

Table A.4a Factors for Calculating One-Sided 95% Statistical Bounds for a Normal Distribution

Number of Given Observations	Factors for Confidence Bounds for the Standard Deviation σ		Factors for Simultaneous Prediction Bounds to exceed (be exceeded by) all $m = n$ Future Observations	
	Factor for Calculating		Factor for Calculating	
n	Lower Bound	Upper Bound	Lower Bound	Upper Bound
4	0.62	2.92	0.33	3.05
5	0.65	2.37	0.40	2.53
6	0.67	2.09	0.44	2.25
7	0.69	1.92	0.48	2.07
8	0.71	1.80	0.51	1.95
9	0.72	1.71	0.54	1.85
10	0.73	1.65	0.56	1.78
12	0.75	1.55	0.60	1.68
15	0.77	1.46	0.63	1.58
20	0.79	1.37	0.68	1.47
25	0.81	1.32	0.71	1.41
30	0.83	1.28	0.73	1.36
40	0.85	1.23	0.77	1.31
60	0.87	1.18	0.81	1.24
∞	1.00	1.00	1.00	1.00

The one-sided 95% statistical bound is $c'_{L(0.95;\,n)}s$ or $c'_{U(0.95;\,n)}s$ where $c'_{L(0.95;\,n)}$ and $c'_{U(0.95;\,n)}$ are the appropriate tabulated factors for lower and upper bounds, respectively, and s is the sample standard deviation of a sample size n. A similar table first appeared in Hahn (1970b). Adapted with permission of the American Society for Quality Control.

Table A.4b Factors for Calculating One-Sided 99% Statistical Bounds for a Normal Distribution

Number of Given Observations	Factors for Confidence Bounds for the Standard Deviation σ		Factors for Simultaneous Prediction Bounds to exceed (be exceeded by) all $m = n$ Future Observations	
	Factor for Calculating		Factor for Calculating	
n	Lower Bound	Upper Bound	Lower Bound	Upper Bound
4	0.51	5.11	0.18	5.43
5	0.55	3.67	0.25	4.00
6	0.58	3.00	0.30	3.31
7	0.60	2.62	0.34	2.91
8	0.62	2.38	0.38	2.64
9	0.63	2.20	0.41	2.46
10	0.64	2.08	0.43	2.31
12	0.67	1.90	0.47	2.11
15	0.69	1.73	0.52	1.92
20	0.72	1.58	0.57	1.74
25	0.75	1.49	0.61	1.63
30	0.76	1.43	0.64	1.56
40	0.79	1.35	0.68	1.46
60	0.82	1.27	0.74	1.36
∞	1.00	1.00	1.00	1.00

The one-sided 99% statistical bound is $c'_{L(0.99;\,n)}$ or $c'_{U(0.99;\,n)}s$ where $c'_{L(0.99;\,n)}$ and $c'_{U(0.99;\,n)}$ are the appropriate tabulated factors for lower and upper bounds, respectively, and s is the sample standard deviation of a sample size n. A similar table first appeared in Hahn (1970b). Adapted with permission of the American Society for Quality Control.

Table A.5a   Standard Normal Cumulative Distribution Probabilities:
$\Pr(z \le z') = \Phi(z')$

z'	.00	.01	.02	.03	.04	.05	.06	.07	.08	.09
0.0	.5000	.4960	.4920	.4880	.4840	.4801	.4761	.4721	.4681	.4641
-0.1	.4602	.4562	.4522	.4483	.4443	.4404	.4364	.4325	.4286	.4247
-0.2	.4207	.4168	.4129	.4090	.4052	.4013	.3974	.3936	.3897	.3859
-0.3	.3821	.3783	.3745	.3707	.3669	.3632	.3594	.3557	.3520	.3483
-0.4	.3446	.3409	.3372	.3336	.3300	.3264	.3228	.3192	.3156	.3121
-0.5	.3085	.3050	.3015	.2981	.2946	.2912	.2877	.2843	.2810	.2776
-0.6	.2743	.2709	.2676	.2643	.2611	.2578	.2546	.2514	.2483	.2451
-0.7	.2420	.2389	.2358	.2327	.2296	.2266	.2236	.2206	.2177	.2148
-0.8	.2119	.2090	.2061	.2033	.2005	.1977	.1949	.1922	.1894	.1867
-0.9	.1841	.1814	.1788	.1762	.1736	.1711	.1685	.1660	.1635	.1611
-1.0	.1587	.1562	.1539	.1515	.1492	.1469	.1446	.1423	.1401	.1379
-1.1	.1357	.1335	.1314	.1292	.1271	.1251	.1230	.1210	.1190	.1170
-1.2	.1151	.1131	.1112	.1093	.1075	.1056	.1038	.1020	.1003	.0985
-1.3	.0968	.0951	.0934	.0918	.0901	.0885	.0869	.0853	.0838	.0823
-1.4	.0808	.0793	.0778	.0764	.0749	.0735	.0721	.0708	.0694	.0681
-1.5	.0668	.0655	.0643	.0630	.0618	.0606	.0594	.0582	.0571	.0559
-1.6	.0548	.0537	.0526	.0516	.0505	.0495	.0485	.0475	.0465	.0455
-1.7	.0446	.0436	.0427	.0418	.0409	.0401	.0392	.0384	.0375	.0367
1.8	.0359	.0351	.0344	.0336	.0329	.0322	.0314	.0307	.0301	.0294
-1.9	.0287	.0281	.0274	.0268	.0262	.0256	.0250	.0244	.0239	.0233
-2.0	.0228	.0222	.0217	.0212	.0207	.0202	.0197	.0192	.0188	.0183
-2.1	.0179	.0174	.0170	.0166	.0162	.0158	.0154	.0150	.0146	.0143
-2.2	.0139	.0136	.0132	.0129	.0125	.0122	.0119	.0116	.0113	.0110
-2.3	.0107	.0104	.0102	$.0^2990$	$.0^2964$	$.0^2939$	$.0^2914$	$.0^2889$	$.0^2866$	$.0^2842$
-2.4	$.0^2820$	$.0^2798$	$.0^2776$	$.0^2755$	$.0^2734$	$.0^2714$	$.0^2695$	$.0^2676$	$.0^2657$	$.0^2639$
-2.5	$.0^2621$	$.0^2604$	$.0^2587$	$.0^2570$	$.0^2554$	$.0^2539$	$.0^2523$	$.0^2508$	$.0^2494$	$.0^2480$
-2.6	$.0^2466$	$.0^2453$	$.0^2440$	$.0^2427$	$.0^2415$	$.0^2402$	$.0^2391$	$.0^2379$	$.0^2368$	$.0^2357$
-2.7	$.0^2347$	$.0^2336$	$.0^2326$	$.0^2317$	$.0^2307$	$.0^2298$	$.0^2289$	$.0^2280$	$.0^2272$	$.0^2264$
-2.8	$.0^2256$	$.0^2248$	$.0^2240$	$.0^2233$	$.0^2226$	$.0^2219$	$.0^2212$	$.0^2205$	$.0^2199$	$.0^2193$
-2.9	$.0^2187$	$.0^2181$	$.0^2175$	$.0^2169$	$.0^2164$	$.0^2159$	$.0^2154$	$.0^2149$	$.0^2144$	$.0^2139$
-3.0	$.0^2135$	$.0^2131$	$.0^2126$	$.0^2122$	$.0^2118$	$.0^2114$	$.0^2111$	$.0^2107$	$.0^2104$	$.0^2100$
-3.1	$.0^3968$	$.0^3935$	$.0^3904$	$.0^3874$	$.0^3845$	$.0^3816$	$.0^3789$	$.0^3762$	$.0^3736$	$.0^3711$
-3.2	$.0^3687$	$.0^3664$	$.0^3641$	$.0^3619$	$.0^3598$	$.0^3577$	$.0^3557$	$.0^3538$	$.0^3519$	$.0^3501$
-3.3	$.0^3483$	$.0^3466$	$.0^3450$	$.0^3434$	$.0^3419$	$.0^3404$	$.0^3390$	$.0^3376$	$.0^3362$	$.0^3349$
-3.4	$.0^3337$	$.0^3325$	$.0^3313$	$.0^3302$	$.0^3291$	$.0^3280$	$.0^3270$	$.0^3260$	$.0^3251$	$.0^3242$
-3.5	$.0^3233$	$.0^3224$	$.0^3216$	$.0^3208$	$.0^3200$	$.0^3193$	$.0^3185$	$.0^3178$	$.0^3172$	$.0^3165$
-3.6	$.0^3159$	$.0^3153$	$.0^3147$	$.0^3142$	$.0^3136$	$.0^3131$	$.0^3126$	$.0^3121$	$.0^3117$	$.0^3112$
-3.7	$.0^3108$	$.0^3104$	$.0^4996$	$.0^4957$	$.0^4920$	$.0^4884$	$.0^4850$	$.0^4816$	$.0^4784$	$.0^4753$
-3.8	$.0^4723$	$.0^4695$	$.0^4667$	$.0^4641$	$.0^4615$	$.0^4591$	$.0^4567$	$.0^4544$	$.0^4522$	$.0^4501$
-3.9	$.0^4481$	$.0^4461$	$.0^4443$	$.0^4425$	$.0^4407$	$.0^4391$	$.0^4375$	$.0^4359$	$.0^4345$	$.0^4330$
-4.0	$.0^4317$	$.0^4304$	$.0^4291$	$.0^4279$	$.0^4267$	$.0^4256$	$.0^4245$	$.0^4235$	$.0^4225$	$.0^4216$

A similar table first appeared in Hald (1952). Adapted with permission of the publisher.

Table A.5b Standard Normal Cumulative Distribution Probabilities:
$\Pr(z \le z') = \Phi(z')$

z'	.00	.01	.02	.03	.04	.05	.06	.07	.08	.09
0.0	.5000	.5040	.5080	.5120	.5160	.5199	.5239	.5279	.5319	.5359
0.1	.5398	.5438	.5478	.5517	.5557	.5596	.5636	.5675	.5714	.5753
0.2	.5793	.5832	.5871	.5910	.5948	.5987	.6026	.6064	.6103	.6141
0.3	.6179	.6217	.6255	.6293	.6331	.6368	.6406	.6443	.6480	.6517
0.4	.6554	.6591	.6628	.6664	.6700	.6736	.6772	.6808	.6844	.6879
0.5	.6915	.6950	.6985	.7019	.7054	.7088	.7123	.7157	.7190	.7224
0.6	.7257	.7291	.7324	.7357	.7389	.7422	.7454	.7486	.7517	.7549
0.7	.7580	.7611	.7642	.7673	.7704	.7734	.7764	.7794	.7823	.7852
0.8	.7881	.7910	.7939	.7967	.7995	.8023	.8051	.8078	.8106	.8133
0.9	.8159	.8186	.8212	.8238	.8264	.8289	.8315	.8340	.8365	.8389
1.0	.8413	.8438	.8461	.8485	.8508	.8531	.8554	.8577	.8599	.8621
1.1	.8643	.8665	.8686	.8708	.8729	.8749	.8770	.8790	.8810	.8830
1.2	.8849	.8869	.8888	.8907	.8925	.8944	.8962	.8980	.8997	.9015
1.3	.9032	.9049	.9066	.9082	.9099	.9115	.9131	.9147	.9162	.9177
1.4	.9192	.9207	.9222	.9236	.9251	.9265	.9279	.9292	.9306	.9319
1.5	.9332	.9345	.9357	.9370	.9382	.9394	.9406	.9418	.9429	.9441
1.6	.9452	.9463	.9474	.9484	.9495	.9505	.9515	.9525	.9535	.9545
1.7	.9554	.9564	.9573	.9582	.9591	.9599	.9608	.9616	.9625	.9633
1.8	.9641	.9649	.9656	.9664	.9671	.9678	.9686	.9693	.9699	.9706
1.9	.9713	.9719	.9726	.9732	.9738	.9744	.9750	.9756	.9761	.9767
2.0	.9772	.9778	.9783	.9788	.9793	.9798	.9803	.9808	.9812	.9817
2.1	.9821	.9826	.9830	.9834	.9838	.9842	.9846	.9850	.9854	.9857
2.2	.9861	.9864	.9868	.9871	.9875	.9878	.9881	.9884	.9887	.9890
2.3	.9893	.9896	.9898	$.9^2010$	$.9^2036$	$.9^2061$	$.9^2086$	$.9^2111$	$.9^2134$	$.9^2158$
2.4	$.9^2180$	$.9^2202$	$.9^2224$	$.9^2245$	$.9^2266$	$.9^2286$	$.9^2305$	$.9^2324$	$.9^2343$	$.9^2361$
2.5	$.9^2379$	$.9^2396$	$.9^2413$	$.9^2430$	$.9^2446$	$.9^2461$	$.9^2477$	$.9^2492$	$.9^2506$	$.9^2520$
2.6	$.9^2534$	$.9^2547$	$.9^2560$	$.9^2573$	$.9^2585$	$.9^2598$	$.9^2609$	$.9^2621$	$.9^2632$	$.9^2643$
2.7	$.9^2653$	$.9^2664$	$.9^2674$	$.9^2683$	$.9^2693$	$.9^2702$	$.9^2711$	$.9^2720$	$.9^2728$	$.9^2736$
2.8	$.9^2744$	$.9^2752$	$.9^2760$	$.9^2767$	$.9^2774$	$.9^2781$	$.9^2788$	$.9^2795$	$.9^2801$	$.9^2807$
2.9	$.9^2813$	$.9^2819$	$.9^2825$	$.9^2831$	$.9^2836$	$.9^2841$	$.9^2846$	$.9^2851$	$.9^2856$	$.9^2861$
3.0	$.9^2865$	$.9^2869$	$.9^2874$	$.9^2878$	$.9^2882$	$.9^2886$	$.9^2889$	$.9^2893$	$.9^2896$	$.9^2900$
3.1	$.9^3032$	$.9^3065$	$.9^3096$	$.9^3126$	$.9^3155$	$.9^3184$	$.9^3211$	$.9^3238$	$.9^3264$	$.9^3289$
3.2	$.9^3313$	$.9^3336$	$.9^3359$	$.9^3381$	$.9^3402$	$.9^3423$	$.9^3443$	$.9^3462$	$.9^3481$	$.9^3499$
3.3	$.9^3517$	$.9^3534$	$.9^3550$	$.9^3566$	$.9^3581$	$.9^3596$	$.9^3610$	$.9^3624$	$.9^3638$	$.9^3651$
3.4	$.9^3663$	$.9^3675$	$.9^3687$	$.9^3698$	$.9^3709$	$.9^3720$	$.9^3730$	$.9^3740$	$.9^3749$	$.9^3758$
3.5	$.9^3767$	$.9^3776$	$.9^3784$	$.9^3792$	$.9^3800$	$.9^3807$	$.9^3815$	$.9^3822$	$.9^3828$	$.9^3835$
3.6	$.9^3841$	$.9^3847$	$.9^3853$	$.9^3858$	$.9^3864$	$.9^3869$	$.9^3874$	$.9^3879$	$.9^3883$	$.9^3888$
3.7	$.9^3892$	$.9^3896$	$.9^4004$	$.9^4043$	$.9^4080$	$.9^4116$	$.9^4150$	$.9^4184$	$.9^4216$	$.9^4247$
3.8	$.9^4277$	$.9^4305$	$.9^4333$	$.9^4359$	$.9^4385$	$.9^4409$	$.9^4433$	$.9^4456$	$.9^4478$	$.9^4499$
3.9	$.9^4519$	$.9^4539$	$.9^4557$	$.9^4575$	$.9^4593$	$.9^4609$	$.9^4625$	$.9^4641$	$.9^4655$	$.9^4670$
4.0	$.9^4683$	$.9^4696$	$.9^4709$	$.9^4721$	$.9^4733$	$.9^4744$	$.9^4755$	$.9^4765$	$.9^4775$	$.9^4784$

A similar table first appeared in Hald (1952). Adapted with permission of the publisher.

Table A.6 Standard Normal Distribution Percentiles

100(1 − α) Percentiles (One-sided Factors)

1 − α	$z_{(1-\alpha)}$	1 − α	$z_{(1-\alpha)}$
0.000001	−4.753	0.50	0.000
0.000005	−4.417	0.55	0.126
0.00001	−4.265	0.60	0.253
0.00005	−3.891	0.65	0.385
0.0001	−3.719	0.70	0.524
0.0005	−3.291	0.75	0.674
0.001	−3.090	0.80	0.842
0.005	−2.576	0.85	1.036
0.01	−2.326	0.90	1.282
0.02	−2.054	0.92	1.405
0.03	−1.881	0.94	1.555
0.04	−1.751	0.95	1.645
0.05	−1.645	0.96	1.751
0.06	−1.555	0.97	1.881
0.08	−1.405	0.98	2.054
0.10	−1.282	0.99	2.326
0.15	−1.036	0.995	2.576
0.20	−0.842	0.999	3.090
0.25	−0.674	0.9995	3.291
0.30	−0.524	0.9999	3.719
0.35	−0.385	0.99995	3.891
0.40	−0.253	0.99999	4.265
0.45	−0.126	0.999995	4.417
0.50	0.000	0.999999	4.753

100(1 − α/2) Percentiles (Two-sided Factors)

1 − α	$z_{(1-\alpha/2)}$	1 − α	$z_{(1-\alpha/2)}$
0.000001	−4.892	0.50	0.674
0.000005	−4.565	0.55	0.755
0.00001	−4.417	0.60	0.842
0.00005	−4.056	0.65	0.935
0.0001	−3.891	0.70	1.036
0.0005	−3.481	0.75	1.150
0.001	−3.291	0.80	1.282
0.005	−2.807	0.85	1.440
0.01	−2.576	0.90	1.645
0.02	−2.326	0.92	1.751
0.03	−2.170	0.94	1.881
0.04	−2.054	0.95	1.960
0.05	−1.960	0.96	2.054
0.06	−1.881	0.97	2.170
0.08	−1.751	0.98	2.326
0.10	−1.645	0.99	2.576
0.15	−1.440	0.995	2.807
0.20	−1.282	0.999	3.291
0.25	−1.150	0.9995	3.481
0.30	−1.036	0.9999	3.891
0.35	−0.935	0.99995	4.056
0.40	−0.842	0.99999	4.417
0.45	−0.755	0.999995	4.565
0.50	−0.674	0.999999	4.892

Table A.7 100γ Percentiles of the Student's _t_ Distribution with _r_ Degrees of Freedom: $t_{(\gamma;\,r)}$

r	0.750	0.800	0.850	0.900	0.950	0.975	0.980	0.990	0.995	0.999
1	1.000	1.376	1.963	3.078	6.314	12.71	15.89	31.82	63.66	318.3
2	0.816	1.061	1.386	1.886	2.920	4.303	4.849	6.965	9.925	22.33
3	0.765	0.978	1.250	1.638	2.353	3.182	3.482	4.541	5.841	10.21
4	0.741	0.941	1.190	1.533	2.132	2.776	2.999	3.747	4.604	7.173
5	0.727	0.920	1.156	1.476	2.015	2.571	2.757	3.365	4.032	5.893
6	0.718	0.906	1.134	1.440	1.943	2.447	2.612	3.143	3.707	5.208
7	0.711	0.896	1.119	1.415	1.895	2.365	2.517	2.998	3.499	4.785
8	0.706	0.889	1.108	1.397	1.860	2.306	2.449	2.896	3.355	4.501
9	0.703	0.883	1.100	1.383	1.833	2.262	2.398	2.821	3.250	4.297
10	0.700	0.879	1.093	1.372	1.812	2.228	2.359	2.764	3.169	4.144
11	0.697	0.876	1.088	1.363	1.796	2.201	2.328	2.718	3.106	4.025
12	0.695	0.873	1.083	1.356	1.782	2.179	2.303	2.681	3.055	3.930
13	0.694	0.870	1.079	1.350	1.771	2.160	2.282	2.650	3.012	3.852
14	0.692	0.868	1.076	1.345	1.761	2.145	2.264	2.624	2.977	3.787
15	0.691	0.866	1.074	1.341	1.753	2.131	2.249	2.602	2.947	3.733
16	0.690	0.865	1.071	1.337	1.746	2.120	2.235	2.583	2.921	3.686
17	0.689	0.863	1.069	1.333	1.740	2.110	2.224	2.567	2.898	3.646
18	0.688	0.862	1.067	1.330	1.734	2.101	2.214	2.552	2.878	3.610
19	0.688	0.861	1.066	1.328	1.729	2.093	2.205	2.539	2.861	3.579
20	0.687	0.860	1.064	1.325	1.725	2.086	2.197	2.528	2.845	3.552
21	0.686	0.859	1.063	1.323	1.721	2.080	2.189	2.518	2.831	3.527
22	0.686	0.858	1.061	1.321	1.717	2.074	2.183	2.508	2.819	3.505
23	0.685	0.858	1.060	1.319	1.714	2.069	2.177	2.500	2.807	3.485
24	0.685	0.857	1.059	1.318	1.711	2.064	2.172	2.492	2.797	3.467
25	0.684	0.856	1.058	1.316	1.708	2.060	2.167	2.485	2.787	3.450
26	0.684	0.856	1.058	1.315	1.706	2.056	2.162	2.479	2.779	3.435
27	0.684	0.855	1.057	1.314	1.703	2.052	2.158	2.473	2.771	3.421
28	0.683	0.855	1.056	1.313	1.701	2.048	2.154	2.467	2.763	3.408
29	0.683	0.854	1.055	1.311	1.699	2.045	2.150	2.462	2.756	3.396
30	0.683	0.854	1.055	1.310	1.697	2.042	2.147	2.457	2.750	3.385
35	0.682	0.852	1.052	1.306	1.690	2.030	2.133	2.438	2.724	3.340
40	0.681	0.851	1.050	1.303	1.684	2.021	2.123	2.423	2.704	3.307
50	0.679	0.849	1.047	1.299	1.676	2.009	2.109	2.403	2.678	3.261
60	0.679	0.848	1.045	1.296	1.671	2.000	2.099	2.390	2.660	3.232
70	0.678	0.847	1.044	1.294	1.667	1.994	2.093	2.381	2.648	3.211
80	0.678	0.846	1.043	1.292	1.664	1.990	2.088	2.374	2.639	3.195
90	0.677	0.846	1.042	1.291	1.662	1.987	2.084	2.368	2.632	3.183
100	0.677	0.845	1.042	1.290	1.660	1.984	2.081	2.364	2.626	3.174
120	0.677	0.845	1.041	1.289	1.658	1.980	2.076	2.358	2.617	3.160
∞	0.675	0.842	1.036	1.282	1.645	1.960	2.054	2.327	2.576	3.091
Two-sided Confidence Level	0.500	0.600	0.700	0.800	0.900	0.950	0.960	0.980	0.990	0.998

Table A.7 gives $t_{(\gamma;\,r)}$ for $\gamma > 0.50$. For $\gamma < 0.50$, use the relationship $t_{(1-\gamma;\,r)} = -t_{(\gamma;\,r)}$.

Table A.8 100γ Percentiles of the Chi-Square Distribution with r Degrees of Freedom: $\chi^2_{(\gamma;\,r)}$

r	0.001	0.005	0.010	0.020	0.025	0.050	0.100	0.250	0.500	0.750	0.900	0.950	0.975	0.980	0.990	0.995	0.999
1	$^{0}016$	$^{0}039$	$^{0}016$	$^{0}062$	$^{0}098$	$^{0}039$	0.016	0.102	0.455	1.323	2.706	3.841	5.024	5.412	6.635	7.880	10.83
2	$^{0}020$	0.010	0.020	0.040	0.051	0.103	0.211	0.575	1.386	2.773	4.605	5.991	7.378	7.824	9.210	10.60	13.82
3	0.024	0.072	0.115	0.185	0.216	0.352	0.584	1.213	2.366	4.108	6.251	7.815	9.348	9.838	11.34	12.84	16.27
4	0.091	0.207	0.297	0.429	0.484	0.711	1.064	1.923	3.357	5.385	7.779	9.488	11.14	11.67	13.28	14.86	18.47
5	0.210	0.412	0.554	0.752	0.831	1.145	1.610	2.675	4.351	6.626	9.236	11.07	12.83	13.39	15.09	16.75	20.51
6	0.381	0.676	0.872	1.134	1.237	1.635	2.204	3.455	5.348	7.841	10.64	12.59	14.45	15.03	16.81	18.55	22.46
7	0.598	0.989	1.239	1.564	1.690	2.167	2.833	4.255	6.346	9.037	12.02	14.07	16.01	16.62	18.48	20.28	24.32
8	0.857	1.344	1.646	2.032	2.180	2.733	3.490	5.071	7.344	10.22	13.36	15.51	17.53	18.17	20.09	21.96	26.13
9	1.152	1.735	2.088	2.532	2.700	3.325	4.168	5.899	8.343	11.39	14.68	16.92	19.02	19.68	21.67	23.59	27.88
10	1.479	2.156	2.558	3.059	3.247	3.940	4.865	6.737	9.342	12.55	15.99	18.31	20.48	21.16	23.21	25.19	29.59
11	1.834	2.603	3.053	3.609	3.816	4.575	5.578	7.584	10.34	13.70	17.28	19.68	21.92	22.62	24.73	26.76	31.26
12	2.214	3.074	3.570	4.178	4.404	5.226	6.304	8.438	11.34	14.85	18.55	21.03	23.34	24.05	26.22	28.30	32.91
13	2.617	3.565	4.107	4.765	5.009	5.892	7.041	9.299	12.34	15.98	19.81	22.36	24.74	25.47	27.69	29.82	34.53
14	3.040	4.075	4.660	5.368	5.629	6.571	7.790	10.17	13.34	17.12	21.06	23.68	26.12	26.87	29.14	31.32	36.12
15	3.482	4.601	5.229	5.985	6.262	7.261	8.547	11.04	14.34	18.25	22.31	25.00	27.49	28.26	30.58	32.80	37.70
16	3.941	5.142	5.812	6.614	6.908	7.962	9.312	11.91	15.34	19.37	23.54	26.30	28.85	29.63	32.00	34.27	39.25
17	4.416	5.697	6.408	7.255	7.564	8.672	10.09	12.79	16.34	20.49	24.77	27.59	30.19	31.00	33.41	35.72	40.79
18	4.904	6.265	7.015	7.906	8.231	9.390	10.86	13.68	17.34	21.60	25.99	28.87	31.53	32.35	34.81	37.16	42.31
19	5.406	6.844	7.633	8.567	8.906	10.12	11.65	14.56	18.34	22.72	27.20	30.14	32.85	33.69	36.19	38.58	43.82
20	5.920	7.434	8.260	9.237	9.591	10.85	12.44	15.45	19.34	23.83	28.41	31.41	34.17	35.02	37.57	40.00	45.31
21	6.446	8.033	8.897	9.914	10.28	11.59	13.24	16.34	20.34	24.93	29.62	32.67	35.48	36.34	38.93	41.40	46.80
22	6.982	8.643	9.542	10.60	10.98	12.34	14.04	17.24	21.34	26.04	30.81	33.92	36.78	37.66	40.29	42.80	48.27
23	7.529	9.260	10.20	11.29	11.69	13.09	14.85	18.14	22.34	27.14	32.01	35.17	38.08	38.97	41.64	44.18	49.73
24	8.084	9.886	10.86	11.99	12.40	13.85	15.66	19.04	23.34	28.24	33.20	36.42	39.36	40.27	42.98	45.56	51.18
25	8.649	10.52	11.52	12.70	13.12	14.61	16.47	19.94	24.34	29.34	34.38	37.65	40.65	41.57	44.31	46.93	52.62
26	9.221	11.16	12.20	13.41	13.84	15.38	17.29	20.84	25.34	30.43	35.56	38.89	41.92	42.86	45.64	48.29	54.05
27	9.802	11.81	12.88	14.13	14.57	16.15	18.11	21.75	26.34	31.53	36.74	40.11	43.19	44.14	46.96	49.64	55.48
28	10.39	12.46	13.56	14.85	15.31	16.93	18.94	22.66	27.34	32.62	37.92	41.34	44.46	45.42	48.28	50.99	56.89
29	10.99	13.12	14.26	15.57	16.05	17.71	19.77	23.57	28.34	33.71	39.09	42.56	45.72	46.69	49.59	52.34	58.30
30	11.59	13.79	14.95	16.31	16.79	18.49	20.60	24.48	29.34	34.80	40.26	43.77	46.98	47.96	50.89	53.67	59.70
31	12.20	14.46	15.66	17.04	17.54	19.28	21.43	25.39	30.34	35.89	41.42	44.99	48.23	49.23	52.19	55.00	61.10
32	12.81	15.13	16.36	17.78	18.29	20.07	22.27	26.30	31.34	36.97	42.58	46.19	49.48	50.49	53.49	56.33	62.49
33	13.43	15.81	17.07	18.53	19.05	20.87	23.11	27.22	32.34	38.06	43.75	47.40	50.73	51.74	54.78	57.65	63.87
34	14.06	16.50	17.79	19.28	19.81	21.66	23.95	28.14	33.34	39.14	44.90	48.60	51.97	53.00	56.06	58.96	65.25
35	14.69	17.19	18.51	20.03	20.57	22.46	24.80	29.05	34.34	40.22	46.06	49.80	53.20	54.24	57.34	60.27	66.62
40	17.89	20.70	22.16	23.84	24.43	26.51	29.05	33.66	39.33	45.62	51.80	55.76	59.34	60.44	63.69	66.76	73.39
45	21.23	24.31	25.90	27.72	28.37	30.61	33.35	38.29	44.33	50.98	57.51	61.66	65.41	66.56	69.96	73.16	80.07
50	24.65	27.99	29.71	31.66	32.36	34.77	37.69	42.94	49.33	56.33	63.17	67.50	71.42	72.61	76.15	79.49	86.65
55	28.15	31.73	33.57	35.66	36.40	38.96	42.06	47.61	54.33	61.66	68.80	73.31	77.38	78.62	82.29	85.75	93.16
60	31.72	35.53	37.48	39.70	40.48	43.19	46.46	52.29	59.33	66.98	74.40	79.08	83.30	84.58	88.38	91.95	99.60
70	39.02	43.27	45.44	47.89	48.76	51.74	55.33	61.70	69.33	77.58	85.53	90.53	95.02	96.39	100.4	104.2	112.3
80	46.50	51.17	53.54	56.21	57.15	60.39	64.28	71.14	79.33	88.13	96.58	101.9	106.6	108.1	112.3	116.3	124.8
90	54.14	59.20	61.76	64.64	65.65	69.13	73.29	80.62	89.33	98.65	107.6	113.1	118.1	119.6	124.1	128.3	137.2
100	61.87	67.30	70.05	73.13	74.22	77.93	82.36	90.14	99.33	109.1	118.5	124.3	129.6	131.1	135.8	140.2	149.5

Table A.9a 50th Percentiles of the F Distribution with r_1 Numerator and r_2 Denominator Degrees of Freedom: $F_{(0.50; r_1, r_2)}$

r_2 \ r_1:	1	2	3	4	5	6	7	8	9	10	12	15	20	25	30	40	60	120	∞
1	1.000	1.500	1.709	1.823	1.894	1.942	1.977	2.004	2.025	2.042	2.068	2.092	2.119	2.133	2.145	2.157	2.172	2.185	2.201
2	0.667	1.000	1.135	1.207	1.252	1.282	1.305	1.321	1.334	1.345	1.361	1.377	1.393	1.403	1.410	1.418	1.426	1.434	1.443
3	0.585	0.881	1.000	1.063	1.102	1.129	1.148	1.163	1.174	1.183	1.197	1.211	1.225	1.234	1.239	1.246	1.254	1.261	1.268
4	0.549	0.828	0.941	1.000	1.037	1.062	1.080	1.093	1.104	1.113	1.126	1.139	1.152	1.160	1.165	1.172	1.178	1.185	1.192
5	0.528	0.799	0.907	0.965	1.000	1.024	1.041	1.054	1.065	1.073	1.085	1.098	1.111	1.118	1.123	1.130	1.136	1.143	1.149
6	0.515	0.780	0.886	0.942	0.977	1.000	1.017	1.030	1.040	1.048	1.060	1.072	1.085	1.092	1.097	1.103	1.109	1.116	1.122
7	0.506	0.767	0.871	0.926	0.960	0.983	1.000	1.013	1.022	1.030	1.042	1.054	1.066	1.074	1.079	1.085	1.091	1.097	1.103
8	0.499	0.757	0.860	0.915	0.948	0.971	0.988	1.000	1.010	1.018	1.029	1.041	1.053	1.060	1.065	1.071	1.077	1.083	1.089
9	0.494	0.749	0.852	0.906	0.939	0.962	0.978	0.990	1.000	1.008	1.019	1.031	1.043	1.050	1.055	1.061	1.067	1.073	1.079
10	0.490	0.743	0.845	0.899	0.932	0.954	0.971	0.983	0.992	1.000	1.012	1.023	1.035	1.042	1.047	1.053	1.058	1.064	1.070
11	0.486	0.739	0.840	0.893	0.926	0.948	0.964	0.977	0.986	0.994	1.005	1.017	1.028	1.035	1.040	1.046	1.052	1.058	1.064
12	0.484	0.735	0.835	0.888	0.921	0.943	0.959	0.972	0.981	0.989	1.000	1.011	1.023	1.030	1.035	1.041	1.046	1.052	1.058
13	0.481	0.731	0.832	0.885	0.917	0.939	0.955	0.967	0.977	0.984	0.996	1.007	1.019	1.025	1.030	1.036	1.042	1.048	1.053
14	0.479	0.729	0.828	0.881	0.914	0.936	0.952	0.964	0.973	0.980	0.992	1.003	1.015	1.022	1.026	1.032	1.038	1.044	1.049
15	0.478	0.726	0.826	0.878	0.911	0.933	0.948	0.960	0.970	0.977	0.989	1.000	1.011	1.018	1.023	1.029	1.034	1.040	1.046
16	0.476	0.724	0.823	0.876	0.908	0.930	0.946	0.958	0.967	0.975	0.986	0.997	1.009	1.015	1.020	1.026	1.032	1.037	1.043
17	0.475	0.722	0.821	0.874	0.906	0.928	0.943	0.955	0.965	0.972	0.983	0.995	1.006	1.013	1.017	1.023	1.029	1.035	1.040
18	0.474	0.721	0.819	0.872	0.904	0.926	0.941	0.953	0.962	0.970	0.981	0.992	1.004	1.011	1.015	1.021	1.027	1.032	1.038
19	0.473	0.719	0.818	0.870	0.902	0.924	0.939	0.951	0.961	0.968	0.979	0.990	1.002	1.009	1.013	1.019	1.025	1.030	1.036
20	0.472	0.718	0.816	0.868	0.900	0.922	0.938	0.950	0.959	0.966	0.977	0.989	1.000	1.007	1.011	1.017	1.023	1.029	1.034
21	0.471	0.717	0.815	0.867	0.899	0.921	0.936	0.948	0.957	0.965	0.976	0.987	0.998	1.005	1.010	1.015	1.021	1.027	1.033
22	0.470	0.715	0.814	0.866	0.898	0.919	0.935	0.947	0.956	0.963	0.974	0.986	0.997	1.004	1.008	1.014	1.020	1.025	1.031
23	0.470	0.714	0.813	0.864	0.896	0.918	0.934	0.945	0.955	0.962	0.973	0.984	0.996	1.002	1.007	1.013	1.018	1.024	1.030
24	0.469	0.714	0.811	0.863	0.895	0.917	0.932	0.944	0.953	0.961	0.972	0.983	0.994	1.001	1.006	1.011	1.017	1.023	1.028
25	0.468	0.713	0.811	0.862	0.894	0.916	0.931	0.943	0.952	0.960	0.971	0.982	0.993	1.000	1.005	1.010	1.016	1.022	1.027
26	0.468	0.712	0.810	0.861	0.893	0.915	0.930	0.942	0.951	0.959	0.970	0.981	0.992	0.999	1.003	1.009	1.015	1.020	1.026
27	0.467	0.711	0.809	0.861	0.892	0.914	0.930	0.941	0.950	0.958	0.969	0.980	0.991	0.998	1.003	1.008	1.014	1.020	1.025
28	0.467	0.711	0.808	0.860	0.892	0.913	0.929	0.940	0.950	0.957	0.968	0.979	0.990	0.997	1.002	1.007	1.013	1.019	1.024
29	0.467	0.710	0.808	0.859	0.891	0.913	0.928	0.940	0.949	0.956	0.967	0.978	0.990	0.996	1.001	1.006	1.012	1.018	1.023
30	0.466	0.709	0.807	0.858	0.890	0.912	0.927	0.939	0.948	0.955	0.966	0.978	0.989	0.995	1.000	1.006	1.011	1.017	1.023
35	0.465	0.707	0.804	0.856	0.887	0.909	0.924	0.936	0.945	0.952	0.963	0.974	0.986	0.992	0.997	1.002	1.008	1.014	1.019
40	0.463	0.705	0.802	0.854	0.885	0.906	0.922	0.934	0.943	0.950	0.961	0.972	0.983	0.990	0.994	1.000	1.006	1.011	1.017
60	0.460	0.701	0.798	0.849	0.880	0.901	0.917	0.928	0.937	0.945	0.956	0.967	0.978	0.984	0.989	0.994	1.000	1.006	1.011
120	0.458	0.697	0.793	0.844	0.875	0.896	0.912	0.923	0.932	0.939	0.950	0.961	0.972	0.979	0.983	0.989	0.994	1.000	1.006
∞	0.455	0.693	0.789	0.839	0.870	0.891	0.907	0.918	0.927	0.934	0.945	0.956	0.967	0.974	0.978	0.983	0.989	0.995	1.000

Table A.9b 75th Percentiles of the F Distribution with r_1 Numerator and r_2 Denominator Degrees of Freedom: $F_{(0.75;\,r_1,\,r_2)}$

r_2 \ r_1	1	2	3	4	5	6	7	8	9	10	12	15	20	25	30	40	60	120	∞
1	5.828	7.500	8.200	8.581	8.320	8.983	9.102	9.192	9.263	9.320	9.406	9.493	9.581	9.634	9.670	9.714	9.759	9.804	9.849
2	2.571	3.000	3.153	3.232	3.280	3.312	3.335	3.353	3.366	3.377	3.393	3.410	3.426	3.436	3.443	3.451	3.459	3.468	3.476
3	2.024	2.280	2.356	2.390	2.410	2.422	2.430	2.436	2.441	2.445	2.450	2.455	2.460	2.463	2.465	2.467	2.470	2.472	2.474
4	1.807	2.000	2.047	2.064	2.072	2.077	2.079	2.080	2.081	2.082	2.083	2.083	2.083	2.083	2.082	2.082	2.082	2.081	2.081
5	1.692	1.853	1.884	1.893	1.895	1.895	1.894	1.892	1.891	1.890	1.888	1.885	1.882	1.880	1.878	1.876	1.874	1.872	1.869
6	1.621	1.762	1.784	1.787	1.785	1.782	1.779	1.776	1.773	1.771	1.767	1.762	1.757	1.753	1.751	1.748	1.744	1.741	1.737
7	1.573	1.701	1.717	1.716	1.711	1.706	1.701	1.697	1.693	1.690	1.684	1.678	1.671	1.667	1.664	1.659	1.655	1.650	1.645
8	1.538	1.657	1.668	1.664	1.658	1.651	1.645	1.640	1.635	1.631	1.624	1.617	1.609	1.603	1.600	1.595	1.589	1.584	1.578
9	1.512	1.624	1.632	1.625	1.617	1.609	1.602	1.596	1.591	1.586	1.579	1.570	1.561	1.555	1.551	1.545	1.539	1.533	1.526
10	1.491	1.598	1.603	1.595	1.585	1.576	1.569	1.562	1.556	1.551	1.543	1.534	1.524	1.517	1.512	1.506	1.499	1.492	1.484
11	1.475	1.577	1.580	1.570	1.560	1.550	1.542	1.535	1.528	1.523	1.514	1.504	1.493	1.486	1.480	1.474	1.467	1.459	1.451
12	1.461	1.560	1.561	1.550	1.539	1.529	1.520	1.512	1.505	1.500	1.490	1.480	1.468	1.460	1.454	1.447	1.439	1.431	1.422
13	1.450	1.545	1.545	1.534	1.521	1.511	1.501	1.493	1.486	1.480	1.470	1.459	1.447	1.438	1.432	1.425	1.416	1.408	1.398
14	1.440	1.533	1.532	1.519	1.507	1.495	1.485	1.477	1.470	1.463	1.453	1.441	1.428	1.420	1.414	1.405	1.397	1.387	1.377
15	1.432	1.523	1.520	1.507	1.494	1.482	1.472	1.463	1.456	1.449	1.438	1.426	1.413	1.404	1.397	1.389	1.380	1.370	1.359
16	1.425	1.514	1.510	1.497	1.483	1.471	1.460	1.451	1.443	1.437	1.425	1.413	1.399	1.390	1.383	1.374	1.365	1.354	1.343
17	1.419	1.506	1.501	1.487	1.473	1.461	1.450	1.441	1.433	1.426	1.414	1.401	1.387	1.377	1.370	1.361	1.351	1.341	1.329
18	1.413	1.499	1.494	1.479	1.464	1.452	1.441	1.431	1.423	1.416	1.404	1.391	1.376	1.366	1.359	1.350	1.340	1.328	1.316
19	1.408	1.493	1.487	1.472	1.457	1.444	1.432	1.423	1.414	1.407	1.395	1.382	1.367	1.356	1.349	1.339	1.329	1.317	1.305
20	1.404	1.487	1.481	1.465	1.450	1.437	1.425	1.415	1.407	1.399	1.387	1.374	1.358	1.348	1.340	1.330	1.319	1.307	1.295
21	1.400	1.482	1.475	1.459	1.444	1.430	1.419	1.409	1.400	1.392	1.380	1.366	1.350	1.340	1.332	1.322	1.311	1.298	1.285
22	1.396	1.477	1.470	1.454	1.438	1.424	1.413	1.402	1.394	1.386	1.374	1.359	1.343	1.332	1.324	1.314	1.303	1.290	1.276
23	1.393	1.473	1.466	1.449	1.433	1.419	1.407	1.397	1.388	1.380	1.368	1.353	1.337	1.326	1.318	1.307	1.295	1.282	1.268
24	1.390	1.470	1.462	1.445	1.428	1.414	1.402	1.392	1.383	1.375	1.362	1.347	1.331	1.319	1.311	1.300	1.288	1.275	1.261
25	1.387	1.466	1.458	1.441	1.424	1.410	1.398	1.387	1.378	1.370	1.357	1.342	1.325	1.314	1.306	1.294	1.282	1.269	1.254
26	1.384	1.463	1.454	1.437	1.420	1.406	1.394	1.383	1.374	1.366	1.352	1.337	1.320	1.309	1.300	1.289	1.277	1.263	1.248
27	1.382	1.460	1.451	1.433	1.417	1.402	1.390	1.379	1.370	1.361	1.348	1.333	1.315	1.304	1.295	1.284	1.271	1.257	1.242
28	1.380	1.457	1.448	1.430	1.413	1.399	1.386	1.375	1.366	1.358	1.344	1.329	1.311	1.299	1.291	1.279	1.266	1.252	1.236
29	1.378	1.455	1.445	1.427	1.410	1.395	1.383	1.372	1.362	1.354	1.340	1.325	1.307	1.295	1.286	1.275	1.262	1.247	1.231
30	1.376	1.452	1.443	1.424	1.407	1.392	1.380	1.369	1.359	1.351	1.337	1.321	1.303	1.291	1.282	1.270	1.257	1.242	1.226
35	1.368	1.443	1.432	1.413	1.395	1.380	1.367	1.355	1.345	1.337	1.323	1.306	1.288	1.275	1.266	1.253	1.239	1.223	1.205
40	1.363	1.435	1.424	1.404	1.386	1.371	1.357	1.345	1.335	1.327	1.312	1.295	1.276	1.263	1.253	1.240	1.225	1.208	1.189
60	1.349	1.419	1.405	1.385	1.366	1.349	1.335	1.323	1.312	1.303	1.287	1.269	1.248	1.234	1.223	1.208	1.191	1.172	1.148
120	1.336	1.402	1.387	1.365	1.345	1.328	1.313	1.300	1.289	1.279	1.262	1.243	1.220	1.204	1.192	1.175	1.156	1.131	1.099
∞	1.323	1.387	1.370	1.347	1.325	1.307	1.291	1.278	1.266	1.255	1.237	1.217	1.192	1.174	1.160	1.141	1.117	1.085	1.014

Table A.9b gives $F_{(\gamma;\,r_1,\,r_2)}$ for $\gamma = 0.75$. For $\gamma = 0.25$ use the relationship $F_{(0.25;\,r_1,\,r_2)} = 1/F_{(0.75;\,r_2,\,r_1)}$.

Table A.9c 90th Percentiles of the F Distribution with r_1 Numerator and r_2 Denominator Degrees of Freedom: $F_{(0.90; r_1, r_2)}$

$r_2 \backslash r_1$	1	2	3	4	5	6	7	8	9	10	12	15	20	25	30	40	60	120	∞
1	39.86	49.50	53.59	55.83	57.24	58.20	58.91	59.44	59.86	60.19	60.71	61.22	61.74	62.05	62.26	62.53	62.79	63.06	63.32
2	8.526	9.000	9.162	9.243	9.293	9.326	9.349	9.367	9.381	9.392	9.408	9.425	9.441	9.451	9.458	9.466	9.475	9.483	9.491
3	5.538	5.462	5.391	5.343	5.309	5.285	5.266	5.252	5.240	5.230	5.216	5.199	5.185	5.174	5.168	5.160	5.151	5.143	5.134
4	4.545	4.325	4.191	4.107	4.051	4.010	3.979	3.955	3.936	3.920	3.896	3.870	3.845	3.828	3.818	3.804	3.790	3.775	3.761
5	4.060	3.780	3.620	3.520	3.453	3.404	3.368	3.339	3.316	3.297	3.268	3.238	3.207	3.187	3.174	3.158	3.140	3.123	3.105
6	3.776	3.463	3.289	3.181	3.107	3.055	3.014	2.983	2.958	2.937	2.905	2.871	2.836	2.815	2.800	2.781	2.762	2.742	2.722
7	3.589	3.257	3.074	2.960	2.883	2.827	2.785	2.751	2.725	2.703	2.668	2.632	2.595	2.571	2.555	2.535	2.514	2.493	2.471
8	3.458	3.113	2.924	2.806	2.726	2.668	2.624	2.589	2.561	2.538	2.502	2.464	2.425	2.400	2.383	2.361	2.339	2.316	2.293
9	3.360	3.006	2.813	2.693	2.611	2.551	2.505	2.469	2.440	2.416	2.379	2.340	2.298	2.272	2.255	2.232	2.208	2.184	2.160
10	3.285	2.924	2.728	2.605	2.522	2.461	2.414	2.377	2.347	2.323	2.284	2.243	2.201	2.174	2.155	2.132	2.107	2.082	2.056
11	3.225	2.860	2.660	2.536	2.451	2.389	2.342	2.304	2.274	2.248	2.209	2.167	2.123	2.095	2.076	2.052	2.026	2.000	1.972
12	3.177	2.807	2.606	2.480	2.394	2.331	2.283	2.245	2.213	2.188	2.147	2.105	2.060	2.031	2.012	1.986	1.960	1.932	1.904
13	3.136	2.763	2.560	2.434	2.347	2.283	2.234	2.195	2.164	2.138	2.097	2.053	2.007	1.978	1.958	1.931	1.904	1.876	1.847
14	3.102	2.726	2.522	2.395	2.307	2.242	2.193	2.154	2.122	2.095	2.054	2.010	1.962	1.933	1.912	1.885	1.857	1.828	1.798
15	3.073	2.695	2.490	2.361	2.273	2.208	2.158	2.119	2.086	2.059	2.017	1.972	1.924	1.894	1.873	1.845	1.817	1.787	1.755
16	3.048	2.668	2.462	2.333	2.244	2.178	2.128	2.088	2.055	2.028	1.985	1.940	1.891	1.860	1.839	1.811	1.782	1.751	1.719
17	3.026	2.645	2.437	2.308	2.218	2.152	2.102	2.061	2.028	2.001	1.958	1.912	1.862	1.831	1.809	1.781	1.751	1.719	1.686
18	3.007	2.624	2.416	2.286	2.196	2.130	2.079	2.038	2.005	1.977	1.933	1.887	1.837	1.805	1.783	1.754	1.723	1.691	1.657
19	2.990	2.606	2.397	2.266	2.176	2.109	2.058	2.017	1.984	1.956	1.912	1.865	1.814	1.782	1.759	1.730	1.699	1.666	1.631
20	2.975	2.589	2.380	2.249	2.158	2.091	2.040	1.999	1.965	1.937	1.892	1.845	1.794	1.761	1.738	1.708	1.677	1.643	1.608
21	2.961	2.575	2.365	2.233	2.142	2.075	2.023	1.982	1.948	1.920	1.875	1.827	1.776	1.742	1.719	1.689	1.657	1.623	1.587
22	2.949	2.561	2.351	2.219	2.128	2.061	2.008	1.967	1.933	1.904	1.859	1.811	1.759	1.726	1.702	1.671	1.639	1.604	1.567
23	2.937	2.549	2.339	2.207	2.115	2.047	1.995	1.953	1.919	1.890	1.845	1.796	1.744	1.710	1.686	1.655	1.622	1.587	1.550
24	2.927	2.538	2.327	2.195	2.103	2.035	1.983	1.941	1.906	1.878	1.832	1.783	1.730	1.696	1.672	1.641	1.607	1.571	1.533
25	2.918	2.528	2.317	2.184	2.092	2.024	1.971	1.929	1.895	1.866	1.820	1.771	1.718	1.683	1.659	1.627	1.593	1.557	1.518
26	2.909	2.519	2.307	2.175	2.082	2.014	1.961	1.919	1.884	1.855	1.809	1.760	1.706	1.671	1.647	1.615	1.581	1.544	1.504
27	2.901	2.511	2.299	2.165	2.073	2.005	1.951	1.909	1.874	1.845	1.799	1.749	1.695	1.660	1.636	1.603	1.569	1.531	1.491
28	2.894	2.503	2.291	2.157	2.064	1.996	1.943	1.900	1.865	1.836	1.790	1.740	1.685	1.650	1.625	1.592	1.558	1.520	1.479
29	2.887	2.495	2.283	2.149	2.057	1.988	1.935	1.892	1.857	1.827	1.781	1.731	1.676	1.641	1.615	1.583	1.547	1.509	1.468
30	2.881	2.489	2.276	2.142	2.049	1.980	1.927	1.884	1.849	1.819	1.773	1.722	1.667	1.632	1.606	1.573	1.538	1.499	1.457
35	2.855	2.461	2.247	2.113	2.019	1.950	1.896	1.852	1.817	1.787	1.739	1.688	1.632	1.595	1.569	1.535	1.497	1.457	1.412
40	2.835	2.440	2.226	2.091	1.997	1.927	1.873	1.829	1.793	1.763	1.715	1.662	1.605	1.568	1.541	1.506	1.467	1.425	1.378
60	2.791	2.393	2.177	2.041	1.946	1.875	1.819	1.775	1.738	1.707	1.657	1.603	1.543	1.504	1.476	1.437	1.395	1.348	1.292
120	2.748	2.347	2.130	1.992	1.896	1.824	1.767	1.722	1.684	1.652	1.601	1.545	1.482	1.440	1.409	1.368	1.320	1.265	1.194
∞	2.706	2.303	2.084	1.945	1.848	1.775	1.717	1.671	1.632	1.599	1.547	1.488	1.421	1.376	1.343	1.296	1.241	1.170	1.027

Table A.9c gives $F_{(\gamma; r_1, r_2)}$ for $\gamma = 0.90$. For $\gamma = 0.10$ use the relationship $F_{(0.10; r_1, r_2)} = 1/F_{(0.90; r_2, r_1)}$.

Table A.9d 95th Percentiles of the F Distribution with r_1 Numerator and r_2 Denominator Degrees of Freedom: $F_{(0.95;r_1,r_2)}$

r_2 \ r_1	1	2	3	4	5	6	7	8	9	10	12	15	20	25	30	40	60	120	∞
1	161.4	199.5	215.7	224.6	230.2	234.0	236.8	238.9	240.5	241.9	243.9	245.9	248.0	249.3	250.1	251.1	252.2	253.3	254.3
2	18.51	19.00	19.16	19.25	19.30	19.33	19.35	19.37	19.38	19.40	19.41	19.43	19.45	19.46	19.46	19.47	19.48	19.49	19.50
3	10.13	9.552	9.276	9.117	9.013	8.942	8.886	8.845	8.813	8.787	8.745	8.701	8.663	8.633	8.617	8.594	8.572	8.549	8.527
4	7.709	6.944	6.591	6.388	6.256	6.163	6.094	6.041	5.999	5.965	5.911	5.858	5.803	5.769	5.745	5.717	5.688	5.658	5.628
5	6.608	5.786	5.410	5.192	5.051	4.950	4.875	4.819	4.772	4.735	4.677	4.619	4.558	4.521	4.496	4.463	4.431	4.398	4.365
6	5.987	5.143	4.757	4.534	4.387	4.284	4.207	4.147	4.099	4.060	4.000	3.938	3.874	3.835	3.808	3.774	3.740	3.705	3.669
7	5.591	4.737	4.347	4.120	3.972	3.866	3.787	3.726	3.676	3.637	3.575	3.511	3.445	3.404	3.376	3.340	3.305	3.267	3.230
8	5.318	4.459	4.066	3.838	3.687	3.581	3.500	3.438	3.388	3.347	3.284	3.218	3.150	3.108	3.080	3.043	3.005	2.967	2.928
9	5.117	4.256	3.863	3.633	3.482	3.374	3.293	3.229	3.179	3.137	3.073	3.006	2.936	2.893	2.864	2.826	2.788	2.748	2.707
10	4.965	4.103	3.708	3.478	3.326	3.217	3.135	3.072	3.020	2.978	2.913	2.845	2.774	2.730	2.700	2.661	2.621	2.580	2.538
11	4.844	3.982	3.588	3.357	3.204	3.095	3.012	2.948	2.896	2.854	2.788	2.719	2.646	2.601	2.570	2.531	2.490	2.448	2.405
12	4.747	3.885	3.490	3.259	3.106	2.996	2.913	2.848	2.796	2.753	2.687	2.617	2.544	2.498	2.466	2.426	2.384	2.341	2.297
13	4.667	3.806	3.411	3.179	3.026	2.915	2.832	2.767	2.714	2.671	2.604	2.533	2.459	2.412	2.380	2.339	2.297	2.252	2.207
14	4.600	3.739	3.344	3.112	2.958	2.848	2.764	2.699	2.646	2.602	2.534	2.463	2.388	2.341	2.308	2.266	2.223	2.178	2.131
15	4.543	3.682	3.287	3.056	2.901	2.791	2.707	2.641	2.588	2.544	2.475	2.403	2.327	2.280	2.247	2.204	2.160	2.114	2.067
16	4.494	3.634	3.239	3.007	2.852	2.741	2.657	2.591	2.538	2.494	2.425	2.352	2.276	2.227	2.194	2.151	2.106	2.059	2.010
17	4.451	3.592	3.197	2.965	2.810	2.699	2.614	2.548	2.494	2.450	2.381	2.308	2.230	2.182	2.148	2.104	2.058	2.011	1.961
18	4.414	3.555	3.160	2.928	2.773	2.661	2.577	2.510	2.456	2.412	2.342	2.269	2.191	2.141	2.107	2.063	2.017	1.968	1.918
19	4.381	3.522	3.127	2.895	2.740	2.628	2.543	2.477	2.423	2.378	2.308	2.234	2.155	2.106	2.071	2.026	1.980	1.930	1.879
20	4.351	3.493	3.098	2.866	2.711	2.599	2.514	2.447	2.393	2.348	2.278	2.203	2.124	2.074	2.039	1.994	1.946	1.896	1.844
21	4.325	3.467	3.072	2.840	2.685	2.573	2.488	2.420	2.366	2.321	2.250	2.176	2.096	2.045	2.010	1.965	1.916	1.866	1.812
22	4.301	3.443	3.049	2.817	2.661	2.549	2.464	2.397	2.342	2.297	2.226	2.151	2.071	2.020	1.984	1.938	1.889	1.838	1.784
23	4.279	3.422	3.028	2.796	2.640	2.528	2.442	2.375	2.320	2.275	2.204	2.128	2.048	1.996	1.961	1.914	1.865	1.813	1.758
24	4.260	3.403	3.009	2.776	2.621	2.508	2.423	2.355	2.300	2.255	2.183	2.108	2.027	1.975	1.939	1.892	1.842	1.790	1.734
25	4.242	3.385	2.991	2.759	2.603	2.490	2.405	2.337	2.282	2.236	2.165	2.089	2.007	1.955	1.919	1.872	1.822	1.768	1.712
26	4.225	3.369	2.975	2.743	2.587	2.474	2.388	2.321	2.265	2.220	2.148	2.072	1.990	1.938	1.901	1.853	1.803	1.749	1.691
27	4.210	3.354	2.960	2.728	2.572	2.459	2.373	2.305	2.250	2.204	2.132	2.056	1.974	1.921	1.884	1.836	1.785	1.731	1.673
28	4.196	3.340	2.947	2.714	2.558	2.445	2.359	2.291	2.236	2.190	2.118	2.041	1.959	1.906	1.869	1.820	1.769	1.714	1.655
29	4.183	3.328	2.934	2.701	2.545	2.432	2.346	2.278	2.223	2.177	2.104	2.027	1.945	1.891	1.854	1.805	1.754	1.698	1.638
30	4.171	3.316	2.922	2.690	2.534	2.421	2.334	2.266	2.211	2.165	2.092	2.015	1.932	1.878	1.841	1.792	1.740	1.684	1.623
35	4.121	3.267	2.874	2.641	2.485	2.372	2.285	2.217	2.161	2.114	2.041	1.963	1.878	1.824	1.786	1.735	1.681	1.623	1.559
40	4.085	3.232	2.839	2.606	2.450	2.336	2.249	2.180	2.124	2.077	2.003	1.924	1.839	1.783	1.744	1.693	1.637	1.577	1.510
60	4.001	3.150	2.758	2.525	2.368	2.254	2.166	2.097	2.040	1.993	1.917	1.836	1.748	1.690	1.649	1.594	1.534	1.467	1.390
120	3.920	3.072	2.680	2.447	2.290	2.175	2.087	2.016	1.959	1.910	1.834	1.750	1.659	1.598	1.554	1.495	1.429	1.352	1.255
∞	3.842	2.997	2.606	2.373	2.215	2.100	2.011	1.939	1.881	1.832	1.753	1.667	1.572	1.507	1.460	1.395	1.320	1.223	1.035

Table A.9d gives $F_{(\gamma;r_1,r_2)}$ for $\gamma = 0.95$. For $\gamma = 0.05$ use the relationship $F_{(0.05;r_1,r_2)} = 1/F_{(0.95;r_2,r_1)}$.

Table A.9e 97.5th Percentiles of the F Distribution with r_1 Numerator and r_2 Denominator Degrees of Freedom: $F_{(0.975; r_1, r_2)}$

r_2 \ r_1:	1	2	3	4	5	6	7	8	9	10	12	15	20	25	30	40	60	120	∞
1	647.8	799.5	864.2	899.6	921.8	937.1	948.2	956.7	963.3	968.6	976.7	984.9	993.1	998.1	1001.	1006.	1010.	1014.	1018.
2	38.51	39.00	39.17	39.25	39.30	39.33	39.36	39.37	39.39	39.40	39.41	39.43	39.45	39.46	39.46	39.47	39.48	39.49	39.50
3	17.44	16.04	15.44	15.10	14.88	14.74	14.63	14.54	14.47	14.42	14.33	14.26	14.17	14.11	14.08	14.04	13.99	13.95	13.90
4	12.22	10.65	9.979	9.604	9.364	9.197	9.074	8.980	8.905	8.845	8.752	8.657	8.559	8.502	8.463	8.411	8.360	8.309	8.258
5	10.01	8.434	7.764	7.388	7.147	6.978	6.854	6.757	6.681	6.620	6.525	6.427	6.329	6.268	6.226	6.174	6.123	6.069	6.016
6	8.813	7.260	6.599	6.227	5.988	5.820	5.696	5.600	5.523	5.461	5.367	5.269	5.168	5.107	5.065	5.012	4.959	4.904	4.850
7	8.073	6.542	5.890	5.523	5.285	5.119	4.995	4.900	4.823	4.761	4.666	4.568	4.466	4.404	4.362	4.309	4.254	4.199	4.143
8	7.571	6.059	5.416	5.053	4.817	4.652	4.528	4.433	4.357	4.295	4.199	4.101	3.999	3.937	3.894	3.840	3.784	3.728	3.671
9	7.209	5.715	5.078	4.718	4.484	4.320	4.197	4.102	4.026	3.964	3.868	3.769	3.667	3.604	3.561	3.506	3.449	3.392	3.334
10	6.937	5.456	4.826	4.468	4.236	4.072	3.950	3.855	3.779	3.717	3.621	3.522	3.418	3.355	3.311	3.256	3.199	3.140	3.081
11	6.724	5.256	4.630	4.275	4.044	3.881	3.759	3.664	3.588	3.526	3.430	3.330	3.226	3.162	3.118	3.062	3.003	2.944	2.884
12	6.554	5.096	4.474	4.121	3.891	3.728	3.607	3.512	3.436	3.373	3.277	3.177	3.073	3.008	2.963	2.907	2.848	2.787	2.726
13	6.414	4.965	4.347	3.996	3.767	3.604	3.483	3.388	3.312	3.250	3.153	3.053	2.948	2.882	2.837	2.780	2.720	2.659	2.596
14	6.298	4.857	4.242	3.892	3.663	3.501	3.380	3.285	3.209	3.147	3.050	2.949	2.844	2.778	2.732	2.674	2.614	2.552	2.488
15	6.200	4.765	4.153	3.804	3.577	3.415	3.293	3.199	3.123	3.060	2.963	2.862	2.756	2.689	2.644	2.585	2.524	2.461	2.396
16	6.115	4.687	4.077	3.729	3.502	3.341	3.219	3.125	3.049	2.986	2.889	2.788	2.681	2.614	2.568	2.509	2.447	2.383	2.317
17	6.042	4.619	4.011	3.665	3.438	3.277	3.156	3.061	2.985	2.922	2.825	2.723	2.616	2.548	2.502	2.442	2.380	2.315	2.248
18	5.978	4.560	3.954	3.608	3.382	3.221	3.100	3.005	2.929	2.866	2.769	2.667	2.559	2.491	2.445	2.384	2.321	2.256	2.188
19	5.922	4.508	3.903	3.559	3.333	3.172	3.051	2.956	2.880	2.817	2.720	2.617	2.509	2.441	2.394	2.333	2.270	2.203	2.134
20	5.871	4.461	3.859	3.515	3.289	3.128	3.007	2.913	2.837	2.774	2.676	2.573	2.464	2.396	2.349	2.287	2.223	2.156	2.086
21	5.827	4.420	3.819	3.475	3.250	3.089	2.969	2.874	2.798	2.735	2.637	2.534	2.425	2.356	2.308	2.246	2.182	2.114	2.043
22	5.786	4.383	3.783	3.440	3.215	3.055	2.934	2.839	2.763	2.700	2.602	2.498	2.389	2.320	2.272	2.210	2.145	2.076	2.004
23	5.750	4.349	3.751	3.408	3.183	3.023	2.902	2.808	2.731	2.668	2.570	2.466	2.357	2.287	2.239	2.176	2.111	2.042	1.969
24	5.717	4.319	3.721	3.379	3.155	2.995	2.874	2.779	2.703	2.640	2.541	2.437	2.327	2.257	2.209	2.146	2.080	2.010	1.936
25	5.686	4.291	3.694	3.353	3.129	2.969	2.848	2.753	2.677	2.614	2.515	2.411	2.300	2.230	2.182	2.118	2.052	1.981	1.907
26	5.659	4.265	3.670	3.329	3.105	2.945	2.824	2.729	2.653	2.590	2.491	2.387	2.276	2.206	2.157	2.093	2.026	1.954	1.879
27	5.633	4.242	3.647	3.307	3.083	2.923	2.802	2.707	2.631	2.568	2.469	2.364	2.253	2.183	2.133	2.069	2.002	1.930	1.854
28	5.610	4.221	3.626	3.286	3.063	2.903	2.782	2.687	2.611	2.547	2.448	2.344	2.232	2.161	2.112	2.048	1.980	1.907	1.830
29	5.588	4.201	3.607	3.267	3.044	2.884	2.763	2.669	2.592	2.529	2.429	2.325	2.213	2.142	2.092	2.028	1.959	1.886	1.808
30	5.568	4.182	3.589	3.250	3.026	2.867	2.746	2.651	2.575	2.511	2.412	2.307	2.195	2.124	2.074	2.009	1.940	1.866	1.788
35	5.485	4.106	3.517	3.178	2.956	2.796	2.676	2.581	2.504	2.440	2.341	2.235	2.122	2.049	1.999	1.932	1.861	1.785	1.703
40	5.424	4.051	3.463	3.126	2.904	2.744	2.624	2.529	2.452	2.388	2.288	2.182	2.068	1.994	1.943	1.875	1.803	1.724	1.638
60	5.286	3.925	3.343	3.008	2.786	2.627	2.507	2.412	2.334	2.270	2.169	2.061	1.944	1.869	1.815	1.744	1.667	1.581	1.484
120	5.152	3.805	3.227	2.894	2.674	2.515	2.395	2.299	2.222	2.157	2.055	1.945	1.825	1.746	1.690	1.614	1.530	1.433	1.312
∞	5.026	3.690	3.118	2.787	2.568	2.410	2.289	2.193	2.115	2.050	1.946	1.834	1.710	1.627	1.568	1.485	1.390	1.271	1.042

Table A.9e gives $F_{(\gamma, r_1, r_2)}$ for $\gamma = 0.975$. For $\gamma = 0.025$ use the relationship $F_{(0.025, r_1, r_2)} = 1/F_{(0.975, r_2, r_1)}$.

Table A.9f 99th Percentiles of the F Distribution with r_1 Numerator and r_2 Denominator Degrees of Freedom: $F_{(0.99; r_1, r_2)}$

r_2 \ r_1:	1	2	3	4	5	6	7	8	9	10	12	15	20	25	30	40	60	120	∞
1	4052.	5000.	5403.	5625.	5764.	5859.	5928.	5981.	6022.	6056.	6106.	6157.	6209.	6240.	6261.	6287.	6313.	6339.	6366.
2	98.50	99.00	99.17	99.25	99.30	99.33	99.35	99.37	99.39	99.40	99.42	99.43	99.45	99.46	99.47	99.47	99.48	99.49	99.50
3	34.12	30.82	29.46	28.71	28.24	27.91	27.67	27.50	27.34	27.22	27.03	26.85	26.67	26.58	26.50	26.41	26.32	26.22	26.13
4	21.20	18.00	16.69	15.98	15.52	15.21	14.93	14.80	14.66	14.55	14.37	14.19	14.02	13.91	13.84	13.75	13.65	13.56	13.46
5	16.26	13.27	12.06	11.39	10.97	10.67	10.46	10.29	10.16	10.05	9.890	9.722	9.552	9.448	9.377	9.297	9.202	9.112	9.022
6	13.75	10.92	9.779	9.149	8.745	8.466	8.260	8.102	7.976	7.874	7.718	7.559	7.397	7.295	7.228	7.145	7.057	6.969	6.881
7	12.25	9.547	8.451	7.847	7.460	7.192	6.993	6.840	6.719	6.620	6.469	6.314	6.156	6.057	5.992	5.908	5.822	5.737	5.651
8	11.26	8.649	7.591	7.006	6.632	6.371	6.178	6.029	5.911	5.814	5.667	5.515	5.360	5.263	5.199	5.116	5.033	4.946	4.860
9	10.56	8.022	6.992	6.422	6.057	5.802	5.613	5.467	5.351	5.257	5.112	4.962	4.808	4.713	4.648	4.566	4.484	4.398	4.312
10	10.04	7.559	6.552	5.994	5.636	5.386	5.200	5.057	4.943	4.849	4.706	4.558	4.406	4.311	4.247	4.166	4.083	3.996	3.910
11	9.646	7.206	6.217	5.668	5.316	5.069	4.886	4.744	4.632	4.539	4.398	4.251	4.099	4.005	3.941	3.860	3.776	3.690	3.604
12	9.330	6.927	5.953	5.412	5.064	4.820	4.640	4.499	4.387	4.296	4.155	4.010	3.858	3.765	3.701	3.619	3.536	3.449	3.362
13	9.074	6.701	5.740	5.205	4.861	4.620	4.441	4.302	4.191	4.100	3.960	3.815	3.664	3.571	3.507	3.425	3.342	3.254	3.167
14	8.862	6.515	5.564	5.035	4.695	4.456	4.278	4.140	4.030	3.939	3.800	3.656	3.505	3.412	3.347	3.266	3.181	3.094	3.005
15	8.683	6.359	5.417	4.893	4.556	4.318	4.142	4.005	3.895	3.805	3.666	3.522	3.372	3.278	3.214	3.132	3.047	2.960	2.870
16	8.531	6.226	5.292	4.772	4.438	4.202	4.026	3.890	3.780	3.691	3.553	3.409	3.259	3.165	3.101	3.018	2.933	2.844	2.754
17	8.400	6.112	5.185	4.669	4.336	4.102	3.927	3.791	3.682	3.593	3.455	3.312	3.161	3.068	3.003	2.920	2.835	2.746	2.654
18	8.285	6.013	5.092	4.579	4.248	4.015	3.841	3.706	3.597	3.508	3.371	3.227	3.077	2.983	2.919	2.835	2.749	2.660	2.567
19	8.185	5.926	5.010	4.500	4.171	3.938	3.765	3.631	3.523	3.434	3.297	3.153	3.003	2.909	2.844	2.761	2.674	2.584	2.491
20	8.096	5.849	4.938	4.431	4.103	3.871	3.699	3.564	3.457	3.368	3.231	3.088	2.938	2.843	2.779	2.695	2.608	2.517	2.423
21	8.017	5.780	4.874	4.369	4.042	3.812	3.640	3.506	3.398	3.310	3.173	3.030	2.880	2.785	2.720	2.636	2.549	2.457	2.362
22	7.945	5.719	4.817	4.313	3.988	3.758	3.587	3.453	3.346	3.258	3.121	2.978	2.827	2.733	2.668	2.583	2.495	2.403	2.307
23	7.881	5.664	4.765	4.263	3.939	3.710	3.539	3.406	3.299	3.211	3.074	2.931	2.781	2.686	2.620	2.536	2.447	2.354	2.257
24	7.823	5.614	4.718	4.218	3.895	3.667	3.496	3.363	3.256	3.168	3.032	2.889	2.738	2.643	2.577	2.492	2.404	2.310	2.212
25	7.770	5.568	4.676	4.177	3.855	3.627	3.457	3.324	3.217	3.129	2.993	2.850	2.699	2.604	2.538	2.453	2.364	2.269	2.171
26	7.721	5.526	4.637	4.140	3.818	3.591	3.421	3.288	3.182	3.094	2.958	2.815	2.664	2.569	2.503	2.417	2.327	2.233	2.133
27	7.677	5.488	4.601	4.106	3.785	3.558	3.388	3.256	3.149	3.062	2.926	2.783	2.632	2.536	2.470	2.384	2.294	2.198	2.098
28	7.636	5.453	4.568	4.074	3.754	3.527	3.358	3.226	3.120	3.032	2.896	2.753	2.602	2.506	2.440	2.354	2.263	2.167	2.066
29	7.598	5.420	4.538	4.045	3.725	3.499	3.330	3.198	3.092	3.005	2.869	2.726	2.574	2.478	2.412	2.325	2.234	2.138	2.036
30	7.562	5.390	4.510	4.018	3.699	3.473	3.304	3.173	3.066	2.979	2.843	2.700	2.549	2.453	2.386	2.299	2.208	2.111	2.008
35	7.419	5.268	4.396	3.908	3.592	3.368	3.200	3.069	2.963	2.876	2.740	2.597	2.445	2.348	2.281	2.193	2.099	2.000	1.893
40	7.314	5.179	4.313	3.828	3.514	3.291	3.124	2.993	2.888	2.801	2.665	2.522	2.369	2.271	2.203	2.114	2.019	1.917	1.806
60	7.077	4.977	4.126	3.649	3.339	3.119	2.953	2.823	2.718	2.632	2.496	2.352	2.198	2.098	2.028	1.936	1.836	1.726	1.602
120	6.851	4.787	3.949	3.480	3.174	2.956	2.792	2.663	2.559	2.472	2.336	2.192	2.035	1.933	1.860	1.763	1.656	1.533	1.383
∞	6.638	4.608	3.784	3.321	3.019	2.804	2.641	2.513	2.409	2.323	2.187	2.041	1.880	1.775	1.699	1.595	1.475	1.328	1.050

Table A.9f gives $F_{(\gamma; r_1, r_2)}$ for $\gamma = 0.99$. For $\gamma = 0.01$ use the relationship $F_{(0.01; r_1, r_2)} = 1/F_{(0.99; r_2, r_1)}$.

Table A.10a Factors $g_{(1-\alpha; p, n)}$ for Calculating Normal Distribution Two-Sided $100(1 - \alpha)\%$ Tolerance Intervals (to Control the Center of the Distribution)

n	p = 0.500					p = 0.700					p = 0.800					n
$1-\alpha$:	0.50	0.80	0.90	0.95	0.99	0.50	0.80	0.90	0.95	0.99	0.50	0.80	0.90	0.95	0.99	
2	1.243	3.369	6.808	13.652	68.316	1.865	5.023	10.142	20.331	101.732	2.275	6.110	12.333	24.722	123.699	2
3	0.942	1.700	2.492	3.585	8.122	1.430	2.562	3.747	5.382	12.181	1.755	3.134	4.577	6.572	14.867	3
4	0.852	1.335	1.766	2.288	4.028	1.300	2.026	2.673	3.456	6.073	1.600	2.486	3.276	4.233	7.431	4
5	0.808	1.173	1.473	1.812	2.824	1.236	1.788	2.239	2.750	4.274	1.523	2.198	2.750	3.375	5.240	5
6	0.782	1.081	1.314	1.566	2.270	1.198	1.651	2.003	2.384	3.446	1.477	2.034	2.464	2.930	4.231	6
7	0.764	1.021	1.213	1.415	1.954	1.172	1.562	1.853	2.159	2.973	1.446	1.925	2.282	2.657	3.656	7
8	0.752	0.979	1.143	1.313	1.750	1.153	1.499	1.749	2.006	2.668	1.423	1.849	2.156	2.472	3.284	8
9	0.742	0.947	1.092	1.239	1.608	1.139	1.451	1.672	1.896	2.455	1.407	1.791	2.062	2.337	3.024	9
10	0.735	0.923	1.053	1.183	1.503	1.128	1.414	1.613	1.811	2.297	1.393	1.746	1.990	2.234	2.831	10
11	0.729	0.903	1.021	1.139	1.422	1.119	1.385	1.566	1.744	2.175	1.382	1.710	1.932	2.152	2.682	11
12	0.724	0.886	0.996	1.103	1.357	1.112	1.360	1.527	1.690	2.078	1.374	1.680	1.885	2.086	2.563	12
13	0.720	0.873	0.974	1.073	1.305	1.106	1.339	1.495	1.645	1.999	1.366	1.654	1.846	2.031	2.466	13
14	0.717	0.861	0.956	1.048	1.261	1.100	1.322	1.467	1.608	1.933	1.360	1.633	1.812	1.985	2.386	14
15	0.714	0.851	0.941	1.027	1.224	1.096	1.307	1.444	1.575	1.877	1.355	1.614	1.783	1.945	2.317	15
16	0.711	0.842	0.927	1.008	1.193	1.092	1.293	1.423	1.547	1.829	1.350	1.598	1.758	1.911	2.259	16
17	0.709	0.835	0.915	0.992	1.165	1.089	1.282	1.405	1.522	1.788	1.346	1.584	1.736	1.881	2.207	17
18	0.707	0.828	0.905	0.978	1.141	1.086	1.271	1.389	1.501	1.751	1.342	1.571	1.717	1.854	2.163	18
19	0.705	0.822	0.895	0.965	1.120	1.083	1.262	1.375	1.481	1.719	1.339	1.559	1.699	1.830	2.123	19
20	0.704	0.816	0.887	0.953	1.101	1.081	1.253	1.362	1.464	1.690	1.336	1.549	1.683	1.809	2.087	20
21	0.702	0.811	0.879	0.943	1.084	1.079	1.246	1.350	1.448	1.664	1.333	1.540	1.669	1.789	2.055	21
22	0.701	0.806	0.872	0.934	1.068	1.077	1.239	1.340	1.434	1.640	1.331	1.531	1.656	1.772	2.026	22
23	0.700	0.802	0.866	0.925	1.054	1.075	1.232	1.330	1.420	1.619	1.329	1.523	1.644	1.755	2.000	23
24	0.699	0.798	0.860	0.917	1.042	1.073	1.226	1.321	1.408	1.599	1.327	1.516	1.633	1.741	1.976	24
25	0.698	0.795	0.855	0.910	1.030	1.072	1.221	1.313	1.397	1.581	1.325	1.509	1.623	1.727	1.954	25
26	0.697	0.791	0.850	0.903	1.019	1.070	1.216	1.305	1.387	1.565	1.323	1.503	1.613	1.714	1.934	26
27	0.696	0.788	0.845	0.897	1.009	1.069	1.211	1.298	1.378	1.550	1.322	1.497	1.604	1.703	1.915	27
28	0.695	0.785	0.841	0.891	1.000	1.068	1.207	1.291	1.369	1.535	1.320	1.492	1.596	1.692	1.898	28
29	0.694	0.783	0.837	0.886	0.991	1.067	1.202	1.285	1.360	1.522	1.319	1.487	1.589	1.682	1.882	29
30	0.694	0.780	0.833	0.881	0.983	1.066	1.199	1.279	1.353	1.510	1.318	1.482	1.581	1.672	1.866	30
35	0.691	0.770	0.817	0.859	0.950	1.061	1.182	1.255	1.320	1.459	1.312	1.462	1.551	1.632	1.803	35
40	0.689	0.761	0.805	0.843	0.925	1.058	1.170	1.236	1.296	1.420	1.308	1.446	1.528	1.602	1.756	40
50	0.686	0.750	0.787	0.820	0.889	1.054	1.152	1.209	1.260	1.365	1.303	1.424	1.495	1.558	1.688	50
60	0.684	0.741	0.775	0.804	0.864	1.051	1.139	1.190	1.235	1.327	1.299	1.408	1.471	1.527	1.641	60
120	0.679	0.718	0.740	0.759	0.797	1.044	1.104	1.137	1.166	1.225	1.290	1.365	1.406	1.442	1.514	120
240	0.677	0.704	0.719	0.731	0.756	1.040	1.082	1.104	1.124	1.162	1.286	1.337	1.365	1.390	1.437	240
480	0.676	0.694	0.705	0.713	0.730	1.038	1.067	1.083	1.096	1.122	1.284	1.320	1.339	1.355	1.387	480
∞	0.674	0.674	0.674	0.674	0.674	1.036	1.036	1.036	1.036	1.036	1.282	1.282	1.282	1.282	1.282	∞

The factors in this table were computed with an algorithm provided by Robert E. Odeh.

Table A.10b Factors $g_{(1-\alpha;\,p,\,n)}$ for Calculating Normal Distribution Two-Sided $100(1-\alpha)\%$ Tolerance Intervals (to Control the Center of the Distribution)

n	p = 0.900					p = 0.950					p = 0.990					n
$1-\alpha$:	0.50	0.80	0.90	0.95	0.99	0.50	0.80	0.90	0.95	0.99	0.50	0.80	0.90	0.95	0.99	
2	2.869	7.688	15.512	31.092	155.569	3.376	9.032	18.221	36.519	182.720	4.348	11.613	23.423	46.944	234.877	2
3	2.229	3.967	5.788	8.306	18.782	2.634	4.679	6.823	9.789	22.131	3.415	6.051	8.819	12.647	28.586	3
4	2.039	3.159	4.157	5.368	9.416	2.416	3.736	4.913	6.341	11.118	3.144	4.850	6.372	8.221	14.405	4
5	1.945	2.801	3.499	4.291	6.655	2.308	3.318	4.142	5.077	7.870	3.010	4.318	5.387	6.598	10.220	5
6	1.888	2.595	3.141	3.733	5.383	2.243	3.078	3.723	4.422	6.373	2.930	4.013	4.850	5.758	8.292	6
7	1.850	2.460	2.913	3.390	4.658	2.199	2.920	3.456	4.020	5.520	2.876	3.813	4.508	5.241	7.191	7
8	1.823	2.364	2.754	3.156	4.189	2.167	2.808	3.270	3.746	4.968	2.836	3.670	4.271	4.889	6.479	8
9	1.802	2.292	2.637	2.986	3.860	2.143	2.723	3.132	3.546	4.581	2.806	3.562	4.094	4.633	5.980	9
10	1.785	2.235	2.546	2.856	3.617	2.124	2.657	3.026	3.393	4.294	2.783	3.478	3.958	4.437	5.610	10
11	1.772	2.189	2.473	2.754	3.429	2.109	2.604	2.941	3.273	4.073	2.764	3.410	3.849	4.282	5.324	11
12	1.761	2.152	2.414	2.670	3.279	2.096	2.560	2.871	3.175	3.896	2.748	3.353	3.759	4.156	5.096	12
13	1.752	2.120	2.364	2.601	3.156	2.085	2.522	2.812	3.093	3.751	2.735	3.306	3.684	4.051	4.909	13
14	1.744	2.093	2.322	2.542	3.054	2.076	2.490	2.762	3.024	3.631	2.723	3.265	3.620	3.962	4.753	14
15	1.737	2.069	2.285	2.492	2.967	2.068	2.463	2.720	2.965	3.529	2.714	3.229	3.565	3.885	4.621	15
16	1.731	2.049	2.254	2.449	2.893	2.061	2.439	2.682	2.913	3.441	2.705	3.198	3.517	3.819	4.507	16
17	1.726	2.030	2.226	2.410	2.828	2.055	2.417	2.649	2.868	3.364	2.698	3.171	3.474	3.761	4.408	17
18	1.721	2.014	2.201	2.376	2.771	2.050	2.398	2.620	2.828	3.297	2.691	3.146	3.436	3.709	4.321	18
19	1.717	2.000	2.178	2.346	2.720	2.045	2.381	2.593	2.793	3.237	2.685	3.124	3.402	3.663	4.244	19
20	1.714	1.987	2.158	2.319	2.675	2.041	2.365	2.570	2.760	3.184	2.680	3.104	3.372	3.621	4.175	20
21	1.710	1.975	2.140	2.294	2.635	2.037	2.351	2.548	2.731	3.136	2.675	3.086	3.344	3.583	4.113	21
22	1.707	1.964	2.123	2.272	2.598	2.034	2.338	2.528	2.705	3.092	2.670	3.070	3.318	3.549	4.056	22
23	1.705	1.954	2.108	2.251	2.564	2.030	2.327	2.510	2.681	3.053	2.666	3.054	3.295	3.518	4.005	23
24	1.702	1.944	2.094	2.232	2.534	2.027	2.316	2.494	2.658	3.017	2.662	3.040	3.274	3.489	3.958	24
25	1.700	1.936	2.081	2.215	2.506	2.025	2.306	2.479	2.638	2.981	2.659	3.027	3.254	3.462	3.915	25
26	1.698	1.928	2.069	2.199	2.480	2.022	2.296	2.464	2.619	2.953	2.656	3.015	3.235	3.437	3.875	26
27	1.696	1.921	2.058	2.184	2.456	2.020	2.288	2.451	2.601	2.925	2.653	3.004	3.218	3.415	3.838	27
28	1.694	1.914	2.048	2.170	2.434	2.018	2.279	2.439	2.585	2.898	2.650	2.993	3.202	3.393	3.804	28
29	1.692	1.907	2.038	2.157	2.413	2.016	2.272	2.427	2.569	2.874	2.648	2.983	3.187	3.373	3.772	29
30	1.691	1.901	2.029	2.145	2.394	2.014	2.265	2.417	2.555	2.851	2.645	2.974	3.173	3.355	3.742	30
35	1.684	1.876	1.991	2.094	2.314	2.006	2.234	2.371	2.495	2.756	2.636	2.935	3.114	3.276	3.618	35
40	1.679	1.856	1.961	2.055	2.253	2.001	2.211	2.336	2.448	2.684	2.628	2.905	3.069	3.216	3.524	40
50	1.672	1.827	1.918	1.999	2.166	1.992	2.177	2.285	2.382	2.580	2.618	2.861	3.003	3.129	3.390	50
60	1.668	1.807	1.888	1.960	2.106	1.987	2.154	2.250	2.335	2.509	2.611	2.830	2.956	3.068	3.297	60
120	1.656	1.752	1.805	1.851	1.943	1.974	2.087	2.151	2.206	2.315	2.594	2.743	2.826	2.899	3.043	120
240	1.651	1.716	1.753	1.784	1.844	1.967	2.045	2.088	2.125	2.197	2.585	2.688	2.744	2.793	2.887	240
480	1.648	1.694	1.718	1.739	1.780	1.963	2.018	2.048	2.073	2.121	2.580	2.652	2.691	2.724	2.787	480
∞	1.645	1.645	1.645	1.645	1.645	1.960	1.960	1.960	1.960	1.960	2.576	2.576	2.576	2.576	2.576	∞

The factors in this table were computed with an algorithm provided by Robert E. Odeh.

Table A.11a Factors $g''_{(1-\alpha; p, n)}$ for Calculating Normal Distribution Two-Sided $100(1-\alpha)\%$ Tolerance Intervals (to Simultaneously Control Each Tail of the Distribution to $100p\%$ or Less)

n	$1-\alpha$: p = 0.250					p = 0.150					p = 0.100					n
	0.600	0.800	0.900	0.950	0.990	0.600	0.800	0.900	0.950	0.990	0.600	0.800	0.900	0.950	0.990	
2	1.784	4.869	9.847	19.748	98.829	2.333	6.302	12.730	25.522	127.707	2.702	7.272	14.682	29.431	147.264	2
3	1.330	2.455	3.617	5.214	11.830	1.773	3.210	4.705	6.765	15.322	2.072	3.724	5.448	7.826	17.714	3
4	1.180	1.914	2.557	3.330	5.892	1.596	2.526	3.348	4.340	7.644	1.876	2.944	3.890	5.035	8.853	4
5	1.100	1.666	2.121	2.631	4.137	1.503	2.217	2.795	3.445	5.378	1.774	2.593	3.257	4.007	6.239	5
6	1.049	1.520	1.879	2.264	3.324	1.444	2.036	2.489	2.978	4.331	1.710	2.388	2.908	3.471	5.032	6
7	1.012	1.421	1.722	2.035	2.856	1.402	1.915	2.292	2.687	3.731	1.665	2.251	2.685	3.139	4.340	7
8	0.984	1.349	1.611	1.877	2.551	1.371	1.827	2.154	2.488	3.341	1.631	2.153	2.527	2.911	3.892	8
9	0.962	1.294	1.527	1.760	2.336	1.346	1.760	2.050	2.342	3.067	1.605	2.078	2.410	2.744	3.577	9
10	0.944	1.250	1.461	1.670	2.175	1.326	1.707	1.969	2.229	2.862	1.583	2.018	2.318	2.616	3.342	10
11	0.929	1.214	1.408	1.598	2.049	1.309	1.663	1.903	2.139	2.703	1.565	1.970	2.244	2.514	3.160	11
12	0.916	1.184	1.364	1.539	1.948	1.295	1.627	1.849	2.066	2.576	1.550	1.929	2.183	2.430	3.014	12
13	0.905	1.158	1.326	1.489	1.865	1.283	1.596	1.803	2.004	2.471	1.537	1.895	2.132	2.360	2.894	13
14	0.895	1.135	1.294	1.446	1.795	1.272	1.569	1.764	1.951	2.383	1.526	1.865	2.087	2.301	2.794	14
15	0.886	1.115	1.266	1.409	1.735	1.262	1.545	1.730	1.906	2.308	1.516	1.839	2.049	2.250	2.708	15
16	0.879	1.098	1.241	1.377	1.683	1.254	1.525	1.700	1.866	2.243	1.507	1.816	2.015	2.205	2.635	16
17	0.872	1.082	1.219	1.349	1.638	1.246	1.506	1.673	1.831	2.187	1.499	1.795	1.986	2.165	2.570	17
18	0.866	1.068	1.200	1.323	1.597	1.240	1.489	1.650	1.800	2.137	1.492	1.777	1.959	2.130	2.513	18
19	0.860	1.056	1.182	1.300	1.562	1.233	1.474	1.628	1.772	2.092	1.485	1.760	1.935	2.099	2.463	19
20	0.854	1.044	1.166	1.279	1.529	1.228	1.461	1.609	1.747	2.053	1.479	1.745	1.913	2.070	2.417	20
21	0.850	1.033	1.151	1.260	1.500	1.222	1.448	1.591	1.724	2.017	1.473	1.731	1.893	2.044	2.377	21
22	0.845	1.024	1.137	1.243	1.474	1.217	1.437	1.575	1.703	1.984	1.468	1.718	1.875	2.020	2.339	22
23	0.841	1.014	1.125	1.227	1.449	1.213	1.426	1.560	1.683	1.954	1.464	1.707	1.858	1.999	2.305	23
24	0.837	1.006	1.113	1.212	1.427	1.209	1.416	1.546	1.666	1.926	1.459	1.696	1.843	1.979	2.274	24
25	0.833	0.998	1.102	1.199	1.406	1.205	1.407	1.533	1.649	1.901	1.455	1.685	1.828	1.960	2.246	25
26	0.830	0.991	1.092	1.186	1.387	1.201	1.398	1.521	1.634	1.878	1.451	1.676	1.815	1.943	2.219	26
27	0.827	0.984	1.083	1.174	1.370	1.198	1.390	1.510	1.619	1.856	1.448	1.667	1.802	1.927	2.194	27
28	0.824	0.978	1.074	1.163	1.353	1.194	1.383	1.499	1.606	1.836	1.444	1.659	1.791	1.912	2.172	28
29	0.821	0.971	1.066	1.153	1.338	1.191	1.376	1.489	1.593	1.817	1.441	1.651	1.780	1.898	2.150	29
30	0.818	0.966	1.058	1.143	1.323	1.188	1.369	1.480	1.582	1.799	1.438	1.644	1.770	1.884	2.130	30
35	0.807	0.941	1.025	1.101	1.262	1.176	1.340	1.441	1.532	1.725	1.425	1.612	1.726	1.828	2.046	35
40	0.798	0.922	0.999	1.069	1.215	1.166	1.318	1.410	1.493	1.668	1.415	1.588	1.692	1.785	1.982	40
50	0.784	0.893	0.960	1.021	1.146	1.151	1.284	1.364	1.436	1.585	1.399	1.551	1.641	1.722	1.889	50
60	0.774	0.872	0.932	0.986	1.098	1.141	1.260	1.332	1.395	1.527	1.388	1.525	1.605	1.677	1.824	60
120	0.743	0.811	0.851	0.887	0.960	1.109	1.190	1.238	1.279	1.364	1.355	1.448	1.501	1.548	1.642	120
240	0.723	0.769	0.797	0.822	0.870	1.087	1.143	1.175	1.203	1.259	1.333	1.397	1.433	1.464	1.526	240
480	0.708	0.741	0.760	0.777	0.810	1.072	1.111	1.133	1.152	1.190	1.317	1.362	1.387	1.408	1.450	480
∞	0.674	0.674	0.674	0.674	0.674	1.036	1.036	1.036	1.036	1.036	1.282	1.282	1.282	1.282	1.282	∞

The factors in this table were provided by Robert E. Odeh.

Table A.11b Factors $g''_{(1-\alpha, p, n)}$ for Calculating Normal Distribution Two-Sided $100(1-\alpha)\%$ Tolerance Intervals (to Simultaneously Control Each Tail of the Distribution to $100p\%$ or Less)

n	p = 0.050					p = 0.025					p = 0.005					n
1 − α :	0.600	0.800	0.900	0.950	0.990	0.600	0.800	0.900	0.950	0.990	0.600	0.800	0.900	0.950	0.990	
2	3.246	8.708	17.574	35.225	176.251	3.717	9.954	20.082	40.251	201.393	4.635	12.386	24.984	50.073	250.531	2
3	2.512	4.487	6.554	9.408	21.281	2.893	5.151	7.516	10.785	24.388	3.636	6.450	9.402	13.485	30.483	3
4	2.288	3.565	4.700	6.074	10.664	2.645	4.105	5.405	6.980	12.246	3.341	5.164	6.788	8.760	15.354	4
5	2.174	3.152	3.948	4.847	7.531	2.520	3.639	4.550	5.582	8.661	3.194	4.592	5.733	7.025	10.887	5
6	2.102	2.912	3.535	4.209	6.085	2.442	3.368	4.082	4.855	7.008	3.104	4.263	5.156	6.125	8.827	6
7	2.052	2.753	3.271	3.815	5.258	2.388	3.190	3.783	4.407	6.063	3.042	4.045	4.789	5.571	7.650	7
8	2.015	2.639	3.086	3.546	4.723	2.348	3.061	3.574	4.100	5.451	2.997	3.889	4.531	5.192	6.888	8
9	1.986	2.551	2.948	3.348	4.346	2.317	2.964	3.418	3.876	5.021	2.961	3.771	4.340	4.915	6.353	9
10	1.963	2.483	2.840	3.197	4.066	2.292	2.887	3.296	3.704	4.702	2.933	3.678	4.192	4.704	5.956	10
11	1.944	2.426	2.754	3.076	3.849	2.271	2.824	3.199	3.568	4.454	2.910	3.602	4.072	4.535	5.649	11
12	1.927	2.380	2.682	2.978	3.675	2.253	2.771	3.118	3.456	4.256	2.890	3.539	3.974	4.398	5.403	12
13	1.913	2.340	2.622	2.895	3.533	2.239	2.727	3.050	3.363	4.094	2.874	3.486	3.891	4.284	5.201	13
14	1.901	2.306	2.571	2.825	3.414	2.226	2.689	2.993	3.284	3.958	2.859	3.440	3.821	4.187	5.033	14
15	1.890	2.276	2.526	2.765	3.312	2.214	2.656	2.942	3.216	3.843	2.847	3.400	3.760	4.103	4.890	15
16	1.881	2.249	2.487	2.713	3.225	2.204	2.626	2.898	3.157	3.743	2.835	3.365	3.706	4.030	4.767	16
17	1.872	2.226	2.452	2.666	3.149	2.195	2.600	2.859	3.104	3.657	2.825	3.334	3.659	3.966	4.659	17
18	1.864	2.204	2.421	2.625	3.031	2.187	2.577	2.825	3.058	3.580	2.816	3.305	3.616	3.909	4.564	18
19	1.857	2.185	2.393	2.588	3.021	2.179	2.555	2.793	3.016	3.512	2.808	3.280	3.578	3.858	4.480	19
20	1.851	2.168	2.368	2.555	2.968	2.173	2.536	2.765	2.978	3.451	2.801	3.257	3.544	3.812	4.405	20
21	1.845	2.152	2.345	2.524	2.919	2.166	2.518	2.739	2.944	3.396	2.794	3.236	3.513	3.770	4.337	21
22	1.839	2.138	2.324	2.497	2.876	2.161	2.502	2.715	2.913	3.346	2.788	3.217	3.484	3.731	4.275	22
23	1.834	2.124	2.304	2.471	2.835	2.155	2.487	2.694	2.884	3.301	2.782	3.199	3.458	3.696	4.219	23
24	1.830	2.112	2.287	2.448	2.799	2.150	2.474	2.674	2.858	3.259	2.776	3.182	3.433	3.664	4.168	24
25	1.825	2.100	2.270	2.426	2.765	2.146	2.461	2.655	2.833	3.221	2.772	3.167	3.411	3.634	4.120	25
26	1.821	2.089	2.254	2.406	2.734	2.142	2.449	2.638	2.811	3.185	2.767	3.153	3.390	3.607	4.076	26
27	1.817	2.079	2.240	2.387	2.705	2.138	2.438	2.621	2.790	3.153	2.763	3.140	3.370	3.581	4.036	27
28	1.814	2.070	2.226	2.370	2.678	2.134	2.427	2.606	2.770	3.122	2.758	3.127	3.352	3.557	3.998	28
29	1.810	2.061	2.214	2.353	2.653	2.130	2.417	2.592	2.752	3.093	2.755	3.115	3.335	3.535	3.963	29
30	1.807	2.053	2.202	2.338	2.629	2.127	2.408	2.579	2.734	3.067	2.751	3.104	3.319	3.513	3.930	30
35	1.793	2.017	2.151	2.273	2.530	2.113	2.369	2.522	2.661	2.955	2.736	3.057	3.250	3.424	3.792	35
40	1.783	1.989	2.112	2.223	2.455	2.101	2.338	2.478	2.605	2.870	2.723	3.021	3.197	3.356	3.687	40
50	1.767	1.947	2.054	2.149	2.346	2.085	2.292	2.414	2.522	2.747	2.705	2.966	3.119	3.255	3.536	50
60	1.755	1.918	2.013	2.097	2.270	2.072	2.259	2.368	2.464	2.661	2.693	2.928	3.064	3.184	3.431	60
120	1.720	1.831	1.894	1.949	2.059	2.037	2.164	2.236	2.298	2.423	2.655	2.815	2.905	2.983	3.139	120
240	1.697	1.773	1.816	1.853	1.925	2.013	2.101	2.149	2.191	2.273	2.630	2.741	2.802	2.854	2.956	240
480	1.681	1.734	1.764	1.788	1.837	1.997	2.058	2.091	2.119	2.175	2.614	2.690	2.732	2.767	2.836	480
∞	1.645	1.645	1.645	1.645	1.645	1.960	1.960	1.960	1.960	1.960	2.576	2.576	2.576	2.576	2.576	∞

The factors in this table were provided by Robert E. Odeh.

311

Table A.12a Factors $g'_{(1-\alpha; p, n)}$ for Calculating Normal Distribution One-Sided $100(1 - \alpha)\%$ Tolerance Bounds

n	p = 0.600 0.010	0.025	0.050	0.100	0.200	p = 0.700 0.010	0.025	0.050	0.100	0.200	p = 0.800 0.010	0.025	0.050	0.100	0.200	n
2	-13.821	-5.508	-2.718	-1.286	-0.493	-7.494	-2.968	-1.431	-0.610	-0.091	-3.204	-1.229	-0.521	-0.084	0.288	2
3	-2.728	-1.657	-1.090	-0.650	-0.276	-1.662	-0.966	-0.580	-0.261	0.043	-0.792	-0.380	-0.127	0.111	0.377	3
4	-1.563	-1.061	-0.746	-0.463	-0.187	-0.949	-0.592	-0.355	-0.130	0.111	-0.405	-0.158	0.021	0.209	0.432	4
5	-1.145	-0.811	-0.583	-0.363	-0.133	-0.670	-0.418	-0.237	-0.052	0.155	-0.227	-0.038	0.110	0.272	0.470	5
6	-0.924	-0.668	-0.483	-0.297	-0.096	-0.513	-0.312	-0.160	0.001	0.186	-0.117	0.043	0.173	0.319	0.499	6
7	-0.784	-0.572	-0.413	-0.249	-0.068	-0.409	-0.237	-0.103	0.041	0.211	-0.040	0.103	0.220	0.355	0.522	7
8	-0.685	-0.501	-0.361	-0.213	-0.046	-0.333	-0.181	-0.060	0.073	0.231	0.020	0.150	0.258	0.384	0.540	8
9	-0.611	-0.447	-0.319	-0.183	-0.028	-0.275	-0.136	-0.025	0.099	0.247	0.067	0.188	0.290	0.408	0.556	9
10	-0.552	-0.403	-0.286	-0.159	-0.013	-0.227	-0.100	0.004	0.121	0.261	0.107	0.220	0.316	0.428	0.569	10
11	-0.504	-0.367	-0.258	-0.138	0.000	-0.188	-0.069	0.029	0.139	0.272	0.140	0.247	0.339	0.446	0.580	11
12	-0.464	-0.336	-0.233	-0.121	0.011	-0.155	-0.043	0.050	0.156	0.283	0.169	0.271	0.359	0.461	0.591	12
13	-0.430	-0.310	-0.213	-0.105	0.020	-0.126	-0.020	0.069	0.170	0.292	0.194	0.292	0.376	0.475	0.599	13
14	-0.401	-0.287	-0.194	-0.092	0.029	-0.101	0.001	0.086	0.182	0.300	0.216	0.310	0.392	0.487	0.608	14
15	-0.375	-0.267	-0.178	-0.080	0.037	-0.079	0.019	0.100	0.194	0.307	0.236	0.327	0.406	0.498	0.615	15
16	-0.352	-0.248	-0.163	-0.069	0.044	-0.059	0.035	0.114	0.204	0.314	0.254	0.342	0.419	0.508	0.621	16
17	-0.332	-0.232	-0.150	-0.059	0.050	-0.041	0.050	0.126	0.213	0.320	0.271	0.356	0.430	0.518	0.627	17
18	-0.313	-0.217	-0.138	-0.050	0.056	-0.024	0.063	0.137	0.222	0.325	0.286	0.369	0.441	0.526	0.633	18
19	-0.296	-0.204	-0.127	-0.041	0.061	-0.009	0.075	0.147	0.230	0.330	0.299	0.380	0.451	0.534	0.638	19
20	-0.281	-0.192	-0.117	-0.034	0.066	0.004	0.086	0.156	0.237	0.335	0.312	0.391	0.460	0.541	0.643	20
21	-0.267	-0.180	-0.108	-0.027	0.070	0.017	0.097	0.165	0.243	0.340	0.324	0.401	0.468	0.548	0.647	21
22	-0.254	-0.170	-0.099	-0.020	0.075	0.029	0.107	0.173	0.250	0.344	0.335	0.410	0.476	0.554	0.651	22
23	-0.242	-0.160	-0.091	-0.014	0.078	0.040	0.116	0.181	0.256	0.347	0.345	0.419	0.484	0.560	0.655	23
24	-0.230	-0.151	-0.084	-0.008	0.082	0.050	0.124	0.188	0.261	0.351	0.355	0.427	0.491	0.565	0.659	24
25	-0.220	-0.142	-0.077	-0.003	0.086	0.059	0.132	0.194	0.266	0.354	0.364	0.435	0.497	0.570	0.662	25
26	-0.210	-0.134	-0.070	0.002	0.089	0.068	0.139	0.200	0.271	0.358	0.373	0.442	0.503	0.575	0.665	26
27	-0.201	-0.127	-0.064	0.007	0.092	0.077	0.147	0.206	0.276	0.361	0.381	0.449	0.509	0.580	0.669	27
28	-0.192	-0.120	-0.058	0.011	0.095	0.085	0.153	0.212	0.280	0.363	0.388	0.456	0.515	0.584	0.671	28
29	-0.184	-0.113	-0.053	0.015	0.098	0.092	0.159	0.217	0.284	0.366	0.396	0.462	0.520	0.588	0.674	29
30	-0.176	-0.106	-0.048	0.020	0.100	0.099	0.165	0.222	0.288	0.369	0.403	0.468	0.525	0.592	0.677	30
35	-0.143	-0.079	-0.025	0.037	0.111	0.130	0.191	0.244	0.305	0.380	0.433	0.494	0.547	0.610	0.688	35
40	-0.116	-0.057	-0.007	0.051	0.121	0.155	0.212	0.261	0.319	0.389	0.457	0.514	0.565	0.624	0.698	40
50	-0.077	-0.024	0.021	0.072	0.134	0.193	0.244	0.288	0.340	0.403	0.495	0.547	0.592	0.645	0.712	50
60	-0.047	0.000	0.041	0.088	0.145	0.221	0.268	0.308	0.355	0.413	0.523	0.571	0.612	0.661	0.723	60
120	0.041	0.074	0.103	0.136	0.176	0.308	0.341	0.370	0.404	0.445	0.611	0.646	0.676	0.712	0.756	120
240	0.103	0.126	0.147	0.170	0.199	0.370	0.393	0.414	0.438	0.468	0.676	0.701	0.723	0.749	0.780	240
480	0.147	0.163	0.178	0.194	0.215	0.414	0.431	0.446	0.463	0.484	0.723	0.741	0.757	0.775	0.798	480
∞	0.253	0.253	0.253	0.253	0.253	0.524	0.524	0.524	0.524	0.524	0.842	0.842	0.842	0.842	0.842	∞

The factors in this table can also be used to compute two-sided confidence intervals and one-sided confidence bounds for normal distribution percentiles; see Section 4.4. The factors in this table were computed with an algorithm provided by Robert E. Odeh.

Table A.12b Factors $g'_{(1-\alpha;\, p,\, n)}$ for Calculating Normal Distribution One-Sided $100(1-\alpha)\%$ Tolerance Bounds

n		p = 0.900					p = 0.950					p = 0.990				n
1 − α :	0.010	0.025	0.050	0.100	0.200	0.010	0.025	0.050	0.100	0.200	0.010	0.025	0.050	0.100	0.200	
2	−0.707	−0.143	0.138	0.403	0.737	0.000	0.273	0.475	0.717	1.077	0.564	0.761	0.954	1.225	1.672	2
3	−0.072	0.159	0.334	0.535	0.799	0.295	0.478	0.639	0.840	1.126	0.782	0.958	1.130	1.361	1.710	3
4	0.123	0.298	0.444	0.617	0.847	0.443	0.601	0.743	0.922	1.172	0.924	1.088	1.246	1.455	1.760	4
5	0.238	0.389	0.519	0.675	0.883	0.543	0.687	0.818	0.982	1.209	1.027	1.182	1.331	1.525	1.801	5
6	0.319	0.455	0.575	0.719	0.911	0.618	0.752	0.875	1.028	1.238	1.108	1.256	1.396	1.578	1.834	6
7	0.381	0.507	0.619	0.755	0.933	0.678	0.804	0.920	1.065	1.261	1.173	1.315	1.449	1.622	1.862	7
8	0.431	0.550	0.655	0.783	0.952	0.727	0.847	0.958	1.096	1.281	1.227	1.364	1.493	1.658	1.885	8
9	0.472	0.585	0.686	0.808	0.968	0.768	0.884	0.990	1.122	1.298	1.273	1.406	1.530	1.688	1.904	9
10	0.508	0.615	0.712	0.828	.981	0.804	0.915	1.017	1.144	1.313	1.314	1.442	1.563	1.715	1.922	10
11	0.538	0.642	0.734	0.847	.993	0.835	0.943	1.041	1.163	1.325	1.349	1.474	1.591	1.738	1.937	11
12	0.565	0.665	0.754	0.863	1.004	0.862	0.967	1.062	1.180	1.337	1.381	1.502	1.616	1.758	1.950	12
13	0.589	0.685	0.772	0.877	1.013	0.887	0.989	1.081	1.196	1.347	1.409	1.528	1.638	1.776	1.962	13
14	0.610	0.704	0.788	0.890	1.022	0.909	1.008	1.098	1.210	1.356	1.434	1.551	1.658	1.793	1.973	14
15	0.629	0.721	0.802	0.901	1.029	0.929	1.026	1.114	1.222	1.364	1.458	1.572	1.677	1.808	1.983	15
16	0.647	0.736	0.815	0.912	1.036	0.948	1.042	1.128	1.234	1.372	1.479	1.591	1.694	1.822	1.992	16
17	0.663	0.750	0.827	0.921	1.043	0.965	1.057	1.141	1.244	1.379	1.499	1.608	1.709	1.834	2.000	17
18	0.678	0.763	0.839	0.930	1.049	0.980	1.071	1.153	1.254	1.385	1.517	1.625	1.724	1.846	2.008	18
19	0.692	0.775	0.849	0.939	1.054	0.995	1.084	1.164	1.263	1.391	1.534	1.640	1.737	1.857	2.015	19
20	0.705	0.786	0.858	0.946	1.059	1.008	1.095	1.175	1.271	1.397	1.550	1.654	1.749	1.867	2.022	20
21	0.716	0.796	0.867	0.953	1.064	1.021	1.107	1.184	1.279	1.402	1.565	1.667	1.761	1.876	2.028	21
22	0.728	0.806	0.876	0.960	1.068	1.033	1.117	1.193	1.286	1.407	1.579	1.680	1.772	1.885	2.034	22
23	0.738	0.815	0.884	0.966	1.073	1.044	1.127	1.202	1.293	1.412	1.592	1.691	1.782	1.893	2.039	23
24	0.748	0.823	0.891	0.972	1.076	1.054	1.136	1.210	1.300	1.416	1.605	1.702	1.791	1.901	2.045	24
25	0.757	0.831	0.898	0.978	1.080	1.064	1.145	1.217	1.306	1.420	1.616	1.713	1.801	1.908	2.049	25
26	0.766	0.839	0.904	0.983	1.084	1.074	1.153	1.225	1.311	1.424	1.628	1.723	1.809	1.915	2.054	26
27	0.774	0.846	0.911	0.988	1.087	1.083	1.161	1.231	1.317	1.427	1.638	1.732	1.817	1.922	2.058	27
28	0.782	0.853	0.917	0.993	1.090	1.091	1.168	1.238	1.322	1.431	1.648	1.741	1.825	1.928	2.063	28
29	0.790	0.860	0.922	0.997	1.093	1.099	1.175	1.244	1.327	1.434	1.658	1.749	1.833	1.934	2.067	29
30	0.797	0.866	0.928	1.002	1.095	1.107	1.182	1.250	1.332	1.437	1.667	1.757	1.840	1.940	2.070	30
35	0.828	0.893	0.951	1.020	1.108	1.141	1.212	1.276	1.352	1.451	1.708	1.793	1.871	1.965	2.087	35
40	0.854	0.916	0.970	1.036	1.119	1.169	1.236	1.297	1.369	1.462	1.741	1.823	1.896	1.986	2.101	40
50	0.894	0.950	1.000	1.059	1.134	1.212	1.274	1.329	1.396	1.480	1.793	1.869	1.936	2.018	2.122	50
60	0.924	0.976	1.022	1.077	1.146	1.245	1.303	1.354	1.415	1.493	1.833	1.903	1.966	2.042	2.138	60
120	1.020	1.059	1.093	1.134	1.184	1.352	1.395	1.433	1.478	1.535	1.963	2.016	2.063	2.119	2.189	120
240	1.092	1.121	1.146	1.175	1.211	1.431	1.463	1.492	1.525	1.565	2.061	2.100	2.135	2.176	2.227	240
480	1.145	1.166	1.184	1.205	1.231	1.491	1.514	1.535	1.558	1.588	2.134	2.163	2.189	2.218	2.255	480
∞	1.282	1.282	1.282	1.282	1.282	1.645	1.645	1.645	1.645	1.645	2.326	2.326	2.326	2.326	2.326	∞

The factors in this table can also be used to compute two-sided confidence intervals and one-sided confidence bouds for normal distribution percentiles; see Section 4.4. The factors in this table were computed with an algorithm provided by Robert E. Odeh.

Table A.12c Factors $g'_{(1-\alpha;\, p, n)}$ for Calculating Normal Distributions One-Sided $100(1-\alpha)\%$ Tolerance Bounds

n	p = 0.600					p = 0.700					p = 0.800				
1 − α:	0.800	0.900	0.950	0.975	0.990	0.800	0.900	0.950	0.975	0.990	0.800	0.900	0.950	0.975	0.990
2	1.577	3.343	6.778	13.602	34.038	2.357	4.881	9.843	19.726	49.344	3.417	6.987	14.051	28.140	70.376
3	0.991	1.602	2.399	3.484	5.593	1.441	2.228	3.277	4.722	7.547	2.016	3.039	4.424	6.343	10.111
4	0.819	1.219	1.672	2.209	3.102	1.199	1.693	2.265	2.954	4.112	1.675	2.295	3.026	3.915	5.417
5	0.729	1.042	1.370	1.732	2.287	1.080	1.456	1.861	2.315	3.020	1.514	1.976	2.483	3.058	3.958
6	0.672	0.935	1.199	1.478	1.884	1.006	1.318	1.638	1.982	2.490	1.417	1.795	2.191	2.621	3.262
7	0.631	0.862	1.087	1.317	1.642	0.955	1.225	1.495	1.775	2.176	1.352	1.676	2.005	2.353	2.854
8	0.600	0.808	1.006	1.205	1.478	0.917	1.158	1.393	1.633	1.967	1.304	1.590	1.875	2.170	2.584
9	0.576	0.766	0.945	1.121	1.358	0.888	1.107	1.317	1.528	1.816	1.266	1.525	1.779	2.036	2.391
10	0.556	0.733	0.896	1.055	1.267	0.864	1.066	1.257	1.446	1.701	1.237	1.474	1.703	1.933	2.246
11	0.540	0.705	0.856	1.002	1.194	0.844	1.032	1.208	1.381	1.610	1.212	1.433	1.643	1.851	2.131
12	0.526	0.681	0.823	0.958	1.134	0.827	1.004	1.168	1.327	1.537	1.192	1.398	1.593	1.784	2.039
13	0.513	0.661	0.794	0.921	1.084	0.813	0.980	1.133	1.282	1.475	1.174	1.368	1.551	1.728	1.963
14	0.502	0.643	0.770	0.889	1.041	0.800	0.959	1.104	1.243	1.424	1.159	1.343	1.514	1.681	1.898
15	0.493	0.628	0.748	0.861	1.005	0.789	0.940	1.078	1.210	1.379	1.145	1.321	1.483	1.639	1.843
16	0.484	0.614	0.729	0.836	0.972	0.779	0.924	1.056	1.180	1.340	1.133	1.301	1.455	1.603	1.795
17	0.477	0.601	0.712	0.814	0.944	0.770	0.910	1.035	1.154	1.306	1.123	1.284	1.431	1.572	1.753
18	0.470	0.590	0.696	0.795	0.918	0.762	0.896	1.017	1.131	1.275	1.113	1.268	1.409	1.543	1.716
19	0.463	0.580	0.682	0.777	0.895	0.755	0.885	1.001	1.110	1.248	1.104	1.254	1.389	1.518	1.682
20	0.458	0.570	0.669	0.761	0.875	0.748	0.874	0.986	1.091	1.223	1.096	1.241	1.371	1.495	1.652
21	0.452	0.562	0.658	0.746	0.856	0.742	0.864	0.972	1.073	1.200	1.089	1.229	1.355	1.474	1.625
22	0.447	0.554	0.647	0.732	0.838	0.736	0.855	0.960	1.057	1.180	1.082	1.218	1.340	1.455	1.600
23	0.443	0.546	0.637	0.720	0.822	0.731	0.846	0.948	1.043	1.161	1.076	1.208	1.326	1.437	1.577
24	0.438	0.540	0.628	0.708	0.808	0.726	0.838	0.937	1.029	1.144	1.070	1.199	1.313	1.421	1.556
25	0.434	0.533	0.619	0.698	0.794	0.722	0.831	0.927	1.017	1.128	1.065	1.190	1.302	1.406	1.537
26	0.430	0.527	0.611	0.687	0.781	0.717	0.824	0.918	1.005	1.113	1.060	1.182	1.291	1.392	1.519
27	0.427	0.522	0.604	0.678	0.769	0.713	0.818	0.910	0.994	1.099	1.055	1.174	1.280	1.379	1.502
28	0.423	0.516	0.596	0.669	0.758	0.710	0.812	0.901	0.984	1.086	1.051	1.167	1.271	1.367	1.486
29	0.420	0.511	0.590	0.661	0.748	0.706	0.806	0.894	0.974	1.073	1.047	1.160	1.262	1.355	1.472
30	0.417	0.506	0.583	0.653	0.738	0.703	0.801	0.886	0.965	1.062	1.043	1.154	1.253	1.344	1.458
35	0.404	0.486	0.556	0.619	0.696	0.688	0.778	0.856	0.927	1.014	1.026	1.127	1.217	1.299	1.400
40	0.394	0.470	0.535	0.593	0.663	0.677	0.760	0.831	0.896	0.976	1.013	1.106	1.188	1.263	1.356
50	0.379	0.446	0.503	0.554	0.615	0.659	0.732	0.795	0.852	0.920	0.993	1.075	1.146	1.211	1.291
60	0.367	0.428	0.480	0.525	0.580	0.647	0.713	0.769	0.820	0.881	0.978	1.052	1.116	1.174	1.245
120	0.333	0.375	0.411	0.442	0.478	0.609	0.655	0.693	0.727	0.767	0.936	0.986	1.029	1.068	1.113
240	0.309	0.339	0.363	0.385	0.410	0.584	0.615	0.642	0.665	0.692	0.907	0.942	0.971	0.997	1.028
480	0.293	0.313	0.331	0.346	0.363	0.566	0.588	0.606	0.622	0.641	0.887	0.912	0.932	0.950	0.971
∞	0.253	0.253	0.253	0.253	0.253	0.524	0.524	0.524	0.524	0.524	0.842	0.842	0.842	0.842	0.842

The factors in this table can also be used to compute two-sided confidence intervals and one-sided confidence bounds for normal distribution percentiles; see Section 4.4. The factors in this table were computed with an algorithm provided by Robert E. Odeh.

Table A.12d Factors $g'_{(1-\alpha; p, n)}$ for Calculating Normal Distribution One-Sided $100(1-\alpha)$% Tolerance Bounds

n	p = 0.900					p = 0.950					p = 0.990				
1 − α :	0.800	0.900	0.950	0.975	0.990	0.800	0.900	0.950	0.975	0.990	0.800	0.900	0.950	0.975	0.990
2	5.049	10.253	20.581	41.201	103.029	6.464	13.090	26.260	52.559	131.426	9.156	18.500	37.094	74.234	185.617
3	2.871	4.258	6.155	8.797	13.995	3.604	5.311	7.656	10.927	17.370	5.010	7.340	10.553	15.043	23.896
4	2.372	3.188	4.162	5.354	7.380	2.968	3.957	5.144	6.602	9.083	4.110	5.438	7.042	9.018	12.387
5	2.145	2.742	3.407	4.166	5.362	2.683	3.400	4.203	5.124	6.578	3.711	4.666	5.741	6.980	8.939
6	2.012	2.494	3.006	3.568	4.411	2.517	3.092	3.708	4.385	5.406	3.482	4.243	5.062	5.967	7.335
7	1.923	2.333	2.755	3.206	3.859	2.407	2.894	3.399	3.940	4.728	3.331	3.972	4.642	5.361	6.412
8	1.859	2.219	2.582	2.960	3.497	2.328	2.754	3.187	3.640	4.285	3.224	3.783	4.354	4.954	5.812
9	1.809	2.133	2.454	2.783	3.240	2.268	2.650	3.031	3.424	3.972	3.142	3.641	4.143	4.662	5.389
10	1.770	2.066	2.355	2.647	3.048	2.220	2.568	2.911	3.259	3.738	3.078	3.532	3.981	4.440	5.074
11	1.738	2.011	2.275	2.540	2.898	2.182	2.503	2.815	3.129	3.556	3.026	3.443	3.852	4.265	4.829
12	1.711	1.966	2.210	2.452	2.777	2.149	2.448	2.736	3.023	3.410	2.982	3.371	3.747	4.124	4.633
13	1.689	1.928	2.155	2.379	2.677	2.122	2.402	2.671	2.936	3.290	2.946	3.309	3.659	4.006	4.472
14	1.669	1.895	2.109	2.317	2.593	2.098	2.363	2.614	2.861	3.189	2.914	3.257	3.585	3.907	4.337
15	1.652	1.867	2.068	2.264	2.521	2.078	2.329	2.566	2.797	3.102	2.887	3.212	3.520	3.822	4.222
16	1.637	1.842	2.033	2.218	2.459	2.059	2.299	2.524	2.742	3.028	2.863	3.172	3.464	3.749	4.123
17	1.623	1.819	2.002	2.177	2.405	2.043	2.272	2.486	2.693	2.963	2.841	3.137	3.414	3.684	4.037
18	1.611	1.800	1.974	2.141	2.357	2.029	2.249	2.453	2.650	2.905	2.822	3.105	3.370	3.627	3.960
19	1.600	1.782	1.949	2.108	2.311	2.016	2.227	2.423	2.611	2.854	2.804	3.077	3.331	3.575	3.892
20	1.590	1.765	1.926	2.079	2.276	2.004	2.208	2.396	2.576	2.808	2.789	3.052	3.295	3.529	3.832
21	1.581	1.750	1.905	2.053	2.241	1.993	2.190	2.371	2.544	2.766	2.774	3.028	3.263	3.487	3.777
22	1.572	1.737	1.886	2.028	2.209	1.983	2.174	2.349	2.515	2.729	2.761	3.007	3.233	3.449	3.727
23	1.564	1.724	1.869	2.006	2.180	1.973	2.159	2.328	2.489	2.694	2.749	2.987	3.206	3.414	3.681
24	1.557	1.712	1.853	1.985	2.154	1.965	2.145	2.309	2.465	2.662	2.738	2.969	3.181	3.382	3.640
25	1.550	1.702	1.838	1.966	2.129	1.957	2.132	2.292	2.442	2.633	2.727	2.952	3.158	3.353	3.601
26	1.544	1.691	1.824	1.949	2.106	1.949	2.120	2.275	2.421	2.606	2.718	2.937	3.136	3.325	3.566
27	1.538	1.682	1.811	1.932	2.085	1.943	2.109	2.260	2.402	2.581	2.708	2.922	3.116	3.300	3.533
28	1.533	1.673	1.799	1.917	2.065	1.936	2.099	2.246	2.384	2.558	2.700	2.909	3.098	3.276	3.502
29	1.528	1.665	1.788	1.903	2.047	1.930	2.089	2.232	2.367	2.536	2.692	2.896	3.080	3.254	3.473
30	1.523	1.657	1.777	1.889	2.030	1.924	2.080	2.220	2.351	2.515	2.684	2.884	3.064	3.233	3.447
35	1.502	1.624	1.732	1.833	1.957	1.900	2.041	2.167	2.284	2.430	2.652	2.833	2.995	3.145	3.334
40	1.486	1.598	1.697	1.789	1.902	1.880	2.010	2.125	2.232	2.364	2.627	2.793	2.941	3.078	3.249
50	1.461	1.559	1.646	1.724	1.821	1.852	1.965	2.065	2.156	2.269	2.590	2.735	2.862	2.980	3.125
60	1.444	1.532	1.609	1.679	1.764	1.832	1.933	2.022	2.103	2.202	2.564	2.694	2.807	2.911	3.038
120	1.393	1.452	1.503	1.549	1.604	1.772	1.841	1.899	1.952	2.015	2.488	2.574	2.649	2.716	2.797
240	1.358	1.399	1.434	1.465	1.501	1.733	1.780	1.819	1.854	1.896	2.437	2.497	2.547	2.591	2.645
480	1.335	1.363	1.387	1.408	1.433	1.706	1.738	1.766	1.790	1.818	2.403	2.444	2.479	2.509	2.545
∞	1.282	1.282	1.282	1.282	1.282	1.645	1.645	1.645	1.645	1.645	2.326	2.326	2.326	2.326	2.326

The factors in this table can also be used to compute two-sided confidence intervals and one-sided confidence bounds for normal distribution percentiles; see Section 4.4. The factors in this table were computed with an algorithm provided by Robert E. Odeh.

Table A.13 Factors $r_{(1-\alpha; m, n)}$ for Calculating Normal Distribution Two-Sided $100(1-\alpha)\%$ Prediction Intervals for m Future Observations Using the Results of a Previous Sample of n Observations

$1-\alpha$	n	m: 1	2	3	4	5	6	7	8	9	10	12	16	20	40	60	80	100
0.90	4	2.631	3.329	3.742	4.032	4.255	4.435	4.585	4.713	4.826	4.925	5.095	5.357	5.555	6.143	6.468	6.692	6.861
	5	2.335	2.909	3.246	3.483	3.665	3.812	3.936	4.041	4.134	4.216	4.356	4.573	4.738	5.229	5.502	5.690	5.833
	6	2.177	2.685	2.982	3.190	3.350	3.480	3.589	3.682	3.764	3.836	3.960	4.153	4.299	4.736	4.980	5.148	5.276
	7	2.077	2.546	2.818	3.008	3.155	3.273	3.373	3.458	3.533	3.599	3.713	3.889	4.023	4.426	4.652	4.807	4.925
	8	2.010	2.452	2.706	2.884	3.021	3.132	3.225	3.305	3.375	3.437	3.543	3.708	3.834	4.213	4.425	4.571	4.683
	9	1.960	2.383	2.625	2.794	2.924	3.029	3.118	3.193	3.259	3.318	3.419	3.576	3.696	4.056	4.258	4.398	4.505
	10	1.923	2.331	2.564	2.726	2.851	2.951	3.036	3.108	3.172	3.228	3.325	3.476	3.591	3.936	4.131	4.265	4.368
	15	1.819	2.188	2.395	2.539	2.648	2.737	2.811	2.875	2.930	2.980	3.065	3.197	3.297	3.601	3.772	3.891	3.982
	20	1.772	2.122	2.318	2.454	2.556	2.639	2.709	2.768	2.820	2.866	2.945	3.068	3.162	3.445	3.604	3.715	3.800
	25	1.745	2.085	2.275	2.405	2.504	2.583	2.650	2.707	2.757	2.801	2.877	2.994	3.084	3.354	3.506	3.612	3.693
	30	1.727	2.061	2.246	2.373	2.470	2.547	2.612	2.667	2.716	2.758	2.832	2.946	3.033	3.294	3.442	3.544	3.623
	60	1.685	2.003	2.178	2.298	2.388	2.460	2.520	2.572	2.617	2.656	2.724	2.829	2.909	3.149	3.283	3.377	3.448
	120	1.665	1.976	2.146	2.262	2.349	2.419	2.477	2.526	2.569	2.607	2.673	2.773	2.849	3.078	3.206	3.294	3.362
	∞	1.645	1.949	2.114	2.226	2.311	2.378	2.434	2.481	2.523	2.560	2.622	2.718	2.791	3.008	3.129	3.212	3.276
0.95	4	3.558	4.412	4.923	5.285	5.564	5.789	5.978	6.140	6.282	6.407	6.622	6.954	7.206	7.954	8.370	8.655	8.872
	5	3.041	3.697	4.087	4.364	4.577	4.751	4.896	5.021	5.131	5.229	5.395	5.655	5.852	6.441	6.771	6.997	7.170
	6	2.777	3.333	3.662	3.896	4.076	4.223	4.346	4.452	4.545	4.628	4.770	4.991	5.159	5.665	5.949	6.145	6.295
	7	2.616	3.114	3.407	3.614	3.774	3.905	4.014	4.108	4.191	4.265	4.391	4.589	4.739	5.194	5.449	5.626	5.761
	8	2.508	2.968	3.236	3.426	3.573	3.692	3.792	3.879	3.954	4.022	4.138	4.319	4.457	4.876	5.112	5.275	5.400
	9	2.431	2.863	3.115	3.292	3.429	3.540	3.634	3.714	3.785	3.848	3.956	4.125	4.255	4.647	4.869	5.022	5.140
	10	2.373	2.785	3.024	3.192	3.321	3.426	3.515	3.591	3.658	3.717	3.820	3.980	4.102	4.474	4.685	4.831	4.942
	15	2.215	2.574	2.778	2.921	3.031	3.120	3.194	3.258	3.314	3.365	3.451	3.585	3.689	4.002	4.180	4.305	4.400
	20	2.145	2.480	2.670	2.801	2.902	2.983	3.052	3.110	3.162	3.208	3.286	3.409	3.503	3.788	3.951	4.064	4.151
	25	2.105	2.427	2.608	2.734	2.829	2.907	2.971	3.027	3.075	3.119	3.193	3.309	3.397	3.666	3.820	3.926	4.008
	30	2.079	2.393	2.569	2.690	2.783	2.857	2.920	2.973	3.020	3.062	3.133	3.244	3.329	3.587	3.734	3.837	3.915
	60	2.018	2.312	2.475	2.587	2.672	2.740	2.797	2.846	2.888	2.926	2.991	3.091	3.167	3.398	3.529	3.619	3.689
	120	1.988	2.274	2.431	2.538	2.619	2.685	2.739	2.785	2.826	2.862	2.923	3.018	3.090	3.308	3.430	3.515	3.580
	∞	1.960	2.236	2.388	2.491	2.569	2.631	2.683	2.727	2.766	2.800	2.858	2.948	3.016	3.220	3.335	3.414	3.474
0.99	4	6.530	7.942	8.800	9.411	9.884	10.268	10.590	10.867	11.110	11.325	11.693	12.264	12.698	13.991	14.711	15.207	15.582
	5	5.044	5.972	6.536	6.940	7.253	7.509	7.725	7.911	8.074	8.219	8.469	8.857	9.154	10.045	10.545	10.891	11.153
	6	4.355	5.071	5.503	5.814	6.055	6.253	6.420	6.564	6.690	6.803	6.998	7.302	7.535	8.239	8.637	8.912	9.122
	7	3.963	4.562	4.922	5.181	5.382	5.546	5.685	5.806	5.912	6.006	6.169	6.425	6.621	7.217	7.556	7.791	7.970
	8	3.712	4.238	4.552	4.778	4.953	5.097	5.218	5.323	5.416	5.499	5.641	5.865	6.038	6.564	6.863	7.072	7.231
	9	3.537	4.014	4.297	4.500	4.657	4.787	4.896	4.990	5.074	5.148	5.277	5.479	5.634	6.110	6.382	6.572	6.717
	10	3.408	3.850	4.111	4.297	4.442	4.560	4.660	4.747	4.824	4.892	5.010	5.196	5.339	5.778	6.029	6.205	6.340
	15	3.074	3.426	3.631	3.776	3.888	3.979	4.056	4.123	4.182	4.234	4.325	4.468	4.578	4.916	5.112	5.249	5.354
	20	2.932	3.247	3.428	3.556	3.655	3.735	3.802	3.860	3.912	3.957	4.036	4.160	4.256	4.550	4.720	4.839	4.931
	25	2.852	3.148	3.317	3.435	3.526	3.600	3.663	3.716	3.763	3.806	3.878	3.992	4.079	4.348	4.503	4.612	4.696
	30	2.802	3.085	3.247	3.359	3.446	3.516	3.575	3.625	3.670	3.710	3.778	3.885	3.968	4.221	4.366	4.468	4.547
	60	2.684	2.939	3.082	3.182	3.258	3.319	3.371	3.415	3.454	3.488	3.547	3.639	3.710	3.925	4.048	4.134	4.200
	120	2.629	2.871	3.006	3.100	3.171	3.229	3.277	3.318	3.354	3.386	3.441	3.526	3.591	3.789	3.901	3.980	4.040
	∞	2.576	2.806	2.934	3.022	3.089	3.143	3.188	3.226	3.260	3.289	3.340	3.419	3.479	3.661	3.764	3.835	3.889

The factors in this table were computed and provided to us by Robert E. Odeh.

Table A.14 Factors $r'_{(1-\alpha;\, m,\, n)}$ for Calculating Normal Distribution One-Sided $100(1-\alpha)\%$ Prediction Bounds for m Future Observations Using the Results of a Previous Sample of n Observations

$1-\alpha$	m	n:1	2	3	4	5	6	7	8	9	10	12	16	20	40	60	80	100
	4	1.831	2.484	2.873	3.150	3.364	3.538	3.684	3.809	3.919	4.017	4.184	4.443	4.640	5.226	5.552	5.776	5.946
	5	1.680	2.240	2.567	2.798	2.975	3.120	3.241	3.345	3.436	3.517	3.656	3.871	4.035	4.526	4.799	4.988	5.131
	6	1.594	2.105	2.399	2.606	2.764	2.893	3.001	3.093	3.174	3.246	3.370	3.562	3.708	4.146	4.391	4.560	4.688
	7	1.539	2.020	2.294	2.485	2.631	2.750	2.849	2.935	3.010	3.076	3.190	3.367	3.502	3.906	4.133	4.290	4.409
	8	1.501	1.961	2.221	2.402	2.540	2.652	2.745	2.826	2.896	2.959	3.066	3.232	3.359	3.741	3.955	4.103	4.215
	9	1.472	1.917	2.167	2.341	2.473	2.580	2.669	2.746	2.813	2.873	2.975	3.134	3.255	3.619	3.823	3.965	4.072
0.90	10	1.451	1.884	2.126	2.294	2.422	2.525	2.611	2.685	2.750	2.807	2.906	3.059	3.175	3.526	3.723	3.859	3.963
	15	1.389	1.792	2.013	2.165	2.280	2.372	2.450	2.516	2.573	2.625	2.712	2.848	2.952	3.263	3.438	3.560	3.652
	20	1.361	1.749	1.961	2.105	2.215	2.302	2.375	2.437	2.492	2.540	2.623	2.750	2.848	3.140	3.304	3.418	3.505
	25	1.344	1.724	1.931	2.071	2.177	2.262	2.332	2.392	2.445	2.491	2.571	2.694	2.787	3.068	3.226	3.335	3.418
	30	1.333	1.708	1.911	2.049	2.152	2.235	2.304	2.363	2.414	2.460	2.537	2.657	2.748	3.021	3.174	3.280	3.361
	60	1.307	1.669	1.864	1.995	2.093	2.171	2.236	2.292	2.340	2.383	2.455	2.567	2.652	2.904	3.046	3.143	3.218
	120	1.294	1.651	1.841	1.969	2.064	2.141	2.204	2.257	2.304	2.345	2.415	2.523	2.605	2.847	2.982	3.075	3.146
	∞	1.282	1.632	1.818	1.943	2.036	2.111	2.172	2.224	2.269	2.309	2.376	2.480	2.559	2.791	2.919	3.008	3.075
	4	2.631	3.401	3.871	4.209	4.472	4.686	4.867	5.023	5.160	5.282	5.491	5.816	6.063	6.805	7.218	7.503	7.719
	5	2.335	2.952	3.320	3.583	3.788	3.955	4.095	4.216	4.323	4.418	4.581	4.835	5.029	5.612	5.940	6.166	6.337
	6	2.177	2.715	3.033	3.259	3.433	3.576	3.696	3.799	3.890	3.971	4.111	4.328	4.495	4.996	5.279	5.475	5.623
	7	2.077	2.570	2.857	3.060	3.217	3.345	3.452	3.545	3.627	3.699	3.824	4.019	4.168	4.620	4.875	5.051	5.186
	8	2.010	2.471	2.738	2.926	3.071	3.189	3.288	3.374	3.449	3.516	3.631	3.811	3.948	4.365	4.601	4.765	4.890
	9	1.960	2.400	2.652	2.830	2.966	3.077	3.170	3.250	3.321	3.384	3.492	3.660	3.790	4.182	4.404	4.558	4.675
0.95	10	1.923	2.346	2.587	2.757	2.887	2.992	3.081	3.157	3.224	3.284	3.387	3.547	3.670	4.042	4.254	4.400	4.512
	15	1.819	2.198	2.411	2.559	2.671	2.762	2.839	2.904	2.962	3.013	3.101	3.237	3.342	3.660	3.841	3.967	4.063
	20	1.772	2.132	2.331	2.469	2.574	2.659	2.730	2.790	2.843	2.891	2.972	3.098	3.194	3.486	3.652	3.767	3.856
	25	1.745	2.094	2.286	2.419	2.519	2.600	2.668	2.726	2.776	2.821	2.898	3.018	3.109	3.386	3.543	3.652	3.736
	30	1.727	2.069	2.257	2.386	2.484	2.562	2.628	2.684	2.733	2.776	2.851	2.966	3.055	3.321	3.472	3.577	3.657
	60	1.685	2.010	2.187	2.308	2.399	2.471	2.532	2.584	2.629	2.669	2.737	2.843	2.923	3.164	3.300	3.395	3.467
	120	1.665	1.982	2.154	2.270	2.358	2.428	2.486	2.536	2.579	2.618	2.683	2.784	2.860	3.089	3.217	3.306	3.374
	∞	1.645	1.955	2.121	2.234	2.319	2.386	2.442	2.490	2.531	2.568	2.630	2.726	2.799	3.016	3.137	3.220	3.283
	4	5.077	6.305	7.073	7.632	8.070	8.430	8.734	8.996	9.228	9.434	9.788	10.341	10.764	12.035	12.747	13.238	13.611
	5	4.105	4.943	5.459	5.833	6.126	6.367	6.571	6.748	6.904	7.043	7.282	7.658	7.946	8.819	9.312	9.653	9.913
	6	3.635	4.298	4.701	4.993	5.221	5.409	5.567	5.705	5.827	5.935	6.122	6.417	6.643	7.332	7.724	7.996	8.204
	7	3.360	3.926	4.267	4.513	4.705	4.862	4.996	5.111	5.213	5.305	5.462	5.710	5.902	6.486	6.819	7.051	7.229
	8	3.180	3.685	3.987	4.203	4.372	4.511	4.628	4.730	4.819	4.900	5.038	5.257	5.425	5.941	6.237	6.443	6.601
	9	3.053	3.517	3.792	3.988	4.141	4.267	4.373	4.465	4.546	4.619	4.744	4.942	5.094	5.562	5.831	6.019	6.163
0.99	10	2.959	3.393	3.648	3.830	3.972	4.087	4.185	4.270	4.345	4.412	4.528	4.710	4.851	5.284	5.532	5.707	5.840
	15	2.711	3.067	3.273	3.418	3.531	3.622	3.699	3.766	3.824	3.877	3.967	4.109	4.219	4.557	4.752	4.889	4.994
	20	2.602	2.927	3.112	3.242	3.342	3.423	3.492	3.551	3.603	3.649	3.728	3.853	3.950	4.246	4.416	4.536	4.628
	25	2.542	2.849	3.023	3.145	3.238	3.314	3.377	3.432	3.480	3.523	3.597	3.712	3.801	4.073	4.230	4.340	4.425
	30	2.503	2.799	2.966	3.083	3.172	3.244	3.305	3.357	3.402	3.443	3.513	3.622	3.707	3.964	4.111	4.215	4.295
	60	2.411	2.682	2.833	2.938	3.017	3.082	3.135	3.181	3.221	3.257	3.318	3.414	3.487	3.709	3.835	3.924	3.991
	120	2.368	2.627	2.771	2.870	2.945	3.006	3.056	3.100	3.137	3.171	3.228	3.317	3.386	3.591	3.707	3.789	3.851
	∞	2.326	2.575	2.712	2.806	2.877	2.934	2.981	3.022	3.057	3.089	3.143	3.226	3.289	3.479	3.587	3.661	3.718

The factors in this table were computed and provided to us by Robert E. Odeh.

Table A.15a   Order Statistics ℓ and u and Actual Confidence Levels for Two-Sided Distribution-Free Confidence Intervals for the 1st Percentile for Various Sample Sizes n and Nominal $100(1 - \alpha)\%$ Confidence Levels

n	0.50	0.60	0.70	0.80	0.90	0.95	0.99
				$1 - \alpha$			
60	1 3	1 3	1 3	1 3	1 3	1 3	1 3
	0.4304*	0.4304*	0.4304*	0.4304*	0.4304*	0.4304*	0.4304*
70	1 3	1 3	1 3	1 3	1 3	1 3	1 3
	0.4718*	0.4718*	0.4718*	0.4718*	0.4718*	0.4718*	0.4718*
80	1 3	1 3	1 3	1 3	1 3	1 3	1 3
	0.5059	0.5059*	0.5059*	0.5059*	0.5059*	0.5059*	0.5059*
90	1 3	1 3	1 3	1 3	1 3	1 3	1 3
	0.5333	0.5333*	0.5333*	0.5333*	0.5333*	0.5333*	0.5333*
100	1 3	1 3	1 3	1 3	1 3	1 3	1 3
	0.5546	0.5546*	0.5546*	0.5546*	0.5546*	0.5546*	0.5546*
150	1 3	1 4	1 4	1 4	1 4	1 4	1 4
	0.5880	0.7139	0.7139	0.7139*	0.7139*	0.7139*	0.7139*
200	1 3	1 4	1 4	1 5	1 5	1 5	1 5
	0.5427	0.7241	0.7241	0.8143	0.8143*	0.8143*	0.8143*
250	2 5	2 5	1 5	1 5	1 6	1 6	1 6
	0.6064	0.6064	0.8111	0.8111	0.8778*	0.8778*	0.8778*
300	2 5	2 5	2 6	1 6	1 7	1 7	1 7
	0.6185	0.6185	0.7194	0.8681	0.9182	0.9182*	0.9182*
400	3 6	3 7	2 7	2 8	1 8	1 9	1 9
	0.5493	0.6537	0.7999	0.8593	0.9318	0.9613	0.9613*
500	4 8	4 8	3 8	3 9	1 9	1 10	1 11
	0.6041	0.6041	0.7443	0.8095	0.9263	0.9623	0.9802*
600	5 9	4 9	4 10	3 10	2 10	1 11	1 13
	0.5646	0.6984	0.7673	0.8560	0.9001	0.9558	0.9891*
700	6 10	5 10	5 11	4 11	3 12	2 13	1 15
	0.5321	0.6599	0.7309	0.8218	0.9185	0.9666	0.9936
800	7 11	6 11	5 11	5 13	3 13	3 14	1 16
	0.5047	0.6270	0.7184	0.8387	0.9237	0.9531	0.9918
900	7 11	7 13	6 13	6 14	5 15	3 15	2 18
	0.5012	0.6715	0.7624	0.8127	0.9052	0.9533	0.9937
1000	8 13	8 14	7 14	6 15	5 16	4 17	3 19
	0.5736	0.6467	0.7367	0.8514	0.9234	0.9635	0.9904

A * indicates that a symmetric confidence interval with the desired confidence cannot be achieved without a larger sample size. With samples of size $n = 50$ or smaller, there is no symmetric confidence interval at any confidence level.

Table A.15b Order Statistics ℓ and u and Actual Confidence Levels for Two-Sided Distribution-Free Confidence Intervals for the 5th Percentile for Various Sample Sizes n and Nominal $100(1 - \alpha)\%$ Confidence Levels

n	$1-\alpha$ 0.50	0.60	0.70	0.80	0.90	0.95	0.99
15	1 3	1 3	1 3	1 3	1 3	1 3	1 3
	0.5005	0.5005*	0.5005*	0.5005*	0.5005*	0.5005*	0.5005*
20	1 3	1 3	1 3	1 3	1 3	1 3	1 3
	0.5660	0.5660*	0.5660*	0.5660*	0.5660*	0.5660*	0.5660*
25	1 3	1 3	1 3	1 3	1 3	1 3	1 3
	0.5955	0.5955*	0.5955*	0.5955*	0.5955*	0.5955*	0.5955*
30	1 3	1 4	1 4	1 4	1 4	1 4	1 4
	0.5975	0.7246	0.7246	0.7246*	0.7246*	0.7246*	0.7246*
35	1 3	1 4	1 4	1 5	1 5	1 5	1 5
	0.5797	0.7382	0.7382	0.8049	0.8049*	0.8049*	0.8049*
40	1 3	1 4	1 4	1 5	1 5	1 5	1 5
	0.5482	0.7333	0.7333	0.8235	0.8235*	0.8235*	0.8235*
45	1 3	1 4	1 4	1 5	1 5	1 5	1 5
	0.5082	0.7139	0.7139	0.8276	0.8276*	0.8276*	0.8276*
50	2 5	2 5	1 5	1 5	1 6	1 6	1 6
	0.6170	0.6170	0.8194	0.8194	0.8853*	0.8853*	0.8853*
60	2 5	2 5	2 6	1 6	1 7	1 7	1 7
	0.6281	0.6281	0.7297	0.8752	0.9242	0.9242*	0.9242*
70	3 6	2 6	2 6	2 7	1 7	1 8	1 8
	0.5490	0.7336	0.7336	0.8104	0.9121	0.9491*	0.9491*
80	3 6	3 7	2 6	2 7	1 8	1 9	1 9
	0.5586	0.6641	0.7032	0.8087	0.9369	0.9651	0.9651*
90	4 7	3 7	3 8	2 8	2 9	1 9	1 10
	0.5003	0.6696	0.7523	0.8620	0.9071	0.9540	0.9756*
100	4 7	4 8	3 8	3 9	1 9	1 10	1 11
	0.5082	0.6142	0.7538	0.8186	0.9310	0.9659	0.9826*
150	6 10	6 11	5 11	5 12	4 13	2 13	1 15
	0.5464	0.6333	0.7422	0.8004	0.9067	0.9574	0.9911
200	8 13	8 14	7 14	6 14	5 16	4 17	3 19
	0.5832	0.6568	0.7464	0.8078	0.9292	0.9672	0.9918
250	11 16	10 16	9 17	9 18	7 19	6 20	4 22
	0.5204	0.6167	0.7564	0.8026	0.9213	0.9598	0.9909
300	13 19	12 19	12 20	11 21	9 22	8 23	6 26
	0.5637	0.6470	0.7030	0.8101	0.9174	0.9548	0.9927
400	18 24	17 25	16 26	15 27	13 28	12 29	9 32
	0.5015	0.6345	0.7436	0.8284	0.9165	0.9502	0.9916
500	22 29	21 30	20 31	19 32	17 34	15 35	12 38
	0.5274	0.6447	0.7419	0.8191	0.9203	0.9589	0.9912
600	27 35	26 35	25 37	24 38	22 40	20 41	16 44
	0.5415	0.6010	0.7355	0.8079	0.9078	0.9517	0.9903
700	32 40	31 41	29 41	28 43	26 45	24 47	21 51
	0.5080	0.6102	0.7028	0.8076	0.9016	0.9546	0.9911
800	36 45	35 46	34 47	33 49	30 51	28 53	25 57
	0.5346	0.6280	0.7088	0.8038	0.9124	0.9580	0.9909
900	41 50	40 52	39 53	37 54	35 57	32 58	28 62
	0.5087	0.6382	0.7131	0.8071	0.9073	0.9528	0.9900
1000	46 56	45 57	43 58	42 60	39 62	37 64	32 68
	0.5290	0.6133	0.7239	0.8070	0.9055	0.9504	0.9904

A * indicates that a symmetric confidence interval with the desired confidence cannot be achieved without a larger sample size. With samples of size $n = 10$ or smaller, there is no symmetric confidence interval at any confidence level.

Table A.15c   Order Statistics ℓ and u and Actual Confidence Levels for Two-Sided Distribution-Free Confidence Intervals for the 10th Percentile for Various Sample Sizes n and Nominal $100(1 - \alpha)\%$ Confidence Levels

n	$1 - \alpha$												
	0.50		0.60		0.70		0.80		0.90		0.95		0.99
10	1	3	1	3	1	3	1	3	1	3	1	3	1 3
	0.5811		0.5811*		0.5811*		0.5811*		0.5811*		0.5811*		0.5811*
15	1	3	1	3	1	4	1	4	1	4	1	4	1 4
	0.6100		0.6100		0.7386		0.7386*		0.7386*		0.7386*		0.7386*
20	1	3	1	4	1	4	1	5	1	5	1	5	1 5
	0.5554		0.7455		0.7455		0.8352		0.8352*		0.8352*		0.8352*
25	2	5	2	5	1	5	1	5	1	6	1	6	1 6
	0.6308		0.6308		0.8302		0.8302		0.8948*		0.8948*		0.8948*
30	2	5	2	5	2	6	1	6	1	7	1	7	1 7
	0.6408		0.6408		0.7431		0.8844		0.9318		0.9318*		0.9318*
35	3	6	2	5	2	6	2	7	1	7	1	8	1 8
	0.5621		0.6084		0.7460		0.8224		0.9198		0.9550		0.9550*
40	3	6	3	7	2	6	2	7	1	8	1	9	1 9
	0.5709		0.6777		0.7133		0.8200		0.9433		0.9697		0.9697*
45	4	7	3	7	3	8	2	8	1	8	1	9	1 10
	0.5126		0.6824		0.7653		0.8719		0.9156		0.9593		0.9792*
50	4	7	4	8	3	8	3	9	2	9	1	10	1 11
	0.5199		0.6276		0.7661		0.8304		0.9083		0.9703		0.9855*
60	5	9	4	8	4	9	3	9	3	11	2	11	1 13
	0.5874		0.6142		0.7210		0.8053		0.9127		0.9520		0.9925
70	6	10	5	10	5	11	4	11	3	12	2	12	1 14
	0.5542		0.6826		0.7539		0.8415		0.9318		0.9504		0.9906
80	7	11	6	11	6	12	5	12	4	13	3	14	2 16
	0.5262		0.6497		0.7226		0.8116		0.9109		0.9626		0.9925
90	8	12	7	12	6	12	6	14	4	14	3	15	2 17
	0.5020		0.6210		0.7103		0.8334		0.9198		0.9621		0.9917
100	8	13	8	14	7	14	7	15	6	16	4	16	3 19
	0.5958		0.6701		0.7590		0.8103		0.9025		0.9523		0.9935
150	13	18	12	19	12	20	11	21	9	22	8	23	6 25
	0.5036		0.6599		0.7161		0.8219		0.9253		0.9603		0.9904
200	18	24	17	25	16	25	15	26	14	28	12	29	10 32
	0.5134		0.6476		0.7120		0.8066		0.9000		0.9561		0.9911
250	22	29	21	29	21	31	19	32	18	34	16	35	13 38
	0.5394		0.6017		0.7034		0.8307		0.9077		0.9558		0.9916
300	27	35	26	35	25	36	24	38	22	40	20	41	17 44
	0.5538		0.6137		0.7107		0.8198		0.9164		0.9575		0.9907
400	36	45	35	46	34	47	33	49	31	51	28	52	25 56
	0.5468		0.6409		0.7218		0.8158		0.9040		0.9540		0.9903
500	46	56	45	57	44	58	42	60	39	62	37	64	33 68
	0.5411		0.6262		0.7009		0.8189		0.9142		0.9563		0.9910
600	56	66	54	67	53	69	51	70	48	73	46	75	42 80
	0.5014		0.6237		0.7218		0.8044		0.9116		0.9520		0.9905
700	65	76	64	78	62	79	60	81	57	84	54	86	50 91
	0.5116		0.6202		0.7160		0.8145		0.9115		0.9559		0.9902
800	75	87	73	88	72	90	70	92	67	95	63	97	58 102
	0.5187		0.6233		0.7097		0.8042		0.9008		0.9546		0.9901
900	84	97	83	99	81	100	79	103	76	106	72	108	67 114
	0.5298		0.6244		0.7090		0.8168		0.9042		0.9543		0.9910
1000	94	107	92	108	91	111	88	113	85	117	81	119	76 125
	0.5067		0.6011		0.7070		0.8127		0.9082		0.9546		0.9902

A * indicates that a symmetric confidence interval with the desired confidence cannot be achieved without a larger sample size.

Table A.15d Order Statistics ℓ and u and Actual Confidence Levels for Two-Sided Distribution-Free Confidence Intervals for the 20th Percentile for Various Sample Sizes n and Nominal $100(1 - \alpha)\%$ Confidence Levels

n	0.50	0.60	0.70	0.80	0.90	0.95	0.99
10	2 4	1 4	1 4	1 5	1 5	1 5	1 5
	0.5033	0.7718	0.7718	0.8598	0.8598*	0.8598*	0.8598*
15	2 5	2 5	2 6	1 5	1 6	1 7	1 7
	0.6686	0.6686	0.7718	0.8006	0.9038	0.9468*	0.9468*
20	3 6	3 7	3 7	2 7	1 7	1 8	1 9
	0.5981	0.7072	0.7072	0.8441	0.9018	0.9563	0.9785*
25	4 7	4 8	3 8	3 9	2 9	2 10	1 11
	0.5460	0.6569	0.7927	0.8550	0.9258	0.9553	0.9907
30	5 8	5 9	4 9	4 10	2 10	2 11	1 13
	0.5056	0.6161	0.7486	0.8162	0.9284	0.9639	0.9957
35	6 10	5 9	5 10	4 11	4 12	2 12	2 14
	0.5822	0.6015	0.7108	0.8648	0.9051	0.9617	0.9908
40	7 11	6 11	6 12	5 12	4 13	4 14	2 15
	0.5533	0.6779	0.7512	0.8366	0.9283	0.9521	0.9906
45	8 12	7 12	7 13	6 13	5 14	4 15	3 17
	0.5284	0.6491	0.7237	0.8103	0.9097	0.9621	0.9923
50	9 13	8 13	7 13	7 15	5 15	5 16	3 18
	0.5066	0.6235	0.7105	0.8359	0.9208	0.9507	0.9925
60	10 15	10 16	9 16	8 16	7 18	6 19	5 21
	0.5802	0.6562	0.7426	0.8024	0.9265	0.9658	0.9912
70	12 17	12 18	11 18	10 19	9 20	7 21	5 23
	0.5448	0.6202	0.7051	0.8229	0.9018	0.9617	0.9914
80	14 19	13 20	13 21	12 22	11 23	9 24	7 26
	0.5151	0.6725	0.7294	0.8334	0.9047	0.9652	0.9924
90	16 22	15 22	15 23	14 24	12 25	11 26	8 28
	0.5630	0.6439	0.7014	0.8078	0.9148	0.9532	0.9904
100	18 24	17 24	16 25	15 26	13 27	12 28	10 31
	0.5397	0.6186	0.7401	0.8321	0.9188	0.9533	0.9916
150	27 34	26 35	25 36	24 37	22 39	20 40	17 43
	0.5249	0.6418	0.7389	0.8162	0.9182	0.9579	0.9913
200	37 45	36 46	35 47	33 48	31 50	29 52	25 55
	0.5170	0.6198	0.7081	0.8157	0.9076	0.9585	0.9915
250	46 55	45 56	44 58	42 59	40 61	38 63	34 67
	0.5232	0.6156	0.7293	0.8215	0.9037	0.9524	0.9910
300	56 66	55 67	53 68	52 70	49 72	46 74	42 78
	0.5272	0.6112	0.7212	0.8047	0.9035	0.9563	0.9903
400	75 86	74 88	72 89	70 91	67 94	64 96	59 101
	0.5082	0.6167	0.7121	0.8109	0.9088	0.9542	0.9911
500	94 107	93 109	91 110	89 112	86 116	82 118	77 124
	0.5326	0.6275	0.7119	0.8017	0.9062	0.9556	0.9915
600	114 128	112 129	110 131	108 134	104 137	101 140	95 146
	0.5239	0.6144	0.7162	0.8148	0.9081	0.9536	0.9908
700	133 148	132 150	130 152	127 155	123 158	119 161	113 168
	0.5215	0.6039	0.7005	0.8135	0.9020	0.9526	0.9907
800	153 169	151 171	149 173	146 175	142 180	138 183	131 190
	0.5196	0.6224	0.7104	0.8002	0.9068	0.9534	0.9909
900	172 189	170 191	168 193	165 196	161 201	156 204	149 211
	0.5212	0.6184	0.7025	0.8036	0.9043	0.9544	0.9901
1000	192 210	190 212	187 214	184 217	180 222	175 225	167 233
	0.5225	0.6148	0.7142	0.8080	0.9030	0.9518	0.9908

* indicates that a symmetric confidence interval with the desired confidence cannot be achieved without a larger sample size.

Table A.15e   Order Statistics ℓ and u and Actual Confidence Levels for Two-Sided Distribution-Free Confidence Intervals for the 30th Percentile for Various Sample Sizes n and Nominal $100(1 - \alpha)\%$ Confidence Levels

n	$1 - \alpha$						
	0.50	0.60	0.70	0.80	0.90	0.95	0.99
10	1 7	1 7	1 7	1 7	1 7	1 7	1 7
	0.3221*	0.3221*	0.3221*	0.3221*	0.3221*	0.3221*	0.3221*
15	1 10	1 10	1 10	1 10	1 10	1 10	1 10
	0.2736*	0.2736*	0.2736*	0.2736*	0.2736*	0.2736*	0.2736*
20	1 13	1 13	1 13	1 13	1 13	1 13	1 13
	0.2269*	0.2269*	0.2269*	0.2269*	0.2269*	0.2269*	0.2269*
25	1 16	1 16	1 16	1 16	1 16	1 16	1 16
	0.1893*	0.1893*	0.1893*	0.1893*	0.1893*	0.1893*	0.1893*
30	1 19	1 19	1 19	1 19	1 19	1 19	1 19
	0.1593*	0.1593*	0.1593*	0.1593*	0.1593*	0.1593*	0.1593*
35	1 22	1 22	1 22	1 22	1 22	1 22	1 22
	0.1350*	0.1350*	0.1350*	0.1350*	0.1350*	0.1350*	0.1350*
40	1 25	1 25	1 25	1 25	1 25	1 25	1 25
	0.1151*	0.1151*	0.1151*	0.1151*	0.1151*	0.1151*	0.1151*
45	1 28	1 28	1 28	1 28	1 28	1 28	1 28
	0.0986*	0.0986*	0.0986*	0.0986*	0.0986*	0.0986*	0.0986*
50	1 31	1 31	1 31	1 31	1 31	1 31	1 31
	0.0848*	0.0848*	0.0848*	0.0848*	0.0848*	0.0848*	0.0848*
60	1 37	1 37	1 37	1 37	1 37	1 37	1 37
	0.0632*	0.0632*	0.0632*	0.0632*	0.0632*	0.0632*	0.0632*
70	1 43	1 43	1 43	1 43	1 43	1 43	1 43
	0.0476*	0.0476*	0.0476*	0.0476*	0.0476*	0.0476*	0.0476*
80	1 49	1 49	1 49	1 49	1 49	1 49	1 49
	0.0360*	0.0360*	0.0360*	0.0360*	0.0360*	0.0360*	0.0360*
90	1 55	1 55	1 55	1 55	1 55	1 55	1 55
	0.0274*	0.0274*	0.0274*	0.0274*	0.0274*	0.0274*	0.0274*
100	1 61	1 61	1 61	1 61	1 61	1 61	1 61
	0.0210*	0.0210*	0.0210*	0.0210*	0.0210*	0.0210*	0.0210*
150	1 91	1 91	1 91	1 91	1 91	1 91	1 91
	0.0057*	0.0057*	0.0057*	0.0057*	0.0057*	0.0057*	0.0057*
200	1 121	1 121	1 121	1 121	1 121	1 121	1 121
	0.0016*	0.0016*	0.0016*	0.0016*	0.0016*	0.0016*	0.0016*
250	1 151	1 151	1 151	1 151	1 151	1 151	1 151
	0.0005*	0.0005*	0.0005*	0.0005*	0.0005*	0.0005*	0.0005*
300	1 181	1 181	1 181	1 181	1 181	1 181	1 181
	0.0001*	0.0001*	0.0001*	0.0001*	0.0001*	0.0001*	0.0001*
400	1 241	1 241	1 241	1 241	1 241	1 241	1 241
	0.0000*	0.0000*	0.0000*	0.0000*	0.0000*	0.0000*	0.0000*
500	1 301	1 301	1 301	1 301	1 301	1 301	1 301
	0.0000*	0.0000*	0.0000*	0.0000*	0.0000*	0.0000*	0.0000*
600	1 361	1 361	1 361	1 361	1 361	1 361	1 361
	0.0000*	0.0000*	0.0000*	0.0000*	0.0000*	0.0000*	0.0000*
700	1 421	1 421	1 421	1 421	1 421	1 421	1 421
	0.0000*	0.0000*	0.0000*	0.0000*	0.0000*	0.0000*	0.0000*
800	1 481	1 481	1 481	1 481	1 481	1 481	1 481
	0.0000*	0.0000*	0.0000*	0.0000*	0.0000*	0.0000*	0.0000*
900	1 541	1 541	1 541	1 541	1 541	1 541	1 541
	0.0000*	0.0000*	0.0000*	0.0000*	0.0000*	0.0000*	0.0000*
1000	1 601	1 601	1 601	1 601	1 601	1 601	1 601
	0.0000*	0.0000*	0.0000*	0.0000*	0.0000*	0.0000*	0.0000*

A * indicates that a symmetric confidence interval with the desired confidence cannot be achieved without a larger sample size.

able A.15f Order Statistics ℓ and u and Actual Confidence Levels for Two-Sided Distribution-Free Confidence Intervals for the 40th Percentile for Various Sample Sizes n and Nominal 100(1 − α)% Confidence Levels

n	0.50	0.60	0.70	0.80	0.90	0.95	0.99
				$1-\alpha$			
10	2 7	1 7	2 8	1 8	1 9	1 9	1 9
	0.5714	0.6117	0.7864	0.8267	0.9476	0.9476*	0.9476*
15	4 10	3 11	3 11	2 12	2 12	1 13	1 13
	0.5063	0.7556	0.7556	0.9043	0.9043	0.9724	0.9724*
20	5 13	4 14	4 14	3 15	2 16	1 17	1 17
	0.5332	0.7340	0.7340	0.8708	0.9485	0.9840	0.9840*
25	6 16	5 17	5 17	4 18	3 19	2 20	1 21
	0.5460	0.7170	0.7170	0.8441	0.9260	0.9706	0.9905
30	7 19	6 20	6 20	5 21	4 22	3 23	1 25
	0.5517	0.7029	0.7029	0.8222	0.9057	0.9564	0.9943
35	8 22	7 23	6 24	6 24	4 26	3 27	1 29
	0.5537	0.6909	0.8039	0.8039	0.9424	0.9740	0.9966
40	9 25	8 26	7 27	6 28	5 29	4 30	2 32
	0.5537	0.6805	0.7882	0.8713	0.9290	0.9648	0.9939
45	10 28	9 29	8 30	7 31	6 32	5 33	3 35
	0.5528	0.6715	0.7747	0.8569	0.9164	0.9554	0.9906
50	11 31	10 32	9 33	8 34	7 35	5 37	3 39
	0.5513	0.6636	0.7629	0.8438	0.9045	0.9720	0.9943
60	13 37	12 38	11 39	10 40	8 42	7 43	4 46
	0.5481	0.6505	0.7431	0.8214	0.9281	0.9587	0.9950
70	15 43	14 44	13 45	12 46	10 48	8 50	6 52
	0.5450	0.6399	0.7271	0.8027	0.9115	0.9682	0.9911
80	17 49	16 50	15 51	13 53	11 55	10 56	7 59
	0.5423	0.6312	0.7139	0.8479	0.9325	0.9583	0.9928
90	19 55	18 56	17 57	15 59	13 61	11 63	8 66
	0.5399	0.6240	0.7028	0.8334	0.9203	0.9679	0.9942
100	21 61	20 62	18 64	17 65	15 67	13 69	10 72
	0.5379	0.6178	0.7614	0.8205	0.9087	0.9602	0.9916
150	31 91	29 93	28 94	26 96	23 99	21 101	17 105
	0.5310	0.6598	0.7189	0.8201	0.9227	0.9611	0.9929
200	41 121	39 123	37 125	35 127	32 130	30 132	25 137
	0.5268	0.6393	0.7410	0.8258	0.9156	0.9525	0.9920
250	51 151	49 153	47 155	44 158	41 161	38 164	33 169
	0.5240	0.6251	0.7184	0.8335	0.9130	0.9602	0.9921
300	61 181	59 183	57 185	54 188	50 192	47 195	41 201
	0.5219	0.6145	0.7010	0.8114	0.9129	0.9571	0.9926
400	81 241	78 244	76 246	73 249	68 254	65 257	58 264
	0.5190	0.6384	0.7119	0.8069	0.9164	0.9546	0.9922
500	101 301	98 304	95 307	92 310	87 315	83 319	76 326
	0.5170	0.6243	0.7228	0.8069	0.9076	0.9550	0.9904
600	121 361	118 364	115 367	111 371	106 376	101 381	93 389
	0.5155	0.6137	0.7053	0.8090	0.9021	0.9568	0.9916
700	141 421	138 424	134 428	130 432	124 438	120 442	111 451
	0.5144	0.6055	0.7180	0.8124	0.9119	0.9519	0.9910
800	161 481	157 485	154 488	149 493	143 499	138 504	129 513
	0.5134	0.6265	0.7052	0.8164	0.9094	0.9555	0.9908
900	181 541	177 545	173 549	169 553	162 560	157 565	147 575
	0.5127	0.6195	0.7180	0.8023	0.9080	0.9527	0.9908
1000	201 601	197 605	193 609	188 614	181 621	176 626	165 637
	0.5120	0.6135	0.7079	0.8081	0.9074	0.9505	0.9910

* indicates that a symmetric confidence interval with the desired confidence cannot be achieved without a larger sample size.

324

Table A.15g Order Statistics ℓ and u and Actual Confidence Levels for Two-Sided Distribution-Free Confidence Intervals for the 50th Percentile for Various Sample Sizes n and Nominal $100(1 - \alpha)\%$ Confidence Levels

n	$1-\alpha$ = 0.50	0.60	0.70	0.80	0.90	0.95	0.99
10	4 7	4 7	3 7	3 8	3 9	2 9	1 10
	0.6563	0.6563	0.7734	0.8906	0.9346	0.9785	0.9980
15	6 9	6 10	5 10	5 11	4 11	4 12	3 13
	0.5455	0.6982	0.7899	0.8815	0.9232	0.9648	0.9926
20	9 13	9 13	8 13	7 13	7 15	6 15	5 17
	0.6167	0.6167	0.7368	0.8108	0.9216	0.9586	0.9928
25	11 15	10 15	10 16	9 16	8 17	8 18	6 19
	0.5756	0.6731	0.7705	0.8314	0.9245	0.9567	0.9906
30	14 18	13 18	13 19	11 19	11 20	10 21	8 23
	0.5269	0.6384	0.7190	0.8504	0.9013	0.9572	0.9948
35	16 20	15 21	14 21	14 22	13 23	12 24	10 26
	0.5004	0.6895	0.7570	0.8245	0.9105	0.9590	0.9940
40	18 23	18 24	17 24	16 25	15 26	14 27	12 29
	0.5704	0.6511	0.7318	0.8461	0.9193	0.9615	0.9936
45	20 25	20 26	19 27	18 27	17 29	16 30	14 32
	0.5386	0.6287	0.7673	0.8161	0.9275	0.9643	0.9934
50	23 28	22 29	22 30	20 30	20 32	19 33	16 35
	0.5201	0.6778	0.7376	0.8392	0.9081	0.9511	0.9934
60	28 34	27 34	26 35	25 36	24 37	23 39	21 41
	0.5574	0.6337	0.7549	0.8450	0.9075	0.9604	0.9901
70	33 39	32 40	31 40	30 41	29 43	27 44	24 46
	0.5233	0.6575	0.7180	0.8118	0.9041	0.9586	0.9914
80	37 44	37 45	36 46	35 47	33 48	32 50	29 52
	0.5660	0.6258	0.7336	0.8179	0.9071	0.9552	0.9903
90	42 49	41 50	41 51	39 52	38 54	36 55	33 58
	0.5392	0.6572	0.7055	0.8298	0.9071	0.9554	0.9920
100	47 54	46 55	45 56	44 57	42 59	41 61	38 64
	0.5159	0.6318	0.7287	0.8067	0.9114	0.9540	0.9907
150	71 80	70 81	69 82	68 84	65 86	63 88	60 92
	0.5375	0.6308	0.7115	0.8073	0.9139	0.9591	0.9910
200	96 106	95 107	93 108	91 110	89 113	87 115	82 119
	0.5193	0.6026	0.7112	0.8210	0.9098	0.9520	0.9913
250	120 131	119 133	117 134	115 136	112 139	110 141	105 146
	0.5133	0.6231	0.7177	0.8160	0.9125	0.9503	0.9906
300	145 157	143 158	142 160	139 162	136 165	134 168	128 173
	0.5108	0.6135	0.7005	0.8159	0.9061	0.9502	0.9907
400	194 208	192 209	190 211	188 214	184 217	181 221	175 227
	0.5155	0.6046	0.7063	0.8059	0.9012	0.9544	0.9907
500	243 259	241 260	239 263	236 265	232 269	229 273	222 280
	0.5252	0.6045	0.7164	0.8054	0.9021	0.9508	0.9905
600	292 309	290 311	288 314	285 317	280 321	276 325	269 333
	0.5123	0.6087	0.7111	0.8083	0.9059	0.9546	0.9910
700	342 360	339 362	337 365	334 368	329 373	325 377	316 385
	0.5034	0.6153	0.7097	0.8009	0.9035	0.9505	0.9909
800	391 411	389 413	386 416	382 419	377 424	373 429	364 437
	0.5202	0.6035	0.7109	0.8092	0.9035	0.9522	0.9902
900	440 461	438 464	435 467	431 470	426 476	421 480	412 490
	0.5161	0.6136	0.7136	0.8064	0.9043	0.9508	0.9907
1000	490 512	487 514	484 517	480 521	474 527	470 532	460 542
	0.5131	0.6068	0.7033	0.8052	0.9063	0.9500	0.9905

Table A.16 Number of Extreme Observations ν to Be Removed from the Ends of a Sample of Size n to Obtain a Two-Sided Distribution-Free Tolerance Interval or to Obtain a One-Sided Distribution-Free Tolerance Bound That Contains at Least 100p% of the Sampled Population with 100(1 − α)% Confidence

In each cell the upper number is the achieved confidence and the lower number is ν.

n	1−α:	p=0.750			p=0.900			p=0.950			p=0.990		
		0.90	0.95	0.99	0.90	0.95	0.99	0.90	0.95	0.99	0.90	0.95	0.99
10		0.9437	0.9437*	0.9437*	0.6513*	0.6513*	0.6513*	0.4013*	0.4013*	0.4013*	0.0956*	0.0956*	0.0956*
	ν	1	1	1	1	1	1	1	1	1	1	1	1
15		0.9198	0.9866	0.9866*	0.7941*	0.7941*	0.7941*	0.5367*	0.5367*	0.5367*	0.1399*	0.1399*	0.1399*
	ν	2	1	1	1	1	1	1	1	1	1	1	1
20		0.9087	0.9757	0.9968	0.8784*	0.8784*	0.8784*	0.6415*	0.6415*	0.6415*	0.1821*	0.1821*	0.1821*
	ν	3	2	1	1	1	1	1	1	1	1	1	1
25		0.9038	0.9679	0.9930	0.9282	0.9282*	0.9282*	0.7226*	0.7226*	0.7226*	0.2222*	0.2222*	0.2222*
	ν	4	3	2	1	1	1	1	1	1	1	1	1
30		0.9021	0.9626	0.9980	0.9576	0.9576	0.9576*	0.7854*	0.7854*	0.7854*	0.2603*	0.2603*	0.2603*
	ν	5	4	2	1	1	1	1	1	1	1	1	1
35		0.9024	0.9590	0.9967	0.9750	0.9750	0.9750*	0.8339*	0.8339*	0.8339*	0.2966*	0.2966*	0.2966*
	ν	6	5	3	1	1	1	1	1	1	1	1	1
40		0.9038	0.9567	0.9953	0.9195	0.9852	0.9852*	0.8715*	0.8715*	0.8715*	0.3310*	0.3310*	0.3310*
	ν	7	6	4	2	1	1	1	1	1	1	1	1
50		0.9084	0.9547	0.9930	0.9662	0.9662	0.9948	0.9231	0.9231*	0.9231*	0.3950*	0.3950*	0.3950*
	ν	9	8	6	2	2	1	1	1	1	1	1	1
60		0.9141	0.9548	0.9912	0.9470	0.9862	0.9982	0.9539	0.9539	0.9539*	0.4528*	0.4528*	0.4528*
	ν	11	10	8	3	2	1	1	1	1	1	1	1
80		0.9260	0.9579	0.9953	0.9120	0.9647	0.9978	0.9139	0.9835	0.9835*	0.5525*	0.5525*	0.5525*
	ν	15	14	11	4	3	2	2	1	1	1	1	1
100		0.9005	0.9624	0.9946	0.9424	0.9763	0.9922	0.9629	0.9629	0.9941	0.6340*	0.6340*	0.6340*
	ν	20	18	15	6	3	2	2	2	1	1	1	1
200		0.9196	0.9595	0.9927	0.9071	0.9680	0.9919	0.9377	0.9736	0.9910	0.8660*	0.8660*	0.8660*
	ν	42	40	36	15	13	11	6	5	4	1	1	1
300		0.9210	0.9544	0.9916	0.9301	0.9542	0.9993	0.9350	0.9659	0.9934	0.9510	0.9510	0.9510*
	ν	65	63	58	23	22	19	10	9	7	1	1	1
400		0.9092	0.9548	0.9922	0.9254	0.9643	0.9908	0.9010	0.9645	0.9906	0.9095	0.9820*	0.9820*
	ν	89	86	80	32	30	27	15	13	11	2	1	1
500		0.9027	0.9575	0.9910	0.9249	0.9607	0.9921	0.9135	0.9657	0.9945	0.9602	0.9602	0.9934
	ν	113	109	103	41	39	35	19	17	14	2	2	1
600		0.9153	0.9520	0.9905	0.9043	0.9591	0.9901	0.9247	0.9680	0.9938	0.9389	0.9830	0.9976
	ν	136	133	126	51	48	44	23	21	18	3	2	1
800		0.9120	0.9542	0.9909	0.9146	0.9593	0.9912	0.9199	0.9606	0.9935	0.9015	0.9583	0.9971
	ν	184	180	172	69	66	61	32	30	26	5	4	2
1000		0.9001	0.9509	0.9901	0.9081	0.9515	0.9901	0.9194	0.9566	0.9907	0.9339	0.9713	0.9973
	ν	233	228	219	88	85	79	41	39	35	6	5	3

* Indicates that a symmetric confidence interval or bound with the desired confidence level cannot be achieved without a larger sample size. Also see Figure 5.3a–d. A table similar to Table A.16 appeared earlier in Somerville (1958). Adapted with permission of the Institute of Mathematical Statistics.

Table A.17 Smallest Sample Size for the Range Formed by the Smallest and Largest Observations to Contain, with 100(1 − α)% Confidence, at Least 100p% of the Sampled Population

	$1-\alpha$						
p	0.50	0.75	0.90	0.95	0.98	0.99	0.999
0.500	3	5	7	8	9	11	14
0.550	4	6	8	9	11	12	16
0.600	4	6	9	10	12	14	19
0.650	5	7	10	12	15	16	22
0.700	6	9	12	14	17	20	27
0.750	7	10	15	18	21	24	33
0.800	9	13	18	22	27	31	42
0.850	11	18	25	30	37	42	58
0.900	17	27	38	46	56	64	89
0.950	34	53	77	93	115	130	181
0.960	42	67	96	117	144	164	227
0.970	56	89	129	157	193	219	304
0.980	84	134	194	236	290	330	458
0.990	168	269	388	473	581	662	920
0.995	336	538	777	947	1165	1325	1843
0.999	1679	2692	3889	4742	5832	6636	9230

A table similar to Table A.17 appeared earlier in Dixon and Massey (1969). Adapted with permission of McGraw-Hill, Inc.

Table A.18 Smallest Sample Size for the Largest (Smallest) Observation to Exceed (Be Exceeded by), with 100(1 − α)% Confidence, at Least 100p% of the Sampled Population

	$1-\alpha$						
p	0.50	0.75	0.90	0.95	0.98	0.99	0.999
0.500	1	2	4	5	6	7	10
0.550	2	3	4	6	7	8	12
0.600	2	3	5	6	8	10	14
0.650	2	4	6	7	10	11	17
0.700	2	4	7	9	11	13	20
0.750	3	5	9	11	14	17	25
0.800	4	7	11	14	18	21	31
0.850	5	9	15	19	25	29	43
0.900	7	14	22	29	38	44	66
0.950	14	28	45	59	77	90	135
0.960	17	34	57	74	96	113	170
0.970	23	46	76	99	129	152	227
0.980	35	69	114	149	194	228	342
0.990	69	138	230	299	390	459	688
0.995	139	277	460	598	781	919	1379
0.999	693	1386	2302	2995	3911	4603	6905

A table similar to Table A.18 appeared earlier in Dixon and Massey (1969). Adapted with permission of McGraw-Hill, Inc.

Table A.19a Largest Number k of m Future Observations That Will Be Contained in (Bounded by) the Distribution-Free $100(1 - \alpha)\%$ Prediction Interval (or Bound) Obtained by Removing ν Blocks from the End(s) of a Previous Sample of Size n

$1-\alpha$	n	$\nu=1$ m: 5	10	25	50	75	100	∞	$\nu=2$ 5	10	25	50	75	100	∞	$\nu=3$ 5	10	25	50	75	100	∞	$\nu=4$ 5	10	25	50	75	100	∞
0.50	5	5	9	22	44	65	87	0.871	4	7	17	34	52	69	0.686	3	5	13	25	38	50	0.500	2	3	8	16	23	31	0.314
	10	5	9	23	47	70	93	0.933	4	9	21	42	63	84	0.838	4	8	19	37	56	74	0.741	3	7	16	32	48	65	0.645
	25	5	10	25	49	73	97	0.973	5	10	24	47	70	94	0.934	5	9	23	45	67	90	0.894	4	9	22	43	64	86	0.855
	50	5	10	25	50	74	99	0.986	5	10	24	49	73	97	0.967	5	10	24	48	71	95	0.947	5	9	23	47	70	93	0.927
	75	5	10	25	50	75	99	0.991	5	10	25	49	74	98	0.978	5	10	24	48	73	97	0.965	5	10	24	48	72	95	0.951
	100	5	10	25	50	75	99	0.993	5	10	25	49	74	98	0.983	5	10	25	49	73	97	0.973	5	10	24	48	72	96	0.963
0.75	5	4	7	19	38	57	76	0.758	2	5	13	27	41	54	0.546	1	3	9	18	27	36	0.359	1	2	4	9	14	19	0.194
	10	4	9	22	43	65	87	0.871	3	6	18	37	56	75	0.753	3	6	16	32	48	64	0.645	2	5	13	27	40	54	0.542
	25	5	9	24	47	71	94	0.946	4	9	22	44	67	89	0.896	4	8	21	42	63	84	0.849	4	8	20	40	60	80	0.804
	50	5	10	24	49	73	97	0.973	5	9	23	47	71	94	0.947	4	9	23	46	69	92	0.923	4	9	22	45	67	90	0.900
	75	5	10	25	49	74	98	0.982	5	10	24	48	72	96	0.965	5	9	23	47	71	94	0.948	5	9	23	46	70	93	0.933
	100	5	10	25	49	74	99	0.986	5	10	24	48	73	97	0.973	5	9	24	48	72	96	0.961	5	9	23	47	71	95	0.949
0.90	5	3	6	15	31	47	62	0.631	1	3	10	20	30	41	0.416	1	2	5	12	18	24	0.247	0	1	2	5	8	11	0.112
	10	4	7	19	39	59	79	0.794	3	6	16	32	49	65	0.663	2	5	13	27	40	54	0.550	1	4	10	21	33	44	0.448
	25	4	9	22	45	68	91	0.912	4	8	21	42	63	84	0.853	3	7	19	39	59	79	0.801	3	7	18	37	55	74	0.752
	50	5	9	23	47	71	95	0.955	4	9	22	45	69	92	0.924	4	8	22	44	66	89	0.897	4	8	21	43	64	86	0.871
	75	5	9	24	48	72	96	0.970	4	9	23	47	70	94	0.949	4	9	23	46	69	92	0.931	4	8	22	45	67	90	0.913
	100	5	10	24	48	73	97	0.977	5	10	23	47	71	95	0.962	4	9	23	47	70	94	0.948	4	9	23	46	69	92	0.934
0.95	5	2	5	13	27	40	54	0.549	0	3	8	16	25	33	0.343	0	1	3	9	13	18	0.189	0	0	1	3	5	7	0.076
	10	3	7	18	36	55	73	0.741	2	5	14	29	44	59	0.606	2	4	11	23	36	48	0.493	1	3	9	18	28	38	0.393
	25	4	8	21	43	66	88	0.887	3	7	20	40	61	81	0.824	3	7	18	37	56	75	0.769	3	6	17	34	52	70	0.718
	50	4	9	23	46	70	93	0.942	4	8	22	44	67	90	0.909	4	8	21	43	65	87	0.879	3	7	20	41	62	84	0.852
	75	4	9	23	47	71	95	0.961	4	9	23	46	69	93	0.938	4	8	22	45	68	90	0.918	4	8	21	44	66	88	0.900
	100	5	9	24	48	72	96	0.970	4	9	23	47	70	94	0.953	4	9	22	46	69	92	0.938	4	8	22	45	68	91	0.924
0.99	5	1	3	9	19	29	39	0.398	0	1	4	10	15	21	0.222	0	0	2	4	7	9	0.106	0	0	0	1	2	2	0.033
	10	2	5	14	30	46	62	0.631	1	4	11	23	35	48	0.496	1	2	8	18	27	37	0.388	0	2	6	13	20	28	0.297
	25	3	7	19	40	61	82	0.832	3	6	17	36	55	74	0.763	2	5	16	33	51	68	0.704	2	5	14	30	46	63	0.651
	50	4	8	21	44	67	90	0.912	3	7	20	42	64	85	0.874	3	7	19	40	61	82	0.842	3	6	18	38	59	79	0.813
	75	4	8	22	46	69	92	0.940	4	8	21	44	67	90	0.915	3	7	20	43	65	87	0.893	3	7	20	41·	63	85	0.872
	100	4	9	23	46	70	94	0.955	4	8	22	45	68	92	0.935	3	8	21	44	67	90	0.919	3	8	21	43	65	88	0.903

Table A.19a is similar to tables that first appeared in Danziger and Davis (1964). Adapted with permission of the Institute of Mathematical Statistics.

Table A.19b Largest Number k of m Future Observations That Will Be Contained in (Bounded by) the Distribution-Free $100(1-\alpha)\%$ Prediction Interval (or Bound) Obtained by Removing ν Blocks from the End(s) of a Previous Sample of Size n

$1-\alpha$	n	$\nu=5$							$\nu=6$							$\nu=7$							$\nu=8$						
$m:$		5	10	25	50	75	100	∞	5	10	25	50	75	100	∞	5	10	25	50	75	100	∞	5	10	25	50	75	100	∞
0.50	5	1	1	3	6	10	13	0.129																					
	10	3	6	14	27	41	55	0.548	2	5	11	23	34	45	0.452	2	4	9	18	27	35	0.355	1	3	6	13	19	26	0.259
	25	4	8	21	41	61	82	0.816	4	8	20	39	58	78	0.776	4	7	19	37	55	74	0.737	4	7	18	35	52	70	0.697
	50	5	9	23	46	68	91	0.907	5	9	22	45	67	89	0.887	4	9	22	44	65	87	0.868	4	9	21	43	64	85	0.848
	75	5	10	24	47	71	94	0.938	5	9	23	46	70	93	0.925	5	9	23	46	69	91	0.911	5	9	23	45	67	90	0.898
	100	5	10	24	48	72	95	0.953	5	10	24	47	71	94	0.943	5	10	23	47	70	93	0.934	5	9	23	46	69	92	0.924
0.75	5	0	0	1	2	4	5	0.056																					
	10	2	4	11	22	33	44	0.445	1	3	8	17	26	35	0.351	1	2	6	13	19	26	0.261	0	1	4	8	13	17	0.176
	25	3	7	19	37	56	75	0.760	3	7	17	35	53	71	0.718	3	6	16	33	50	67	0.675	3	6	15	31	47	63	0.634
	50	4	8	22	43	65	87	0.877	4	8	21	42	64	85	0.855	4	8	20	41	62	83	0.833	4	8	20	40	60	80	0.812
	75	4	9	23	45	68	91	0.918	4	9	22	45	67	90	0.903	4	9	22	44	66	88	0.888	4	8	21	43	65	87	0.873
	100	5	9	23	46	70	93	0.938	4	9	23	46	69	92	0.927	4	9	22	45	68	91	0.916	4	9	22	45	67	90	0.904
0.90	5	0	0	0	1	1	2	0.021																					
	10	1	3	8	17	26	34	0.354	1	2	6	12	19	26	0.267	0	1	4	8	13	18	0.188	0	0	2	5	8	11	0.116
	25	3	6	17	34	52	69	0.705	2	6	15	32	48	65	0.660	2	5	14	30	45	60	0.617	2	5	13	27	42	56	0.574
	50	4	8	20	41	62	83	0.846	3	7	19	40	60	81	0.822	3	7	19	39	58	78	0.799	3	7	18	37	57	76	0.776
	75	4	8	21	44	66	88	0.896	4	8	21	43	65	87	0.880	4	8	20	42	63	85	0.864	4	8	20	41	62	83	0.848
	100	4	9	22	45	68	91	0.922	4	8	22	44	67	90	0.909	4	8	21	44	66	88	0.897	4	8	21	43	65	87	0.885
0.95	5	0	0	0	0	0	1	0.010																					
	10	1	2	6	14	21	29	0.304	0	1	4	10	15	21	0.222	0	1	3	6	10	14	0.150	0	0	1	3	6	8	0.087
	25	2	5	15	32	49	65	0.670	2	5	14	30	45	61	0.625	2	4	13	27	42	56	0.580	2	4	12	25	38	52	0.538
	50	3	7	19	40	60	81	0.826	3	7	18	38	58	78	0.801	3	6	18	37	56	76	0.777	3	6	17	36	54	73	0.753
	75	4	8	21	43	65	86	0.882	3	8	20	42	63	85	0.865	3	7	20	41	62	83	0.848	3	7	19	40	60	81	0.832
	100	4	8	21	44	67	89	0.911	4	8	21	43	66	88	0.898	4	8	21	43	65	87	0.885	3	8	20	42	63	85	0.873
0.99	5	0	0	0	0	0	0	0.002																					
	10	0	1	4	9	15	20	0.218	0	1	4	10	15	13	0.150	0	1	3	6	10	8	0.093	0	0	0	1	2	4	0.048
	25	1	4	13	28	43	58	0.602	1	4	12	25	39	53	0.556	1	3	10	23	36	48	0.512	1	3	9	21	33	44	0.469
	50	2	6	17	37	56	76	0.785	2	6	17	35	54	73	0.758	2	5	16	34	52	70	0.733	2	5	15	33	50	68	0.708
	75	3	7	19	40	61	83	0.853	3	7	19	39	60	81	0.834	3	6	18	38	58	79	0.817	2	6	17	37	57	77	0.799
	100	3	7	20	42	64	86	0.888	3	7	20	41	63	85	0.874	3	7	19	40	62	83	0.860	3	7	19	40	61	82	0.847

Table A.19b is similar to tables that first appeared in Danziger and Davis (1964). Adapted with permission of the Institute of Mathematical Statistics.

Table A.19c Largest Number k of m Future Observations That Will Be Contained in (Bounded by) the Distribution-Free 100(1 − α)% Prediction Interval (or Bound) Obtained by Removing ν Blocks from the End(s) of a Previous Sample of Size n

1 − α	n	ν = 9							ν = 10							ν = 11							ν = 12						
	m:	5	10	25	50	75	100	∞	5	10	25	50	75	100	∞	5	10	25	50	75	100	∞	5	10	25	50	75	100	∞
0.50	5																												
	10	1	2	4	8	12	16	0.162	0	1	2	3	5	7	0.067														
	25	3	7	17	33	49	66	0.658	3	6	16	31	46	62	0.618	3	6	15	29	43	58	0.579	3	5	14	27	40	54	0.539
	50	4	8	21	42	62	83	0.828	4	8	20	41	61	81	0.808	4	8	20	40	59	79	0.788	4	8	19	38	58	77	0.768
	75	5	9	22	44	66	89	0.885	5	9	22	44	65	87	0.872	4	9	22	43	65	86	0.858	4	9	21	42	64	85	0.845
	100	5	9	23	46	69	91	0.914	5	9	23	45	68	90	0.904	5	9	22	45	67	89	0.894	5	9	22	44	66	88	0.884
0.75	5																												
	10	0	1	2	4	7	9	0.096	0	0	1	2	2	2	0.028														
	25	3	5	14	29	44	59	0.593	3	5	13	27	41	54	0.552	2	5	12	25	38	50	0.512	2	4	11	23	35	46	0.473
	50	4	7	19	39	59	78	0.790	3	7	19	38	57	76	0.769	3	7	18	37	55	74	0.748	3	7	17	36	54	72	0.727
	75	4	8	21	42	64	85	0.859	4	8	21	42	63	84	0.844	4	8	20	41	61	82	0.830	4	8	20	40	60	81	0.816
	100	4	9	22	44	66	89	0.893	4	8	22	43	65	88	0.883	4	8	21	43	65	86	0.872	4	8	21	42	64	85	0.861
0.90	5																												
	10	0	0	1	2	3	5	0.055	0	0	0	0	0	1	0.010														
	25	2	4	12	25	38	52	0.533	2	4	11	23	35	48	0.492	1	3	10	21	32	44	0.452	1	3	9	19	29	40	0.413
	50	3	6	18	36	55	74	0.753	3	6	17	35	53	71	0.731	3	6	16	34	51	69	0.709	3	6	16	33	50	67	0.687
	75	3	7	20	40	61	82	0.833	3	7	19	39	60	80	0.817	3	7	19	39	58	78	0.802	3	7	18	38	57	77	0.787
	100	4	8	21	42	64	86	0.873	4	8	20	42	63	85	0.862	4	8	20	41	62	83	0.850	3	7	20	40	61	82	0.839
0.95	5																												
	10	0	0	0	1	2	3	0.037	0	0	0	0	0	0	0.005														
	25	1	4	11	23	35	48	0.496	1	3	10	21	32	44	0.456	1	3	9	19	29	40	0.417	1	2	8	17	26	36	0.379
	50	3	6	16	34	53	71	0.730	2	5	16	33	51	68	0.707	2	5	15	32	49	66	0.684	2	5	15	31	47	64	0.662
	75	3	7	19	39	59	79	0.816	3	7	18	38	58	78	0.800	3	6	18	37	57	76	0.784	3	6	17	36	55	74	0.769
	100	3	7	20	41	62	84	0.860	3	7	20	40	61	83	0.848	3	7	19	40	61	81	0.836	3	7	19	39	60	80	0.825
0.99	5																												
	10	0	0	0	0	0	1	0.016	0	0	0	0	0	0	0.001														
	25	1	2	8	19	29	40	0.429	0	2	7	17	27	36	0.390	0	2	6	15	24	33	0.352	0	1	6	13	21	29	0.316
	50	2	5	14	31	48	65	0.684	2	4	14	30	46	63	0.660	1	4	13	29	45	60	0.637	1	4	13	28	43	58	0.615
	75	2	6	17	36	55	75	0.782	2	6	16	35	54	73	0.766	2	5	16	34	53	71	0.749	2	5	15	33	52	70	0.733
	100	3	6	18	39	59	80	0.834	2	6	18	38	58	79	0.821	2	6	18	37	57	77	0.809	2	6	17	37	56	76	0.796

Table A.19c is similar to tables that first appeared in Danziger and Davis (1964). Adapted with permission of the Institute of Mathematical Statistics.

329

Table A.20a Smallest Sample Size for the Range Formed by the Smallest and Largest Observations to Contain, with $100(1 - \alpha)\%$ Confidence, All m Future Observations from the Previously Sampled Population

	$1 - \alpha$						
m	0.50	0.75	0.90	0.95	0.98	0.99	0.999
1	3	7	19	39	99	199	1999
2	6	14	38	78	198	398	3998
3	8	20	56	116	296	596	5996
4	11	27	75	155	395	795	7995
5	13	33	93	193	493	993	9993
6	15	40	112	232	592	1192	11992
7	18	46	130	270	690	1390	13990
8	20	53	149	309	789	1589	15989
9	23	59	167	347	887	1787	17987
10	25	66	186	386	986	1986	19986
11	28	72	204	424	1084	2184	21984
12	30	79	223	463	1183	2383	23983
13	32	85	241	501	1281	2581	25981
14	35	91	260	540	1380	2780	27980
15	37	98	278	578	1478	2978	29978
16	40	104	297	617	1577	3177	31977
17	42	111	315	655	1675	3375	33975
18	44	117	334	694	1774	3574	35974
19	47	124	352	732	1872	3772	37972
20	49	130	371	771	1971	3971	39971
21	52	137	389	809	2069	4169	41969
22	54	143	408	848	2168	4368	43968
23	57	150	426	886	2266	4566	45966
24	59	156	445	925	2365	4765	47965
25	61	163	463	963	2463	4963	49963
30	73	195	556	1156	2956	5956	59956
35	85	227	648	1348	3448	6948	69948
40	98	260	740	1541	3941	7941	79941
45	110	292	833	1733	4433	8933	89933
50	122	324	925	1926	4926	9926	99926
60	146	389	1110	2311	5911	11911	119911
70	170	453	1295	2696	6896	13896	139896
80	194	518	1480	3080	7881	15881	159881
90	218	583	1665	3465	8866	17866	179866
100	242	647	1850	3850	9851	19851	199851

Table A.20b Smallest Sample Size for the Range Formed by the Smallest and Largest Observations to Contain, with $100(1 - \alpha)$% Confidence, at Least $m - 1$ of m Future Observations from the Previously Sampled Population

	$1 - \alpha$						
m	0.50	0.75	0.90	0.95	0.98	0.99	0.999
2	2	4	7	10	16	23	76
3	3	6	11	17	28	40	132
4	4	8	15	23	39	56	186
5	5	10	19	30	50	72	240
6	6	12	23	36	61	88	294
7	7	14	28	42	72	105	347
8	8	16	32	49	83	121	401
9	9	19	36	55	93	137	454
10	10	21	40	62	104	153	508
11	11	23	44	68	115	169	561
12	12	25	48	74	126	185	615
13	13	27	52	81	137	201	668
14	14	29	56	87	148	217	722
15	15	31	61	94	159	233	775
16	16	33	65	100	170	248	829
17	17	35	69	106	181	264	882
18	18	37	73	113	192	280	936
19	19	39	77	119	203	296	989
20	20	41	81	126	213	312	1043
21	21	43	85	132	224	328	1096
22	22	45	89	138	235	344	1150
23	23	47	93	145	246	360	1203
24	24	49	97	151	257	376	1256
25	25	52	102	157	268	392	1310
30	30	62	122	189	322	472	1577
35	35	72	143	221	377	552	1844
40	40	83	163	253	431	632	2112
45	45	93	184	285	486	712	2379
50	50	103	204	317	540	792	2646
60	60	124	245	381	649	952	3180
70	70	144	286	445	758	1111	3715
80	80	165	328	509	867	1271	4249
90	90	186	369	573	976	1431	4783
100	100	206	410	637	1085	1591	5318

Table A.20c Smallest Sample Size for the Range Formed by the Smallest and Largest Observations to Contain, with $100(1 - \alpha)\%$ Confidence, at Least $m - 2$ of m Future Observations from the Previously Sampled Population

m	$1 - \alpha$						
	0.50	0.75	0.90	0.95	0.98	0.99	0.999
3	2	3	5	6	9	12	27
4	3	4	7	10	14	18	43
5	3	6	9	13	19	25	58
6	4	7	11	16	23	31	73
7	5	8	14	19	28	37	88
8	5	9	16	22	33	43	103
9	6	10	18	25	37	49	117
10	7	12	20	28	42	56	132
11	7	13	22	31	47	62	147
12	8	14	24	34	51	68	161
13	9	15	26	37	56	74	176
14	9	16	28	40	60	80	191
15	10	18	31	43	65	86	205
16	10	19	33	46	69	92	220
17	11	20	35	49	74	98	235
18	12	21	37	52	79	105	249
19	12	22	39	55	83	111	264
20	13	24	41	58	88	117	278
21	14	25	43	61	92	123	293
22	14	26	45	64	97	129	308
23	15	27	48	67	102	135	322
24	15	28	50	70	106	141	337
25	16	30	52	73	111	147	352
30	19	36	62	89	134	178	425
35	22	41	73	104	156	208	498
40	25	47	84	119	179	239	571
45	29	53	94	134	202	269	644
50	32	59	105	149	225	300	717
60	38	71	126	179	271	361	863
70	44	83	147	210	316	422	1009
80	51	95	168	240	362	483	1156
90	57	107	190	270	408	544	1302
100	63	119	211	300	454	605	1448

Table A.21a   Smallest Sample Size for the Largest (Smallest) Observation to Exceed (Be Exceeded by), with $100(1 - \alpha)\%$ Confidence, All m Future Observations from the Previously Sampled Population

m	$1 - \alpha$						
	0.50	0.75	0.90	0.95	0.98	0.99	0.999
1	1	3	9	19	49	99	999
2	2	6	18	38	98	198	1998
3	3	9	27	57	147	297	2997
4	4	12	36	76	196	396	3996
5	5	15	45	95	245	495	4995
6	6	18	54	114	294	594	5994
7	7	21	63	133	343	693	6993
8	8	24	72	152	392	792	7992
9	9	27	81	171	441	891	8991
10	10	30	90	190	490	990	9990
11	11	33	99	209	539	1089	10989
12	12	36	108	228	588	1188	11988
13	13	39	117	247	637	1287	12987
14	14	42	126	266	686	1386	13986
15	15	45	135	285	735	1485	14985
16	16	48	144	304	784	1584	15984
17	17	51	153	323	833	1683	16983
18	18	54	162	342	882	1782	17982
19	19	57	171	361	931	1881	18981
20	20	60	180	380	980	1980	19980
21	21	63	189	399	1029	2079	20979
22	22	66	198	418	1078	2178	21978
23	23	69	207	437	1127	2277	22977
24	24	72	216	456	1176	2376	23976
25	25	75	225	475	1225	2475	24975
30	30	90	270	570	1470	2970	29970
35	35	105	315	665	1715	3465	34965
40	40	120	360	760	1960	3960	39960
45	45	135	405	855	2205	4455	44955
50	50	150	450	950	2450	4950	49950
60	60	180	540	1140	2940	5940	59940
70	70	210	630	1330	3430	6930	69930
80	80	240	720	1520	3920	7920	79920
90	90	270	810	1710	4410	8910	89910
100	100	300	900	1900	4900	9900	99900

Table A.21b   Smallest Sample Size for the Largest (Smallest) Observation to Exceed (Be Exceeded by), with $100(1 - \alpha)\%$ Confidence, at Least $m - 1$ of m Future Observations from the Previously Sampled Population

m	$1 - \alpha$						
	0.50	0.75	0.90	0.95	0.98	0.99	0.999
2	1	2	3	5	9	13	44
3	1	3	6	9	15	22	75
4	2	4	8	12	21	32	107
5	2	5	10	16	28	41	137
6	3	6	12	19	34	50	168
7	3	7	14	23	40	59	199
8	4	8	17	26	46	68	230
9	4	9	19	30	52	77	260
10	4	10	21	33	58	86	291
11	5	11	23	37	64	95	322
12	5	12	25	40	70	104	352
13	6	13	27	44	76	113	383
14	6	14	30	47	82	122	414
15	6	15	32	51	88	131	444
16	7	16	34	54	95	140	475
17	7	17	36	58	101	149	506
18	8	18	38	61	107	158	536
19	8	19	40	65	113	167	567
20	9	20	43	68	119	176	597
21	9	21	45	72	125	185	628
22	9	22	47	75	131	194	659
23	10	23	49	79	137	203	689
24	10	24	51	82	143	212	720
25	11	25	53	86	149	221	751
30	13	30	64	103	180	266	904
35	15	35	75	120	210	311	1057
40	17	40	86	138	240	356	1210
45	19	45	97	155	271	401	1363
50	21	50	108	172	301	446	1516
60	25	60	129	207	362	536	1822
70	29	70	151	242	422	626	2129
80	33	80	172	277	483	716	2435
90	38	90	194	311	544	806	2741
100	42	100	216	346	605	896	3047

Table A.21c Smallest Sample Size for the Largest (Smallest) Observation to Exceed (Be Exceeded by), with $100(1 - \alpha)\%$ Confidence, at Least $m - 2$ of m Future Observations from the Previously Sampled Population

m	$1 - \alpha$						
	0.50	0.75	0.90	0.95	0.98	0.99	0.999
3	1	1	2	3	5	7	17
4	1	2	4	5	8	11	26
5	1	3	5	7	11	15	36
6	2	3	6	9	14	18	45
7	2	4	7	11	16	22	54
8	2	5	8	12	19	26	63
9	3	5	10	14	22	29	72
10	3	6	11	16	25	33	81
11	3	6	12	18	27	37	90
12	3	7	13	19	30	40	99
13	4	8	14	21	33	44	108
14	4	8	15	23	35	48	117
15	4	9	17	24	38	51	126
16	4	9	18	26	41	55	135
17	5	10	19	28	43	59	144
18	5	10	20	30	46	62	153
19	5	11	21	31	49	66	162
20	5	12	22	33	51	70	171
21	6	12	24	35	54	73	180
22	6	13	25	36	57	77	189
23	6	13	26	38	59	81	198
24	6	14	27	40	62	84	207
25	7	15	28	42	65	88	216
30	8	18	34	50	78	106	261
35	9	20	40	59	92	124	306
40	11	23	45	67	105	142	351
45	12	26	51	76	119	161	396
50	13	29	57	84	132	179	441
60	16	35	69	102	159	215	531
70	18	41	80	119	186	252	621
80	21	47	92	136	213	288	711
90	24	53	103	153	239	325	801
100	26	59	115	170	266	361	891

Table A.22a Binomial Cumulative Distribution Probabilities:

$$\Pr(x \le x') = B(x'; n, p) = \sum_{i=0}^{x'} \binom{n}{i} p^i (1-p)^{n-i}$$

n	r'	.01	.02	.03	.04	.05	.06	.08	.10	.15	.20	.25	.30	.35	.40	.45	.50
2	0	.9801	.9604	.9409	.9216	.9025	.8836	.8464	.8100	.7225	.6400	.5625	.5000	.4225	.3600	.3025	.2500
	1	$.9^4000$	$.9^3600$	$.9^3100$	$.9^2840$	$.9^2750$	$.9^2640$	$.9^2360$	$.9^2000$	.9775	.9600	.9375	.9100	.8775	.8400	.7975	.7500
3	0	.9703	.9412	.9127	.8847	.8574	.8306	.7787	.7290	.6141	.5120	.4219	.3430	.2746	.2160	.1664	.1250
	1	$.9^3702$	$.9^2882$	$.9^2735$	$.9^2533$	$.9^2275$	.9896	.9818	.9720	.9393	.8960	.8438	.7840	.7183	.6480	.5748	.5000
	2	$.9^6000$	$.9^5200$	$.9^4730$	$.9^4360$	$.9^3875$	$.9^3784$	$.9^3488$	$.9^3000$	$.9^2663$	$.9^2200$	.9844	.9730	.9571	.9360	.9089	.8750
4	0	.9606	.9224	.8853	.8493	.8145	.7807	.7164	.6561	.5220	.4096	.3164	.2401	.1785	.1296	.0915	.0625
	1	$.9^3408$	$.9^2766$	$.9^2481$	$.9^2090$	.9860	.9801	.9656	.9477	.8905	.8192	.7383	.6517	.5630	.4752	.3910	.3125
	2	$.9^5603$	$.9^4685$	$.9^3894$	$.9^3752$	$.9^3519$	$.9^3175$	$.9^2807$	$.9^2630$	.9880	.9728	.9492	.9163	.8735	.8208	.7585	.6875
	3	$.9^8000$	$.9^6840$	$.9^6190$	$.9^5744$	$.9^5375$	$.9^4870$	$.9^4590$	$.9^4000$	$.9^3494$	$.9^2840$	$.9^2609$	$.9^2190$	.9850	.9744	.9590	.9375
5	0	.9510	.9039	.8587	.8154	.7738	.7339	.6591	.5905	.4437	.3277	.2373	.1681	.1160	.0778	.0503	.0313
	1	$.9^3020$	$.9^2616$	$.9^2153$	.9852	.9774	.9681	.9456	.9185	.8352	.7373	.6328	.5282	.4284	.3370	.2562	.1875
	2	$.9^5015$	$.9^4224$	$.9^3742$	$.9^3398$	$.9^2884$	$.9^2803$	$.9^2547$	$.9^2144$	.9734	.9421	.8965	.8369	.7648	.6826	.5931	.5000
	3	$.9^7504$	$.9^6213$	$.9^5605$	$.9^4876$	$.9^4700$	$.9^3383$	$.9^3808$	$.9^3540$	$.9^2777$	$.9^2328$	.9844	.9692	.9460	.9130	.8688	.8125
	4	$.9^{10}000$	$.9^8680$	$.9^7757$	$.9^6898$	$.9^6688$	$.9^6222$	$.9^5672$	$.9^5000$	$.9^4241$	$.9^3680$	$.9^3023$	$.9^2757$	$.9^2475$	.9898	.9815	.9688
6	0	.9415	.8858	.8330	.7828	.7351	.6899	.6064	.5314	.3771	.2621	.1780	.1176	.0754	.0467	.0277	.0156
	1	$.9^2854$	$.9^2431$	.9875	.9784	.9672	.9541	.9227	.8857	.7765	.6554	.5339	.4202	.3191	.2333	.1636	.1094
	2	$.9^4804$	$.9^3847$	$.9^3496$	$.9^2883$	$.9^2777$	$.9^2624$	$.9^2149$	.9842	.9527	.9011	.8306	.7443	.6471	.5443	.4415	.3438
	3	$.9^6852$	$.9^5768$	$.9^4884$	$.9^4640$	$.9^4136$	$.9^3824$	$.9^3462$	$.9^2873$	$.9^2411$	.9830	.9624	.9295	.8826	.8208	.7447	.6563
	4	$.9^9405$	$.9^7811$	$.9^6858$	$.9^6406$	$.9^5820$	$.9^5557$	$.9^4816$	$.9^4450$	$.9^3601$	$.9^2840$	$.9^2536$	.9891	.9777	.9590	.9308	.8906
	5	$.9^{12}000$	$.9^9936$	$.9^9271$	$.9^8590$	$.9^7844$	$.9^7533$	$.9^6738$	$.9^6000$	$.9^4886$	$.9^3936$	$.9^3756$	$.9^3271$	$.9^2816$	$.9^2590$	$.9^2170$	.9844
7	0	.9321	.8681	.8080	.7514	.6983	.6485	.5578	.4783	.3206	.2097	.1335	.0824	.0490	.0280	.0152	.0078
	1	$.9^2797$	$.9^2214$	.9829	.9706	.9556	.9382	.8974	.8503	.7166	.5767	.4449	.3294	.2338	.1586	.1024	.0625
	2	$.9^4660$	$.9^3736$	$.9^3137$	$.9^2802$	$.9^2624$	$.9^2371$	.9860	.9743	.9262	.8520	.7564	.6471	.5323	.4199	.3164	.2266
	3	$.9^6658$	$.9^5466$	$.9^4736$	$.9^4187$	$.9^3806$	$.9^3609$	$.9^2882$	$.9^2727$	.9879	.9667	.9294	.8740	.8002	.7102	.6083	.5000
	4	$.9^8979$	$.9^7350$	$.9^6515$	$.9^5799$	$.9^5397$	$.9^4853$	$.9^4401$	$.9^3824$	$.9^2878$	$.9^2533$	$.9^2871$	$.9^2714$	.9444	.9037	.8471	.7734
	5	$.9^{11}306$	$.9^9560$	$.9^8503$	$.9^7723$	$.9^6895$	$.9^6690$	$.9^5829$	$.9^5360$	$.9^4305$	$.9^3629$	$.9^2866$	$.9^2621$	$.9^2099$	.9812	.9643	.9375
	6	$.9^{14}000$	$.9^{11}872$	$.9^{10}781$	$.9^9836$	$.9^9219$	$.9^8720$	$.9^7790$	$.9^7000$	$.9^5829$	$.9^4872$	$.9^4390$	$.9^3781$	$.9^3357$	$.9^2836$	$.9^2626$	$.9^2219$
8	0	.9227	.8508	.7837	.7214	.6634	.6096	.5132	.4305	.2725	.1678	.1001	.0576	.0319	.0168	$.0^2837$	$.0^2391$
	1	$.9^2731$	.9897	.9777	.9619	.9428	.9208	.8702	.8131	.6572	.5033	.3671	.2553	.1691	.1064	.0632	.0352
	2	$.9^4461$	$.9^3585$	$.9^2865$	$.9^2692$	$.9^2421$	$.9^2038$	.9789	.9619	.8948	.7969	.6785	.5518	.4278	.3154	.2201	.1445
	3	$.9^6322$	$.9^4895$	$.9^4486$	$.9^3843$	$.9^3628$	$.9^3254$	$.9^2797$	$.9^2498$	.9786	.9437	.8862	.8059	.7064	.5941	.4770	.3633
	4	$.9^8450$	$.9^6830$	$.9^5882$	$.9^4948$	$.9^4848$	$.9^4626$	$.9^3851$	$.9^3568$	$.9^2713$	.9896	.9727	.9420	.8939	.8263	.7396	.6367
	5	$.9^{10}725$	$.9^8826$	$.9^7806$	$.9^6893$	$.9^6599$	$.9^5824$	$.9^5363$	$.9^4766$	$.9^3758$	$.9^2877$	$.9^2577$	$.9^2887$	.9745	.9502	.9115	.8555
	6	$.9^{13}207$	$.9^{10}899$	$.9^9830$	$.9^8874$	$.9^8402$	$.9^7788$	$.9^6844$	$.9^6270$	$.9^4881$	$.9^4155$	$.9^3619$	$.9^2871$	$.9^2643$	$.9^2148$	.9819	.9648
	7	$.9^{16}000$	$.9^{13}744$	$.9^{12}344$	$.9^{11}344$	$.9^{10}961$	$.9^9832$	$.9^8832$	$.9^8000$	$.9^6744$	$.9^5744$	$.9^4847$	$.9^4344$	$.9^3775$	$.9^3345$	$.9^2832$	$.9^2609$

Table A.22b Binomial Cumulative Distribution Probabilities:

$$\Pr(x \le x') = B(x'; n, p) = \sum_{i=0}^{x'} \binom{n}{i} p^{i}(1-p)^{n-i}$$

n	x'	p: .01	.02	.03	.04	.05	.06	.08	.10	.15	.20	.25	.30	.35	.40	.45	.50
9	0	.9135	.8337	.7602	.6925	.6302	.5730	.4722	.3874	.2316	.1342	.0751	.0404	.0207	.0101	$.0^{2}461$	$.0^{2}195$
	1	$.9^{3}656$	.9869	.9718	.9522	.9288	.9022	.8417	.7748	.5995	.4362	.3003	.1960	.1211	.0705	.0385	.0195
	2	$.9^{4}197$	$.9^{3}386$	$.9^{2}802$	$.9^{2}552$	$.9^{2}164$	.9862	.9702	.9470	.8591	.7382	.6007	.4628	.3373	.2318	.1495	.0898
	3	$.9^{4}879$	$.9^{4}814$	$.9^{4}096$	$.9^{3}726$	$.9^{3}357$	$.9^{2}872$	$.9^{2}628$	$.9^{2}167$	.9661	.9144	.8343	.7297	.6089	.4826	.3614	.2539
	4		$.9^{6}623$	$.9^{4}887$	$.9^{4}668$	$.9^{3}686$	$.9^{3}109$	$.9^{2}437$	.9804	.9511	.9012	.8283	.7334	.6214	.6000		
	5		$.9^{6}623$	$.9^{5}723$			$.9^{4}202$	$.9^{3}686$	$.9^{3}358$	$.9^{3}109$?	$.9^{2}437$	.9804	.9511	.9012	.8283	.7334	.6214
	6		$.9^{6}690$	$.9^{5}885$	$.9^{4}668$	$.9^{4}887$	$.9^{5}665$		$.9^{6}347$								
	7								$.9^{6}700$	$.9^{4}536$	$.9^{3}686$	$.9^{3}866$	$.9^{3}567$	$.9^{2}860$	$.9^{2}620$	$.9^{2}092$	.9805
	8									$.9^{5}800$	$.9^{4}811$	$.9^{4}893$	$.9^{4}803$	$.9^{4}212$	$.9^{3}738$	$.9^{3}243$	$.9^{2}805$
10	0	.9044	.8171	.7374	.6648	.5987	.5386	.4344	.3487	.1969	.1074	.0563	.0282	.0135	$.0^{2}605$	$.0^{2}253$	$.0^{3}977$
	1	$.9^{2}573$	.9838	.9655	.9418	.9139	.8824	.8121	.7361	.5443	.3758	.2440	.1493	.0860	.0464	.0233	.0107
	2	$.9^{3}886$	$.9^{3}136$	$.9^{2}724$	$.9^{3}379$	.9885	.9812	.9599	.9298	.8202	.6778	.5256	.3828	.2616	.1673	.0996	.0547
	3	$.9^{5}800$	$.9^{3}853$	$.9^{3}853$	$.9^{3}557$	$.9^{2}897$	$.9^{2}797$	$.9^{2}420$	.9872	.9500	.8791	.7759	.6496	.5138	.3823	.2660	.1719
	4		$.9^{5}259$	$.9^{5}460$	$.9^{4}782$	$.9^{3}363$	$.9^{3}848$	$.9^{2}013$	$.9^{2}837$	$.9^{2}862$	.9672	.9219	.8497	.7515	.6331	.5044	.3770
	5		$.9^{6}259$	$.9^{6}862$	$.9^{6}252$	$.9^{5}363$	$.9^{5}206$	$.9^{4}013$	$.9^{3}853$	$.9^{3}865$	$.9^{2}363$	.9803	.9527	.9051	.8338	.7384	.6230
	6					$.9^{5}725$	$.9^{6}714$	$.9^{5}585$	$.9^{4}088$	$.9^{5}133$	$.9^{3}136$	$.9^{3}649$	.9894	.9740	.9452	.8980	.8281
	7							$.9^{6}798$	$.9^{6}626$	$.9^{5}865$	$.9^{4}221$	$.9^{3}584$	$.9^{2}649$	$.9^{2}518$	.9877	.9726	.9453
	8									$.9^{6}667$	$.9^{5}580$	$.9^{4}704$	$.9^{4}856$	$.9^{3}460$	$.9^{3}832$	$.9^{2}550$	.9893
	9										$.9^{6}898$	$.9^{6}046$	$.9^{5}410$	$.9^{4}724$	$.9^{3}895$	$.9^{3}659$	$.9^{3}023$
11	0	.8953	.8007	.7153	.6382	.5688	.5063	.3996	.3138	.1673	.0859	.0422	.0198	$.0^{2}875$	$.0^{2}363$	$.0^{2}139$	$.0^{3}488$
	1	$.9^{2}482$	.9805	.9587	.9308	.8981	.8618	.7819	.6974	.4922	.3221	.1971	.1130	.0606	.0302	.0139	$.0^{2}586$
	2	$.9^{3}845$	$.9^{2}883$	$.9^{2}628$	$.9^{2}171$	.9848	.9752	.9481	.9104	.7788	.6174	.4552	.3127	.2001	.1189	.0652	.0327
	3	$.9^{5}688$	$.9^{4}528$	$.9^{3}774$	$.9^{3}327$	$.9^{3}845$	$.9^{2}696$	$.9^{2}146$	.9815	.9306	.8389	.7133	.5696	.4256	.2963	.1911	.1133
	4		$.9^{5}866$	$.9^{5}035$	$.9^{4}614$	$.9^{3}888$	$.9^{3}736$	$.9^{2}900$	$.9^{2}725$	.9841	.9496	.8854	.7897	.6683	.5328	.3971	.2744
	5			$.9^{6}704$	$.9^{5}841$	$.9^{5}420$	$.9^{4}834$	$.9^{3}150$	$.9^{3}704$	$.9^{2}734$	.9883	.9657	.9218	.8513	.7535	.6331	.5000
	6					$.9^{6}784$	$.9^{5}255$	$.9^{4}482$	$.9^{4}771$	$.9^{3}678$	$.9^{2}803$	$.9^{2}244$	.9784	.9499	.9006	.8262	.7256
	7							$.9^{6}778$	$.9^{5}482$	$.9^{4}724$	$.9^{3}765$	$.9^{3}803$	$.9^{2}571$	.9878	.9707	.9390	.8867
	8								$.9^{6}875$	$.9^{5}724$	$.9^{4}765$	$.9^{4}189$	$.9^{3}422$	$.9^{2}796$	$.9^{2}408$	.9852	.9673
	9									$.9^{6}842$	$.9^{5}811$	$.9^{5}528$	$.9^{4}528$	$.9^{3}793$	$.9^{3}266$	$.9^{2}779$	$.9^{2}414$
	10										$.9^{6}078$	$.9^{6}762$	$.9^{5}823$	$.9^{5}035$	$.9^{4}581$	$.9^{3}847$	$.9^{3}512$

Table A.22c Binomial Cumulative Distribution Probabilities:

$$\Pr(x \le x') = B(x'; n, p) = \sum_{i=0}^{x'} \binom{n}{i} p^i (1 - p)^{n-i}$$

n = 12

x'	0.01	0.02	0.03	0.04	0.05	0.06	0.08	0.10	0.15	0.20	0.25	0.30	0.35	0.40	0.45	0.50
0	.8864	.7847	.6938	.6127	.5404	.4759	.3677	.2824	.1422	.0687	.0317	.0138	$.0^2569$	$.0^2218$	$.0^2766$	$.0^3244$
1	$.9^3383$	.9769	.9514	.9191	.8816	.8405	.7513	.6590	.4435	.2749	.1584	.0850	.0424	.0196	$.0^2829$	$.0^2317$
2	$.9^3794$	$.9^2846$	$.9^2515$	.9893	.9804	.9684	.9348	.8891	.7358	.5583	.3907	.2528	.1513	.0834	.0421	.0193
3	$.9^5536$	$.9^4304$	$.9^3670$	$.9^3022$	$.9^2776$	$.9^2566$	$.9^2839$	.9744	.9078	.7946	.6488	.4925	.3467	.2253	.1345	.0730
4		$.9^5775$	$.9^4839$	$.9^4360$	$.9^3816$	$.9^3569$	$.9^3842$	$.9^2567$	.9761	.9274	.8424	.7237	.5833	.4382	.3044	.1938
5			$.9^6424$	$.9^5693$	$.9^4889$	$.9^4686$	$.9^4884$	$.9^3459$	$.9^2536$	.9806	.9456	.8822	.7873	.6652	.5269	.3872
6				$.9^6891$	$.9^5505$	$.9^5831$	$.9^6381$	$.9^4498$	$.9^3328$	$.9^2610$	.9857	.9614	.9154	.8418	.7393	.6128
7								$.9^5659$	$.9^4452$	$.9^3419$	$.9^2722$	$.9^2051$	.9745	.9427	.8883	.8062
8								$.9^6834$	$.9^5191$	$.9^4378$	$.9^3608$	$.9^2831$	$.9^2439$	.9847	.9644	.9270
9										$.9^5547$	$.9^4624$	$.9^3794$	$.9^3152$	$.9^2719$	$.9^2212$	.9807
10										$.9^6799$	$.9^5779$	$.9^4831$	$.9^4213$	$.9^3681$	$.9^2892$	$.9^2683$
11												$.9^5662$	$.9^5662$	$.9^4832$	$.9^3310$	$.9^3756$

n = 13

x'	0.01	0.02	0.03	0.04	0.05	0.06	0.08	0.10	0.15	0.20	0.25	0.30	0.35	0.40	0.45	0.50
0	.8775	.7690	.6730	.5882	.5133	.4474	.3383	.2542	.1209	.0550	.0238	$.0^2969$	$.0^2370$	$.0^2131$	$.0^3421$	$.0^3122$
1	$.9^2275$	.9730	.9436	.9068	.8646	.8186	.7206	.6213	.3983	.2336	.1267	.0637	.0296	.0126	$.0^2490$	$.0^2171$
2	$.9^3735$	$.9^2803$	$.9^2384$	.9865	.9755	.9608	.9201	.8661	.6920	.5017	.3326	.2025	.1132	.0579	.0269	.0112
3	$.9^5335$	$.9^4010$	$.9^3534$	$.9^2863$	$.9^2690$	$.9^2402$	.9756	.9658	.8820	.7473	.5843	.4206	.2783	.1686	.0929	.0461
4	$.9^6880$	$.9^5640$	$.9^4744$	$.9^4899$	$.9^3713$	$.9^3335$	$.9^2726$	$.9^2354$	.9658	.9009	.7940	.6543	.5005	.3530	.2279	.1334
5			$.9^6896$	$.9^5449$	$.9^4803$	$.9^4446$	$.9^3726$	$.9^3080$	.9925	.9700	.9198	.8346	.7159	.5744	.4268	.2905
6				$.9^6773$	$.9^5669$	$.9^5652$	$.9^4767$	$.9^4007$	$.9^2873$	$.9^2300$	.9757	.9376	.8705	.7712	.6437	.5000
7					$.9^6804$	$.9^6341$	$.9^5767$	$.9^5007$	$.9^3191$	$.9^2873$	$.9^2435$	.9818	.9538	.9023	.8212	.7095
8							$.9^6851$	$.9^6509$	$.9^4846$	$.9^3834$	$.9^3011$	$.9^2597$	.9874	.9679	.9302	.8666
9									$.9^4191$	$.9^4839$	$.9^3874$	$.9^3348$	$.9^2749$	$.9^2221$	.9797	.9539
10									$.9^5846$	$.9^5893$	$.9^4889$	$.9^4273$	$.9^3652$	$.9^2868$	$.9^2586$	.9888
11									$.9^6894$		$.9^5404$	$.9^5500$	$.9^4703$	$.9^3862$	$.9^3476$	$.9^2829$
12												$.9^5841$	$.9^5882$	$.9^5329$	$.9^4690$	$.9^3878$

n = 14

x'	0.01	0.02	0.03	0.04	0.05	0.06	0.08	0.10	0.15	0.20	0.25	0.30	0.35	0.40	0.45	0.50
0	.8687	.7536	.6528	.5647	.4877	.4205	.3112	.2288	.1028	.0440	.0178	$.0^2678$	$.0^2240$	$.0^3784$	$.0^3232$	$.0^4610$
1	$.9^2160$	.9690	.9355	.8941	.8470	.7963	.6900	.5846	.3567	.1979	.1010	.0475	.0205	$.0^2810$	$.0^2289$	$.0^3916$
2	$.9^3665$	$.9^2753$	$.9^2233$	.9833	.9699	.9522	.9042	.8416	.6479	.4481	.2811	.1608	.0839	.0398	.0170	$.0^2647$
3	$.9^5076$	$.9^3864$	$.9^3363$	$.9^2815$	$.9^2583$	$.9^2203$	.9786	.9559	.8535	.6982	.5213	.3552	.2205	.1243	.0632	.0287
4	$.9^6814$	$.9^5449$	$.9^4612$	$.9^3849$	$.9^3573$	$.9^3016$	$.9^2646$	$.9^2077$	.9533	.8702	.7415	.5842	.4227	.2793	.1672	.0898
5		$.9^6833$	$.9^5822$	$.9^5069$	$.9^4669$	$.9^4100$	$.9^3553$	$.9^2853$	.9885	.9561	.8883	.7805	.6405	.4859	.3373	.2120
6				$.9^6562$	$.9^5804$	$.9^4341$	$.9^4566$	$.9^3819$	$.9^2672$	$.9^2779$	.9617	.9067	.8164	.6925	.5461	.3953
7						$.9^5637$	$.9^5676$	$.9^4828$	$.9^3626$	$.9^2760$	.9897	.9685	.9247	.8499	.7414	.6047
8							$.9^6815$	$.9^5875$	$.9^4540$	$.9^3618$	$.9^2785$	$.9^2171$	.9757	.9417	.8811	.7880
9									$.9^5678$	$.9^4540$	$.9^3658$	$.9^2833$	$.9^2396$	.9825	.9574	.9102
10									$.9^6798$	$.9^5594$	$.9^4602$	$.9^3754$	$.9^2889$	$.9^2609$	.9886	.9713
11										$.9^6752$	$.9^5594$	$.9^4602$	$.9^3859$	$.9^3391$	$.9^2785$	$.9^2353$
12											$.9^5679$	$.9^4747$	$.9^4888$	$.9^4409$	$.9^3747$	$.9^3084$
13											$.9^6840$	$.9^5839$	$.9^5586$	$.9^5732$	$.9^4860$	$.9^4390$

338

Table A.22d Binomial Cumulative Distribution Probabilities:

$$\Pr(x \le x') = B(x'; n, p) = \sum_{i=0}^{x'} \binom{n}{i} p^i (1-p)^{n-i}$$

Note on notation: a superscript preceding a group of digits gives the number of repeated leading 9's or 0's. E.g. $.9^3584 = 0.999584$ and $.0^2475 = 0.00475$.

n = 15

x'	.01	.02	.03	.04	.05	.06	.08	.10	.15	.20	.25	.30	.35	.40	.45	.50
0	.8601	.7386	.6333	.5421	.4633	.3953	.2863	.2059	.0874	.0352	.0134	$.0^2475$	$.0^2156$	$.0^3470$	$.0^3127$	$.0^4305$
1	$.9^2037$	.9647	.9270	.8809	.8290	.7738	.6597	.5490	.3186	.1671	.0802	.0353	.0142	$.0^2517$	$.0^2169$	$.0^3488$
2	$.9^3584$	$.9^2696$	$.9^2063$	.9797	.9638	.9429	.8870	.8159	.6042	.3980	.2361	.1268	.0617	.0271	.0107	$.0^2369$
3	$.9^4875$	$.9^3817$	$.9^3152$	$.9^2755$	$.9^2453$	.9896	.9727	.9444	.8227	.6482	.4613	.2969	.1727	.0905	.0424	.0176
4	$.9^6724$	$.9^5187$	$.9^4433$	$.9^3781$	$.9^3385$	$.9^2860$	$.9^2503$	.9873	.9383	.8358	.6865	.5155	.3519	.2173	.1204	.0592
5		$.9^6726$	$.9^5711$	$.9^4850$	$.9^4472$	$.9^3854$	$.9^3305$	$.9^2775$	.9832	.9389	.8516	.7216	.5643	.4032	.2608	.1509
6			$.9^6886$	$.9^5648$	$.9^5402$	$.9^4883$	$.9^4243$	$.9^3689$	$.9^2639$	.9819	.9434	.8689	.7548	.6098	.4522	.3036
7					$.9^6817$	$.9^5779$	$.9^5355$	$.9^4715$	$.9^3390$	$.9^2576$	.9827	.9500	.8868	.7869	.6535	.5000
8						$.9^6263$	$.9^6243$	$.9^5813$	$.9^4664$	$.9^3215$	$.9^2581$	.9848	.9578	.9050	.8182	.6964
9								$.9^6355$	$.9^5355$	$.9^3885$	$.9^3205$	$.9^2635$	.9876	.9662	.9231	.8491
10									$.9^5166$	$.9^4860$	$.9^3885$	$.9^3328$	$.9^2717$	$.9^2065$	.9745	.9408
11									$.9^6346$	$.9^5876$	$.9^4876$	$.9^4083$	$.9^3521$	$.9^2807$	$.9^2367$	.9824
12										$.9^6899$	$.9^6077$	$.9^5128$	$.9^4434$	$.9^3721$	$.9^2889$	$.9^2631$
13												$.9^6483$	$.9^5582$	$.9^4748$	$.9^3879$	$.9^3512$
14													$.9^6855$	$.9^5893$	$.9^5372$	$.9^4695$

n = 16

x'	.01	.02	.03	.04	.05	.06	.08	.10	.15	.20	.25	.30	.35	.40	.45	.50
0	.8515	.7238	.6143	.5204	.4401	.3716	.2634	.1853	.0743	.0281	.0100	$.0^2332$	$.0^2102$	$.0^3282$	$.0^4701$	$.0^4153$
1	.9891	.9601	.9182	.8673	.8108	.7511	.6299	.5147	.2839	.1407	.0635	.0261	$.0^2976$	$.0^2329$	$.0^3988$	$.0^3259$
2	$.9^4492$	$.9^2631$	.9887	.9758	.9571	.9327	.8689	.7892	.5614	.3518	.1971	.0994	.0451	.0183	$.0^2662$	$.0^2209$
3	$.9^4835$	$.9^3760$	$.9^2890$	$.9^2684$	$.9^2300$	.9868	.9658	.9316	.7899	.5981	.4050	.2459	.1339	.0651	.0281	.0106
4	$.9^6602$	$.9^4884$	$.9^4196$	$.9^3691$	$.9^3143$	$.9^2806$	$.9^2324$	.9830	.9209	.7982	.6302	.4499	.2892	.1666	.0853	.0384
5	$.9^6569$	$.9^5550$	$.9^4768$	$.9^4191$	$.9^3779$	$.9^2896$	$.9^2670$	.9765	.9765	.8103	.6598	.4900	.3288	.1976	.1051	.1051
6		$.9^6569$	$.9^5803$	$.9^4864$	$.9^5402$	$.9^3779$	$.9^3495$	$.9^2875$	.9880	.9733	.9204	.8247	.6881	.5272	.3660	.2272
7			$.9^6803$	$.9^5768$	$.9^5191$	$.9^3779$	$.9^4387$	$.9^3880$	$.9^3387$	$.9^2300$	.9729	.9256	.8406	.7161	.5629	.4018
8					$.9^5402$	$.9^4670$	$.9^5408$	$.9^4875$	$.9^3894$	$.9^2852$	$.9^2253$	.9743	.9329	.8577	.7441	.5982
9					$.9^6650$	$.9^3860$	$.9^6089$	$.9^5547$	$.9^3840$	$.9^2752$	$.9^2836$	$.9^2287$	.9771	.9417	.8759	.7728
10									$.9^3808$	$.9^3674$	$.9^3715$	$.9^2843$	$.9^2380$	.9809	.9514	.8949
11									$.9^5819$	$.9^3670$	$.9^4619$	$.9^3734$	$.9^3870$	$.9^2510$	.9851	.9616
12									$.9^6870$	$.9^3752$	$.9^5622$	$.9^4664$	$.9^3796$	$.9^3062$	$.9^2654$	.9894
13											$.9^6737$	$.9^5702$	$.9^4775$	$.9^3873$	$.9^2435$	$.9^2791$
14												$.9^6835$	$.9^5844$	$.9^3893$	$.9^3419$	$.9^3741$
15														$.9^5571$	$.9^5717$	$.9^4847$

339

Table A.22e Binomial Cumulative Distribution Probabilities:

$$\Pr(x \le x') = B(x'; n, p) = \sum_{i=0}^{x'} \binom{n}{i} p^i (1 - p)^{n-i}$$

n = 17

x' \ p	0.01	0.02	0.03	0.04	0.05	0.06	0.08	0.10	0.15	0.20	0.25	0.30	0.35	0.40	0.45	0.50
0	.8429	.7093	.5958	.4996	.4181	.3493	.2423	.1668	.0631	.0225	$.0^2752$	$.0^2233$	$.0^3660$	$.0^3169$	$.0^4386$	$.0^5763$
1	.9877	.9554	.9091	.8535	.7922	.7283	.6005	.4818	.2525	.1182	.0501	.0193	$.0^2670$	$.0^2209$	$.0^3575$	$.0^3137$
2	$.9^3388$	$.9^2559$	.9866	.9714	.9497	.9218	.8497	.7618	.5198	.3096	.1637	.0774	.0327	.0123	$.0^2409$	$.0^2117$
3	$.9^4786$	$.9^3691$	$.9^2859$	$.9^2599$	$.9^2120$	.9836	.9581	.9174	.7556	.5489	.3530	.2019	.1028	.0464	.0184	$.0^2636$
4	$.9^6440$	$.9^4838$	$.9^3889$	$.9^3577$	$.9^2884$	$.9^2739$	$.9^2105$	.9779	.9013	.7582	.5739	.3887	.2348	.1260	.0596	.0245
5			$.9^5322$	$.9^4654$	$.9^3880$	$.9^3676$	$.9^2851$	$.9^2533$	.9681	.8943	.7653	.5968	.4197	.2639	.1471	.0717
6				$.9^6777$	$.9^5027$	$.9^4682$	$.9^3802$	$.9^3216$	$.9^2172$	.9623	.8929	.7752	.6188	.4478	.2902	.1662
7				$.9^6674$	$.9^5750$	$.9^4789$	$.9^3894$	$.9^3826$	$.9^2826$	.9891	.9598	.8954	.7872	.6405	.4743	.3145
8					$.9^6369$	$.9^5750$	$.9^4789$	$.9^3894$	$.9^3705$	$.9^2742$	.9876	.9597	.9006	.8011	.6626	.5000
9									$.9^4558$	$.9^3918$	$.9^2901$	.9873	.9617	.9081	.8166	.6855
10									$.9^6618$	$.9^4244$	$.9^3375$	$.9^2676$	.9880	.9652	.9174	.8338
11										$.9^5084$	$.9^4001$	$.9^2897$	$.9^2699$	.9894	.9699	.9283
12											$.9^4375$	$.9^3001$	$.9^3411$	$.9^2748$	$.9^2138$	.9755
13											$.9^4876$	$.9^4001$	$.9^4138$	$.9^3549$	$.9^2813$	$.9^2364$
14											$.9^5886$	$.9^4876$	$.9^5110$	$.9^4429$	$.9^3714$	$.9^2883$
15												$.9^5886$	$.9^6422$	$.9^4545$	$.9^4723$	$.9^3863$
16													$.9^6899$	$.9^6828$	$.9^5873$	$.9^5237$

n = 18

x' \ p	0.01	0.02	0.03	0.04	0.05	0.06	0.08	0.10	0.15	0.20	0.25	0.30	0.35	0.40	0.45	0.50
0	.8345	.6951	.5780	.4796	.3972	.3283	.2229	.1501	.0536	.0180	$.0^2564$	$.0^2163$	$.0^3429$	$.0^3102$	$.0^4212$	$.0^5381$
1	.9862	.9505	.8997	.8393	.7735	.7055	.5719	.4503	.2241	.0991	.0395	.0142	$.0^2459$	$.0^2132$	$.0^3334$	$.0^4725$
2	$.9^3271$	$.9^2479$	.9843	.9667	.9419	.9102	.8298	.7338	.4797	.2713	.1353	.0600	.0236	$.0^2823$	$.0^2251$	$.0^3656$
3	$.9^4726$	$.9^3609$	$.9^2823$	$.9^2501$	.9891	.9799	.9494	.9018	.7202	.5010	.3057	.1646	.0783	.0328	$.0^2823$	$.0^2377$
4	$.9^6231$	$.9^4779$	$.9^3850$	$.9^3434$	$.9^2845$	$.9^2656$	.9884	.9718	.8794	.7164	.5187	.3327	.1886	.0942	.0411	.0154
5		$.9^6034$	$.9^5009$	$.9^4499$	$.9^3828$	$.9^3538$	$.9^2791$	$.9^2358$	.9581	.8671	.7175	.5344	.3550	.2088	.1077	.0481
6			$.9^6480$	$.9^5647$	$.9^4828$	$.9^4506$	$.9^3698$	$.9^2883$	.9882	.9487	.8610	.7217	.5491	.3743	.2258	.1189
7				$.9^6800$	$.9^5848$	$.9^4574$	$.9^3648$	$.9^3827$	$.9^2728$	.9837	.9431	.8593	.7283	.5634	.3915	.2403
8					$.9^5891$	$.9^4702$	$.9^4666$	$.9^3791$	$.9^3489$	$.9^2575$	.9807	.9404	.8609	.7368	.5778	.4073
9							$.9^3743$	$.9^3698$	$.9^3827$	$.9^3089$	$.9^3214$	.9790	.9403	.8653	.7473	.5927
10								$.9^3837$	$.9^3827$	$.9^3841$	$.9^2876$	.9788	.9393	.8720	.7597
11									$.9^4214$	$.9^3019$	$.9^3393$	$.9^2383$	.9788	.9424	.8811
12									$.9^4795$	$.9^5019$	$.9^3731$	$.9^2856$	.9797	.9519
13									$.9^6013$	$.9^4605$	$.9^3738$	$.9^3425$	.9846
14										$.9^5564$	$.9^4640$	$.9^3785$	$.9^2872$	$.9^2623$
15										$.9^5659$	$.9^5651$	$.9^4744$	$.9^3003$	$.9^3344$
16												$.9^5786$	$.9^4856$	$.9^3623$	$.9^4275$
17													$.9^6808$	$.9^4868$	$.9^5619$

340

Table A.22f Binomial Cumulative Distribution Probabilities:

$$\Pr(x \le x') = B(x'; n, p) = \sum_{i=0}^{x'} \binom{n}{i} p^i (1-p)^{n-i}$$

n	x'	0.01	0.02	0.03	0.04	0.05	0.06	0.08	0.10	0.15	0.20	0.25	0.30	0.35	0.40	0.45	0.50
19	0	.8262	.6812	.5606	.4604	.3774	.3086	.2051	.1351	.0456	.0144	$.0^2423$	$.0^2114$	$.0^3279$	$.0^4609$	$.0^4117$	$.0^5191$
	1	.9847	.9454	.8900	.8249	.7547	.6829	.5440	.4203	.1985	.0829	.0310	.0104	$.0^2313$	$.0^3833$	$.0^3193$	$.0^4381$
	2	$.9^3141$	$.9^2390$	.9817	.9616	.9335	.8979	.8092	.7054	.4413	.2369	.1113	.0462	.0170	$.0^2546$	$.0^2153$	$.0^3364$
	3	$.9^4656$	$.9^3513$	$.9^2781$	$.9^2388$	.9868	.9757	.9398	.8850	.6841	.4551	.2631	.1332	.0591	.0230	$.0^2772$	$.0^2221$
	4	$.9^5897$	$.9^4706$	$.9^3801$	$.9^3257$	$.9^2799$	$.9^2556$	.9853	.9648	.8556	.6733	.4654	.2822	.1500	.0696	.0280	$.0^2961$
	5		$.9^5861$	$.9^4859$	$.9^4292$	$.9^3759$	$.9^3360$	$.9^2715$	$.9^2141$	.9463	.8369	.6678	.4739	.2968	.1629	.0777	.0318
	6		$.9^6199$	$.9^5461$	$.9^4769$	$.9^4259$	$.9^3555$	$.9^2830$	.9837	.9324	.8251	.6655	.4812	.3081	.1727	.0835
	7			$.9^6667$	$.9^5821$	$.9^5304$	$.9^4434$	$.9^3727$	$.9^2592$	.9767	.9225	.8180	.6656	.4878	.3169	.1796
	8				$.9^6886$	$.9^6465$	$.9^5411$	$.9^4639$	$.9^3157$	$.9^2334$	.9713	.9161	.8145	.6675	.4940	.3238
	9					$.9^6607$	$.9^5649$	$.9^3856$	$.9^2842$	.9674	.9125	.8139	.6710	.5000
	10						$.9^4799$	$.9^3691$	$.9^2110$	.9895	.9653	.9115	.8159	.6762
	11						$.9^5769$	$.9^3771$	$.9^2718$	.9895	.9648	.9129	.8204
	12						$.9^6785$	$.9^5516$	$.9^3383$	$.9^2718$	.9886	.9658	.9165
	13							$.9^4165$	$.9^3892$	$.9^3326$	$.9^2693$	.9891	.9682
	14							$.9^4885$	$.9^4851$	$.9^3885$	$.9^2359$	$.9^2724$	$.9^2039$
	15							$.9^5876$	$.9^5846$	$.9^3899$	$.9^3472$	$.9^2779$
	16							$.9^6899$	$.9^5851$	$.9^3886$	$.9^4279$	$.9^3636$
	17								$.9^6887$	$.9^6189$	$.9^5376$	$.9^4619$
	18									$.9^6742$	$.9^5809$
20	0	.8179	.6676	.5438	.4420	.3585	.2901	.1887	.1216	.0388	.0115	$.0^2317$	$.0^2798$	$.0^3181$	$.0^3366$	$.0^4642$	$.0^4954$
	1	.9831	.9401	.8802	.8103	.7358	.6605	.5169	.3917	.1756	.0692	.0243	$.0^2764$	$.0^2213$	$.0^2524$	$.0^3111$	$.0^3200$
	2	$.9^2900$	$.9^2293$	.9790	.9561	.9245	.8850	.7879	.6769	.4049	.2061	.0913	.0355	.0121	$.0^2361$	$.0^2927$	$.0^2201$
	3	$.9^4574$	$.9^3400$	$.9^2733$	$.9^2259$	.9841	.9710	.9294	.8670	.6477	.4114	.2252	.1071	.0444	.0160	$.0^2493$	$.0^2129$
	4	$.9^5863$	$.9^4614$	$.9^3804$	$.9^3042$	$.9^2743$	$.9^2437$	.9817	.9568	.8298	.6296	.4148	.2375	.1182	.0510	.0189	$.0^2591$
	5		$.9^5805$	$.9^4804$	$.9^4023$	$.9^3671$	$.9^3131$	$.9^2620$	$.9^2361$	.9327	.8042	.6172	.4164	.2454	.1256	.0553	.0207
	6			$.9^5880$	$.9^5199$	$.9^4661$	$.9^3892$	$.9^3362$	$.9^2761$	.9781	.9133	.7858	.6080	.4166	.2500	.1299	.0577
	7				$.9^5714$	$.9^4124$	$.9^4584$	$.9^3584$	$.9^2408$	.9679	.8982	.7723	.6010	.4159	.2520	.1316
	8				$.9^6465$	$.9^5890$	$.9^4124$	$.9^3006$	$.9^3401$	$.9^2867$	.9679	.8982	.8867	.7624	.5956	.4143	.2517
	9					$.9^5802$	$.9^4890$	$.9^3890$	$.9^2006$	$.9^3401$	$.9^2002$	.9591	.8867	.7553	.5914	.4119
	10						$.9^6079$	$.9^4065$	$.9^3006$	$.9^3614$	$.9^2606$	.9861	.9520	.8782	.7553	.5881
	11							$.9^5502$	$.9^4065$	$.9^3752$	$.9^2437$	$.9^2486$	.9829	.9468	.8725	.7507
	12							$.9^6291$	$.9^5470$	$.9^3867$	$.9^2898$	$.9^3065$	$.9^2872$	.9804	.9435	.8692	.7483
	13								$.9^6065$	$.9^4502$	$.9^3848$	$.9^3816$	$.9^2739$	$.9^2398$	.9790	.9420	.8684
	14									$.9^5470$	$.9^4815$	$.9^3705$	$.9^3151$	$.9^2848$	$.9^2353$	.9786	.9423
	15									$.9^6619$	$.9^5815$	$.9^4571$	$.9^3619$	$.9^2689$	$.9^2839$	.9683	.9793
	16										$.9^5820$	$.9^5613$	$.9^4501$	$.9^3683$	$.9^3527$	$.9^2847$	$.9^2409$
	17												$.9^5445$	$.9^4392$	$.9^3723$	$.9^2839$	$.9^2871$
	18												$.9^6457$	$.9^5472$	$.9^4496$	$.9^3641$	$.9^3799$
	19														$.9^5659$	$.9^5705$	$.9^4800$
																$.9^6884$	$.9^6046$

Table A.23a Two-Sided $100(1-\alpha)\%$ Confidence Intervals and One-Sided $100(1-\alpha)\%$ Confidence Bounds for a Binomial Proportion Nonconforming Based on x Nonconforming Units in a Sample of Size n

$1-\alpha$	Lower			Upper			Lower			Upper			Lower			Upper			Lower			Upper		
One-sided:	.99	.95	.90	.90	.95	.99	.99	.95	.90	.90	.95	.99	.99	.95	.90	.90	.95	.99	.99	.95	.90	.90	.95	.99
Two-sided:	.98	.90	.80	.80	.90	.98	.98	.90	.80	.80	.90	.98	.98	.90	.80	.80	.90	.98	.98	.90	.80	.80	.90	.98
x	$n=1$						$n=2$						$n=3$						$n=4$					
0	0	0	0	.900	.950	.990	0	0	0	.684	.776	.900	0	0	0	.536	.632	.785	0	0	0	.438	.527	.684
1	.010	.050	.100	1	1	1	.005	.025	.051	.949	.975	.995	.003	.017	.035	.804	.865	.941	.003	.013	.026	.680	.751	.859
2							.100	.224	.316	1	1	1	.059	.135	.196	.965	.983	.997	.042	.098	.143	.857	.902	.958
3													.215	.368	.464	1	1	1	.141	.249	.320	.974	.987	.997
4																			.316	.473	.562	1	1	1
x	$n=5$						$n=6$						$n=7$						$n=8$					
0	0	0	0	.369	.451	.602	0	0	0	.319	.393	.536	0	0	0	.280	.348	.482	0	0	0	.250	.312	.438
1	.002	.010	.021	.584	.657	.778	.002	.009	.017	.510	.582	.706	.001	.007	.015	.453	.521	.643	.001	.006	.013	.406	.471	.590
2	.033	.076	.112	.753	.811	.894	.027	.063	.093	.667	.729	.827	.023	.053	.079	.596	.659	.764	.020	.046	.069	.538	.600	.707
3	.106	.189	.247	.888	.924	.967	.085	.153	.201	.799	.847	.915	.071	.129	.170	.721	.775	.858	.061	.111	.147	.655	.711	.802
4	.222	.343	.416	.979	.990	.998	.173	.271	.333	.907	.937	.973	.142	.225	.279	.830	.871	.929	.121	.193	.240	.760	.807	.879
5	.398	.549	.631	1	1	1	.294	.418	.490	.983	.991	.998	.236	.341	.404	.921	.947	.977	.198	.289	.345	.853	.889	.939
6							.464	.607	.681	1	1	1	.357	.479	.547	.985	.993	.999	.293	.400	.462	.931	.954	.980
7													.518	.652	.720	1	1	1	.410	.529	.594	.987	.994	.999
8																			.562	.688	.750	1	1	1
x	$n=9$						$n=10$						$n=11$						$n=12$					
0	0	0	0	.226	.283	.401	0	0	0	.206	.259	.369	0	0	0	.189	.238	.342	0	0	0	.175	.221	.319
1	.001	.006	.012	.368	.429	.544	.001	.005	.010	.337	.394	.504	.001	.005	.010	.310	.364	.470	.001	.004	.009	.287	.339	.440
2	.017	.041	.061	.490	.550	.656	.016	.037	.055	.450	.507	.612	.014	.033	.049	.415	.470	.572	.013	.030	.045	.386	.438	.537
3	.053	.098	.129	.599	.655	.750	.048	.087	.116	.552	.607	.703	.043	.079	.105	.511	.564	.660	.039	.072	.096	.475	.527	.622
4	.105	.169	.210	.699	.749	.829	.093	.150	.188	.646	.696	.782	.084	.135	.169	.599	.650	.738	.076	.123	.154	.559	.609	.698
5	.171	.251	.301	.790	.831	.895	.150	.222	.267	.733	.778	.850	.134	.200	.241	.682	.729	.806	.121	.181	.219	.638	.685	.765
6	.250	.345	.401	.871	.902	.947	.218	.304	.354	.812	.850	.907	.194	.271	.318	.759	.800	.866	.175	.245	.288	.712	.755	.825
7	.344	.450	.510	.939	.959	.983	.297	.393	.448	.884	.913	.952	.262	.350	.401	.831	.865	.916	.235	.315	.362	.781	.819	.879
8	.456	.571	.632	.988	.994	.999	.388	.493	.550	.945	.963	.984	.340	.436	.489	.895	.921	.957	.302	.391	.441	.846	.877	.924
9	.599	.717	.774	1	1	1	.496	.606	.663	.990	.995	.999	.428	.530	.585	.951	.967	.986	.378	.473	.525	.904	.928	.961
10							.631	.741	.794	1	1	1	.530	.636	.690	.990	.995	.999	.463	.562	.614	.955	.970	.987
11													.658	.762	.811	1	1	1	.560	.661	.713	.991	.996	.999
12																			.681	.779	.825	1	1	1

Table A.23b Two-Sided 100(1 − α)% Confidence Intervals and One-Sided 100(1 − α)% Confidence Bounds for a Binomial Proportion Nonconforming Based on x Nonconforming Units in a Sample of Size n

Column key for every block:

	Lower			Upper		
1 − α One-sided:	.99	.95	.90	.90	.95	.99
Two-sided:	.98	.90	.80	.80	.90	.98

n = 13

x	.99/.98	.95/.90	.90/.80	.90/.80	.95/.90	.99/.98
0	0	0	0	.162	.206	.298
1	.001	.004	.008	.268	.316	.413
2	.012	.028	.042	.360	.410	.506
3	.036	.066	.088	.444	.495	.588
4	.069	.113	.142	.523	.573	.661
5	.111	.166	.201	.598	.645	.727
6	.159	.224	.264	.669	.713	.787
7	.213	.287	.331	.736	.776	.841
8	.273	.355	.402	.799	.834	.889
9	.339	.427	.477	.858	.887	.931
10	.412	.505	.556	.912	.934	.964
11	.494	.590	.640	.958	.972	.988
12	.587	.684	.732	.992	.996	.999
13	.702	.794	.838	1	1	1

n = 14

x	.99/.98	.95/.90	.90/.80	.90/.80	.95/.90	.99/.98
0	0	0	0	.152	.193	.280
1	.001	.004	.007	.251	.297	.389
2	.011	.026	.039	.337	.385	.478
3	.033	.061	.081	.417	.466	.557
4	.064	.104	.131	.492	.540	.627
5	.102	.153	.185	.563	.610	.692
6	.146	.206	.243	.631	.675	.751
7	.195	.264	.305	.695	.736	.805
8	.249	.325	.369	.757	.794	.854
9	.309	.390	.437	.815	.847	.898
10	.373	.460	.508	.869	.896	.936
11	.443	.534	.583	.919	.939	.967
12	.522	.615	.663	.961	.976	.989
13	.611	.703	.749	.993	.996	.999
14	.720	.807	.848	1	1	1

n = 15

x	.99/.98	.95/.90	.90/.80	.90/.80	.95/.90	.99/.98
0	0	0	0	.142	.181	.264
1	.001	.003	.007	.236	.279	.368
2	.010	.024	.036	.317	.363	.453
3	.031	.057	.076	.393	.440	.529
4	.059	.097	.122	.464	.511	.597
5	.094	.142	.172	.532	.577	.660
6	.135	.191	.226	.596	.640	.718
7	.179	.244	.282	.658	.700	.771
8	.229	.300	.342	.718	.756	.821
9	.282	.360	.404	.774	.809	.865
10	.340	.423	.468	.828	.858	.906
11	.403	.489	.536	.878	.903	.941
12	.471	.560	.607	.924	.943	.969
13	.547	.637	.683	.964	.976	.990
14	.632	.721	.764	.993	.997	.999
15	.736	.819	.858	1	1	1

n = 16

x	.99/.98	.95/.90	.90/.80	.90/.80	.95/.90	.99/.98
0	0	0	0	.134	.171	.250
1	.001	.003	.007	.222	.264	.349
2	.010	.023	.034	.300	.344	.430
3	.029	.053	.071	.371	.417	.503
4	.055	.090	.114	.439	.484	.569
5	.088	.132	.161	.504	.548	.630
6	.125	.178	.210	.565	.609	.687
7	.166	.227	.263	.625	.667	.739
8	.212	.279	.318	.682	.721	.788
9	.261	.333	.375	.737	.773	.834
10	.313	.391	.435	.790	.822	.875
11	.370	.452	.496	.839	.868	.912
12	.431	.516	.561	.886	.910	.945
13	.497	.583	.629	.929	.947	.971
14	.570	.656	.700	.966	.977	.990
15	.651	.736	.778	.993	.997	.999
16	.750	.829	.866	1	1	1

n = 17

x	.99/.98	.95/.90	.90/.80	.90/.80	.95/.90	.99/.98
0	0	0	0	.127	.162	.237
1	.001	.003	.006	.210	.250	.332
2	.009	.021	.032	.284	.326	.410
3	.027	.050	.067	.352	.396	.480
4	.052	.085	.107	.416	.461	.543
5	.082	.124	.151	.478	.522	.603
6	.117	.166	.197	.537	.580	.658
7	.155	.212	.246	.594	.636	.709
8	.197	.260	.297	.650	.689	.758
9	.242	.311	.350	.703	.740	.803
10	.291	.364	.406	.754	.788	.845
11	.342	.420	.463	.803	.834	.883
12	.397	.478	.522	.849	.876	.918
13	.457	.539	.584	.893	.915	.948
14	.520	.604	.648	.933	.950	.973
15	.590	.674	.716	.968	.979	.991
16	.668	.750	.790	.994	.997	.999
17	.763	.838	.873	1	1	1

n = 18

x	.99/.98	.95/.90	.90/.80	.90/.80	.95/.90	.99/.98
0	0	0	0	.120	.153	.226
1	.001	.003	.006	.199	.238	.316
2	.008	.020	.030	.269	.310	.391
3	.025	.047	.063	.334	.377	.458
4	.049	.080	.101	.396	.439	.520
5	.077	.116	.142	.455	.498	.577
6	.110	.155	.185	.512	.554	.631
7	.145	.199	.231	.567	.608	.681
8	.184	.244	.279	.620	.659	.729
9	.226	.291	.329	.671	.709	.774
10	.271	.341	.380	.721	.756	.816
11	.319	.392	.433	.769	.801	.855
12	.369	.446	.488	.815	.844	.890
13	.423	.502	.545	.858	.884	.923
14	.480	.561	.604	.899	.920	.951
15	.542	.623	.666	.937	.953	.975
16	.609	.690	.731	.970	.980	.992
17	.684	.762	.801	.994	.997	.999
18	.774	.847	.880	1	1	1

n = 19

x	.99/.98	.95/.90	.90/.80	.90/.80	.95/.90	.99/.98
0	0	0	0	.114	.146	.215
1	.001	.003	.006	.190	.226	.302
2	.008	.019	.028	.257	.296	.374
3	.024	.044	.059	.319	.359	.439
4	.046	.075	.095	.378	.419	.498
5	.073	.110	.134	.434	.476	.554
6	.103	.147	.175	.489	.530	.606
7	.137	.188	.218	.541	.582	.655
8	.173	.230	.263	.592	.632	.702
9	.212	.274	.310	.642	.680	.746
10	.254	.320	.358	.690	.726	.788
11	.298	.368	.408	.737	.770	.827
12	.345	.418	.459	.782	.812	.863
13	.394	.470	.511	.825	.853	.897
14	.446	.524	.566	.866	.890	.927
15	.502	.581	.622	.905	.925	.954
16	.561	.641	.680	.941	.956	.976
17	.626	.704	.743	.972	.981	.992
18	.698	.774	.810	.994	.997	.999
19	.785	.854	.886	1	1	1

n = 20

x	.99/.98	.95/.90	.90/.80	.90/.80	.95/.90	.99/.98
0	0	0	0	.109	.139	.206
1	.001	.003	.005	.181	.216	.289
2	.008	.018	.027	.245	.283	.358
3	.023	.042	.056	.304	.344	.421
4	.044	.071	.090	.361	.401	.478
5	.069	.104	.127	.415	.456	.532
6	.098	.140	.166	.467	.508	.583
7	.129	.177	.207	.518	.558	.631
8	.163	.217	.249	.567	.606	.677
9	.200	.259	.293	.615	.653	.720
10	.239	.302	.338	.662	.698	.761
11	.280	.347	.385	.707	.741	.800
12	.323	.394	.433	.751	.783	.837
13	.369	.442	.482	.793	.823	.871
14	.417	.492	.533	.834	.860	.902
15	.468	.544	.585	.873	.896	.931
16	.522	.599	.639	.910	.929	.956
17	.579	.656	.696	.944	.958	.977
18	.642	.717	.755	.973	.982	.992
19	.711	.784	.819	.995	.997	.999
20	.794	.861	.891	1	1	1

Table A.24a Poisson Cumulative Distribution Probabilities: $\Pr(x \leq x') = P(x'; \lambda) = \sum_{i=0}^{x'} \exp(-\lambda)\lambda^i/i!$

x' \\ λ:	0.01	0.02	0.03	0.04	0.05	0.06	0.08	0.10	0.12	0.15	0.20	0.25	0.30	0.35	0.40	0.45
0	$.9^2005$	.9802	.9704	.9608	.9512	.9418	.9231	.9048	.8869	.8607	.8187	.7788	.7408	.7047	.6703	.6376
1	$.9^4503$	$.9^3803$	$.9^3559$	$.9^3221$	.9879	.9827	$.9^2697$	$.9^2532$	$.9^2335$	.9898	.9825	.9735	.9631	.9513	.9384	.9246
2	$.9^6835$	$.9^5869$	$.9^5560$	$.9^4896$	$.9^4799$	$.9^4656$	$.9^4196$	$.9^3845$	$.9^3737$	$.9^3497$	$.9^2885$	$.9^2784$	$.9^2640$	$.9^2449$	$.9^2207$	.9891
3				$.9^6897$	$.9^6750$	$.9^6485$	$.9^5840$	$.9^5615$	$.9^5215$	$.9^4813$	$.9^4432$	$.9^3867$	$.9^3734$	$.9^3527$	$.9^3224$	$.9^2880$
4									$.9^6812$	$.9^5441$	$.9^5774$	$.9^5339$	$.9^4842$	$.9^4673$	$.9^4388$	$.9^3894$

x' \\ λ:	0.50	0.55	0.60	0.65	0.70	0.75	0.80	0.85	0.90	0.95	1.00	1.10	1.20	1.30	1.40	1.50
0	.6065	.5769	.5488	.5220	.4966	.4724	.4493	.4274	.4066	.3867	.3679	.3329	.3012	.2725	.2466	.2231
1	.9098	.8943	.8781	.8614	.8442	.8266	.8088	.7907	.7725	.7541	.7358	.6990	.6626	.6268	.5918	.5578
2	.9856	.9815	.9769	.9717	.9659	.9595	.9526	.9451	.9371	.9287	.9197	.9004	.8795	.8571	.8335	.8088
3	$.9^2825$	$.9^2753$	$.9^2664$	$.9^2555$	$.9^2425$	$.9^2271$	$.9^2092$	.9889	.9865	.9839	.9810	.9743	.9662	.9569	.9463	.9344
4	$.9^3828$	$.9^3734$	$.9^3606$	$.9^3435$	$.9^3214$	$.9^2894$	$.9^2859$	$.9^2817$	$.9^2766$	$.9^2705$	$.9^2634$	$.9^2456$	$.9^2225$	.9893	.9857	.9814
5	$.9^4858$	$.9^4759$	$.9^4611$	$.9^4398$	$.9^4100$	$.9^3869$	$.9^3816$	$.9^3746$	$.9^3657$	$.9^3544$	$.9^3406$	$.9^3032$	$.9^2850$	$.9^2777$	$.9^2680$	$.9^2554$
6	$.9^5900$	$.9^5813$	$.9^5671$	$.9^5448$	$.9^5112$	$.9^4862$	$.9^4793$	$.9^4696$	$.9^4566$	$.9^4393$	$.9^4168$	$.9^3851$	$.9^3749$	$.9^3596$	$.9^3378$	$.9^3074$
7		$.9^6872$	$.9^6755$	$.9^6231$	$.9^5872$	$.9^5795$	$.9^5681$	$.9^5518$	$.9^5290$							

x' \\ λ:	1.60	1.70	1.80	1.90	2.00	2.20	2.40	2.60	2.80	3.00	3.20	3.40	3.60	3.80	4.00	4.20
0	.2019	.1827	.1653	.1496	.1353	.1108	.0907	.0743	.0608	.0498	.0408	.0334	.0273	.0224	.0183	.0150
1	.5249	.4932	.4628	.4337	.4060	.3546	.3084	.2674	.2311	.1991	.1712	.1468	.1257	.1074	.0916	.0780
2	.7834	.7572	.7306	.7037	.6767	.6227	.5697	.5184	.4695	.4232	.3799	.3397	.3027	.2689	.2381	.2102
3	.9212	.9068	.8913	.8747	.8571	.8194	.7787	.7360	.6919	.6472	.6025	.5584	.5152	.4735	.4335	.3954
4	.9763	.9704	.9636	.9559	.9473	.9275	.9041	.8774	.8477	.8153	.7806	.7442	.7064	.6678	.6288	.5898
5	$.9^2396$	$.9^2200$	.9896	.9868	.9834	.9751	.9643	.9510	.9349	.9161	.8946	.8705	.8441	.8156	.7851	.7531
6	$.9^3866$	$.9^2812$	$.9^2743$	$.9^2655$	$.9^2547$	$.9^2254$	.9884	.9828	.9756	.9665	.9554	.9421	.9267	.9091	.8893	.8675
7	$.9^3740$	$.9^3612$	$.9^3438$	$.9^3207$	$.9^2890$	$.9^2802$	$.9^2666$	$.9^2467$	$.9^2187$	.9881	.9832	.9769	.9692	.9599	.9489	.9361
8	$.9^4546$	$.9^4283$	$.9^3890$	$.9^3763$	$.9^3530$	$.9^3138$	$.9^2851$	$.9^2757$	$.9^2620$	$.9^2429$	$.9^2171$	.9883	.9840	.9786	.9786	.9721
9	$.9^4286$	$.9^4880$	$.9^4806$	$.9^4696$	$.9^4535$	$.9^3899$	$.9^3798$	$.9^3530$	$.9^3138$	$.9^2890$	$.9^2824$	$.9^2598$	$.9^2420$	$.9^2187$	$.9^2807$	$.9^2716$
10	$.9^5898$	$.9^5818$	$.9^5688$	$.9^4483$	$.9^4169$	$.9^4802$	$.9^4570$	$.9^4133$	$.9^3836$	$.9^3708$	$.9^3503$	$.9^3190$	$.9^3630$	$.9^3408$	$.9^2807$	$.9^2593$
11	$.9^6865$	$.9^6745$	$.9^6537$	$.9^6192$	$.9^5864$	$.9^5643$	$.9^5155$	$.9^5816$	$.9^4626$	$.9^4286$	$.9^3777$	$.9^3871$	$.9^3630$	$.9^3085$	$.9^3408$	$.9^2863$
12							$.9^5209$	$.9^5844$	$.9^5839$	$.9^4689$	$.9^4429$	$.9^4900$	$.9^3832$	$.9^3726$	$.9^3569$
13								$.9^5337$	$.9^5660$	$.9^4839$	$.9^4748$	$.9^4553$	$.9^4237$	$.9^4874$
14									$.9^5844$	$.9^5330$	$.9^5171$	$.9^5407$	$.9^4889$	$.9^4801$	$.9^4655$
15										$.9^6713$	$.9^6876$	$.9^6366$	$.9^5869$	$.9^5741$	$.9^5511$	$.9^5113$

Table A.24b Poisson Cumulative Distribution Probabilities: $\Pr(x \le x') = P(x'; \lambda) = \sum_{i=0}^{x'} \exp(-\lambda)\lambda^i/i!$

In the entries below, a superscript on a leading 0 or 9 denotes the number of times that digit is repeated (e.g. $.0^2823 = .00823$, $.9^3342 = .999342$).

x' \ λ:	4.40	4.60	4.80	5.00	5.20	5.40	5.60	5.80	6.00	6.20	6.40	6.60	6.80	7.00	8.00	9.00
0	.0123	.0101	$.0^2823$	$.0^2674$	$.0^2552$	$.0^2452$	$.0^2370$	$.0^2303$	$.0^2248$	$.0^2203$	$.0^2166$	$.0^2136$	$.0^2111$	$.0^3912$	$.0^3335$	$.0^3123$
1	.0663	.0563	.0477	.0404	.0342	.0289	.0244	.0206	.0174	.0146	.0123	.0103	$.0^2869$	$.0^2730$	$.0^2302$	$.0^2123$
2	.1851	.1626	.1425	.1247	.1088	.0948	.0824	.0715	.0620	.0536	.0463	.0400	.0344	.0296	.0138	$.0^2623$
3	.3594	.3257	.2942	.2650	.2381	.2133	.1906	.1700	.1512	.1342	.1189	.1052	.0928	.0818	.0424	.0212
4	.5512	.5132	.4763	.4405	.4061	.3733	.3422	.3127	.2851	.2592	.2351	.2127	.1920	.1730	.0996	.0550
5	.7199	.6858	.6510	.6160	.5809	.5461	.5119	.4783	.4457	.4141	.3837	.3547	.3270	.3007	.1912	.1157
6	.8436	.8180	.7908	.7622	.7324	.7017	.6703	.6384	.6063	.5742	.5423	.5108	.4799	.4497	.3134	.2068
7	.9214	.9049	.8867	.8666	.8449	.8217	.7970	.7710	.7440	.7160	.6873	.6581	.6285	.5987	.4530	.3239
8	.9642	.9549	.9442	.9319	.9181	.9027	.8857	.8672	.8472	.8259	.8033	.7796	.7548	.7291	.5925	.4557
9	.9851	.9805	.9749	.9682	.9603	.9512	.9409	.9292	.9161	.9016	.8858	.8686	.8502	.8305	.7166	.5874
10	$.9^2431$	$.9^2222$	.9896	.9863	.9823	.9775	.9718	.9651	.9574	.9486	.9386	.9274	.9151	.9015	.8159	.7060
11	$.9^2799$	$.9^2714$	$.9^2601$	$.9^2455$	$.9^2269$	$.9^2037$	.9875	.9841	.9799	.9750	.9693	.9627	.9552	.9467	.8881	.8030
12	$.9^3342$	$.9^3021$	$.9^2858$	$.9^2798$	$.9^2720$	$.9^2625$	$.9^2486$	$.9^2320$	$.9^2117$	.9887	.9857	.9821	.9779	.9730	.9362	.8757
13	$.9^3799$	$.9^3688$	$.9^3529$	$.9^3302$	$.9^3002$	$.9^2866$	$.9^2802$	$.9^2729$	$.9^2637$	$.9^2520$	$.9^2377$	$.9^2197$	.9898	.9872	.9658	.9261
14	$.9^4424$	$.9^4066$	$.9^3853$	$.9^3774$	$.9^3661$	$.9^3502$	$.9^3284$	$.9^2899$	$.9^2860$	$.9^2809$	$.9^2744$	$.9^2661$	$.9^2557$	$.9^2428$	.9827	.9585
15	$.9^4845$	$.9^4737$	$.9^4568$	$.9^4310$	$.9^4028$	$.9^3836$	$.9^3756$	$.9^3644$	$.9^3491$	$.9^3284$	$.9^3008$	$.9^2865$	$.9^2818$	$.9^2759$	$.9^2177$	.9780
16	$.9^5605$	$.9^5302$	$.9^4881$	$.9^4801$	$.9^4678$	$.9^4492$	$.9^4217$	$.9^3882$	$.9^3825$	$.9^3746$	$.9^3638$	$.9^3491$	$.9^3297$	$.9^3042$	$.9^2628$	.9889
17	$.9^6051$	$.9^5825$	$.9^5687$	$.9^5458$	$.9^5039$	$.9^4851$	$.9^4762$	$.9^4628$	$.9^4431$	$.9^4147$	$.9^3874$	$.9^3818$	$.9^3742$	$.9^3638$	$.9^2841$	$.9^2468$
18	$.9^6783$	$.9^6222$	$.9^5860$	$.9^5755$	$.9^5584$	$.9^4889$	$.9^4311$	$.9^4889$	$.9^4824$	$.9^4728$	$.9^4587$	$.9^4099$	$.9^3870$	$.9^3747$	$.9^3350$	$.9^2757$
19		$.9^6582$	$.9^6655$	$.9^6655$	$.9^6373$	$.9^5755$	$.9^5810$	$.9^5683$	$.9^5482$	$.9^5172$	$.9^4870$	$.9^4801$	$.9^4700$	$.9^3556$	$.9^3747$	$.9^2894$
20					$.9^6847$	$.9^6720$	$.9^6502$	$.9^6138$	$.9^5854$	$.9^5760$	$.9^5612$	$.9^5387$	$.9^5049$	$.9^4855$	$.9^4060$	$.9^3561$
21							$.9^6875$	$.9^6776$	$.9^6609$	$.9^6334$	$.9^5889$	$.9^5819$	$.9^5711$	$.9^5547$	$.9^4666$	$.9^3825$
22									$.9^6899$	$.9^6823$	$.9^5696$	$.9^5489$	$.9^5160$	$.9^5865$	$.9^4886$	$.9^4332$
23												$.9^6862$	$.9^6766$	$.9^6611$	$.9^5627$	$.9^4755$
24														$.9^6893$	$.9^5883$	$.9^5135$
25															$.9^6645$	$.9^5706$
26															$.9^6896$	$.9^6036$
27																$.9^6695$
28																
29																
30																
31																

Table A.25 Factors for Computing $100(1 - \alpha)\%$ Two-Sided Confidence Intervals and $100(1 - \alpha)\%$ One-Sided Confidence Bounds for Poisson Distribution Mean Occurrence Rates Based on x Past Occurrences: $G_{L(1-\alpha; x)}$ and $G_{U(1-\alpha; x)}$

| One-sided $1 - \alpha$: | 0.750 | | 0.900 | | 0.950 | | 0.975 | | 0.990 | | 0.995 | | 0.999 | |
| Two-sided $1 - \alpha$: | 0.500 | | 0.800 | | 0.900 | | 0.950 | | 0.980 | | 0.990 | | 0.998 | |
z	Lower	Upper	Lower	Upper	Lower	Upper	Lower	Upper	Lower	Upper	Lower	Upper	Lower	Upper
0	0.000	1.386	0.000	2.303	0.000	2.996	0.000	3.689	0.000	4.605	0.000	5.298	0.000	6.908
1	0.288	2.693	0.105	3.890	0.051	4.744	0.025	5.572	0.010	6.638	0.005	7.430	0.001	9.234
2	0.961	3.920	0.532	5.322	0.355	6.296	0.242	7.225	0.149	8.406	0.103	9.274	0.045	11.23
3	1.727	5.109	1.102	6.681	0.818	7.754	0.619	8.767	0.436	10.04	0.338	10.98	0.190	13.06
4	2.535	6.274	1.745	7.994	1.366	9.154	1.090	10.24	0.823	11.60	0.672	12.59	0.428	14.79
5	3.369	7.423	2.433	9.275	1.970	10.51	1.623	11.67	1.279	13.11	1.078	14.15	0.739	16.45
6	4.219	8.558	3.152	10.53	2.613	11.84	2.202	13.06	1.785	14.57	1.537	15.66	1.107	18.06
7	5.083	9.684	3.895	11.77	3.285	13.15	2.814	14.42	2.330	16.00	2.037	17.13	1.520	19.63
8	5.956	10.80	4.656	12.99	3.981	14.43	3.454	15.76	2.906	17.40	2.571	18.58	1.971	21.15
9	6.838	11.91	5.432	14.21	4.695	15.71	4.115	17.08	3.507	18.78	3.132	20.00	2.452	22.66
10	7.726	13.02	6.221	15.41	5.425	16.96	4.795	18.39	4.130	20.14	3.717	21.40	2.960	24.13
11	8.620	14.12	7.021	16.60	6.169	18.21	5.491	19.68	4.771	21.49	4.321	22.78	3.491	25.59
12	9.519	15.22	7.829	17.78	6.924	19.44	6.201	20.96	5.428	22.82	4.943	24.14	4.042	27.02
13	10.42	16.31	8.646	18.96	7.690	20.67	6.922	22.23	6.099	24.14	5.580	25.50	4.611	28.45
14	11.33	17.40	9.470	20.13	8.464	21.89	7.654	23.49	6.782	25.45	6.231	26.84	5.195	29.85
15	12.24	18.49	10.30	21.29	9.246	23.10	8.395	24.74	7.477	26.74	6.893	28.16	5.793	31.24
16	13.15	19.57	11.14	22.45	10.04	24.30	9.145	25.98	8.181	28.03	7.567	29.48	6.405	32.62
17	14.07	20.65	11.98	23.61	10.83	25.50	9.903	27.22	8.895	29.31	8.251	30.79	7.028	33.99
18	14.99	21.73	12.82	24.76	11.63	26.69	10.67	28.45	9.616	30.58	8.943	32.09	7.662	35.35
19	15.91	22.81	13.67	25.90	12.44	27.88	11.44	29.67	10.35	31.84	9.644	33.38	8.305	36.69
20	16.83	23.88	14.53	27.05	13.26	29.06	12.22	30.89	11.08	33.10	10.35	34.67	8.946	38.04
21	17.75	24.96	15.38	28.18	14.07	30.24	13.00	32.10	11.82	34.35	11.07	35.95	9.607	39.37
22	18.68	26.03	16.24	29.32	14.89	31.41	13.79	33.31	12.57	35.60	11.79	37.22	10.28	40.70
23	19.61	27.10	17.11	30.45	15.72	32.59	14.58	34.51	13.33	36.84	12.52	38.48	10.95	42.01
24	20.54	28.17	17.98	31.58	16.55	33.75	15.38	35.71	14.09	38.08	13.25	39.74	11.64	43.33
25	21.47	29.23	18.85	32.71	17.38	34.92	16.18	36.91	14.85	39.31	13.99	41.00	12.33	44.63
26	22.40	30.30	19.72	33.84	18.22	36.08	16.98	38.10	15.62	40.53	14.74	42.25	13.02	45.93
27	23.34	31.36	20.59	34.96	19.06	37.23	17.79	39.28	16.40	41.76	15.49	43.50	13.72	47.23
28	24.27	32.43	21.47	36.08	19.90	38.39	18.61	40.47	17.17	42.97	16.24	44.74	14.43	48.52
29	25.21	33.49	22.35	37.20	20.75	39.54	19.42	41.65	17.96	44.19	17.00	45.98	15.14	49.80
30	26.15	34.55	23.23	38.32	21.59	40.69	20.24	42.83	18.74	45.40	17.77	47.21	15.86	51.08
40	35.57	45.12	32.14	49.39	30.20	52.07	28.58	54.47	26.77	57.35	25.59	59.36	23.25	63.66
50	45.07	55.62	41.18	60.34	38.96	63.29	37.11	65.92	35.02	69.07	33.65	71.28	30.93	75.97
60	54.61	66.07	50.31	71.20	47.85	74.39	45.78	77.23	43.45	80.63	41.91	83.00	38.86	88.03
100	93.09	107.6	87.42	114.1	84.14	118.1	81.36	121.6	78.21	125.8	76.11	128.8	71.90	134.9

A table similar to Table A.25 appeared earlier in Pearson and Hartley (1954). Adapted with permission of the Biometrika Trustees.

Table A.26a **Smallest Sample Size Such That a Two-Sided $100(1 - \alpha)$% Confidence Interval for the Mean of a Normal Distribution Will Be No Wider Than $\pm k\sigma$ (σ Known)**

k	0.50	0.75	0.80	0.90	0.95	0.98	0.99	0.999
0.01	4550	13234	16424	27056	38415	54119	66349	108109
0.02	1138	3309	4106	6764	9604	13530	16588	27028
0.03	506	1471	1825	3007	4269	6014	7373	12013
0.04	285	828	1027	1691	2401	3383	4147	6757
0.05	182	530	657	1083	1537	2165	2654	4325
0.06	127	368	457	752	1068	1504	1844	3004
0.07	93	271	336	553	784	1105	1355	2207
0.08	72	207	257	423	601	846	1037	1690
0.09	57	164	203	335	475	669	820	1335
0.10	46	133	165	271	385	542	664	1082
0.11	38	110	136	224	318	448	549	894
0.12	32	92	115	188	267	376	461	751
0.13	27	79	98	161	228	321	393	640
0.14	24	68	84	139	196	277	339	552
0.15	21	59	73	121	171	241	295	481
0.16	18	52	65	106	151	212	260	423
0.17	16	46	57	94	133	188	230	375
0.18	15	41	51	84	119	168	205	334
0.19	13	37	46	75	107	150	184	300
0.20	12	34	42	68	97	136	166	271
0.21	11	31	38	62	88	123	151	246
0.22	10	28	34	56	80	112	138	224
0.23	9	26	32	52	73	103	126	205
0.24	8	23	29	47	67	94	116	188
0.25	8	22	27	44	62	87	107	173
0.30	6	15	19	31	43	61	74	121
0.35	4	11	14	23	32	45	55	89
0.40	3	9	11	17	25	34	42	68
0.45	3	7	9	14	19	27	33	54
0.50	2	6	7	11	16	22	27	44
0.60	2	4	5	8	11	16	19	31
0.70	1	3	4	6	8	12	14	23
0.80	1	3	3	5	7	9	11	17
0.90	1	2	3	4	5	7	9	14
1.00	1	2	2	3	4	6	7	11
1.20	1	1	2	2	3	4	5	8
1.40	1	1	1	2	2	3	4	6
1.60	1	1	1	2	2	3	3	5
1.80	1	1	1	1	2	2	3	4
2.00	1	1	1	1	1	2	2	3

A table similar to Table A.26a appeared earlier in Dixon and Massey (1969). Adapted with permission of McGraw-Hill, Inc.

Table A.26b Corrected Sample Size Needed to Find a Confidence Interval for the Mean of a Normal Distribution

$1 - \alpha = 0.75$

n	γ'	0.50	0.60	0.70	0.80	0.90	0.95	0.99
2	0.366	3	4	4	5	6	6	7
3	0.402	4	5	5	6	7	8	9
4	0.420	5	6	7	7	9	10	11
5	0.430	6	7	8	9	10	11	13
6	0.437	7	8	9	10	11	12	15
8	0.447	9	10	11	12	14	15	18
10	0.453	11	12	13	15	17	18	21
15	0.463	16	17	19	21	23	25	28
20	0.468	21	23	24	26	29	31	35
25	0.471	26	28	30	32	35	37	41
30	0.474	31	33	35	37	41	43	48
40	0.478	41	43	46	48	52	55	61
50	0.480	51	54	56	59	63	67	73
60	0.482	61	64	67	70	75	78	85
70	0.483	71	74	77	81	86	90	97
80	0.484	81	84	88	91	97	101	109
90	0.485	91	94	98	102	108	112	121
100	0.486	101	105	108	113	119	124	133

$1 - \alpha = 0.90$

n	γ'	0.50	0.60	0.70	0.80	0.90	0.95	0.99
2	0.206	4	5	5	6	6	7	8
3	0.272	5	6	6	7	8	8	10
4	0.310	6	7	7	8	9	10	12
5	0.334	7	8	9	9	11	12	13
6	0.351	8	9	10	11	12	13	15
8	0.374	10	11	11	13	15	16	18
10	0.389	12	13	14	15	17	19	21
15	0.411	17	18	20	21	23	25	28
20	0.423	22	23	25	27	29	32	35
25	0.432	27	29	30	33	35	38	42
30	0.438	32	34	36	38	41	44	49
40	0.446	42	44	46	49	53	56	61
50	0.452	52	54	57	60	64	67	74
60	0.456	62	64	67	71	75	79	86
70	0.460	72	75	78	81	86	90	98
80	0.462	82	85	88	92	97	102	110
90	0.465	92	95	99	103	108	113	122
100	0.466	102	105	109	113	119	124	133

$1 - \alpha = 0.95$

n	γ'	0.50	0.60	0.70	0.80	0.90	0.95	0.99
2	0.123	5	5	5	6	7	7	8
3	0.187	6	6	6	7	8	9	10
4	0.232	6	7	8	9	10	11	12
5	0.263	7	8	9	10	11	12	14
6	0.286	8	9	10	11	12	13	15
8	0.317	10	11	12	14	15	16	19
10	0.337	12	13	15	16	18	19	22
15	0.369	17	19	20	22	24	26	29
20	0.387	22	24	26	27	30	32	36
25	0.399	27	29	31	33	36	38	43
30	0.408	32	34	36	39	42	44	49
40	0.421	42	45	47	50	53	56	62
50	0.429	52	55	57	60	65	68	74
60	0.436	62	65	68	71	76	80	86
70	0.441	72	75	78	82	87	91	98
80	0.444	82	85	89	93	98	102	110
90	0.448	92	96	99	103	109	114	122
100	0.450	102	106	110	114	120	125	134

$1 - \alpha = 0.99$

n	γ'	0.50	0.60	0.70	0.80	0.90	0.95	0.99
2	0.032	6	6	7	7	8	8	9
3	0.065	7	7	8	8	9	10	11
4	0.100	8	8	9	10	11	12	13
5	0.131	9	9	10	11	12	13	15
6	0.156	10	10	11	12	14	15	17
8	0.197	11	13	14	15	16	17	20
10	0.226	14	15	16	18	19	20	23
15	0.274	19	20	21	23	25	27	30
20	0.303	24	25	27	29	31	33	37
25	0.324	29	30	32	34	37	39	44
30	0.339	34	36	38	40	43	46	50
40	0.360	44	46	48	51	55	58	63
50	0.375	54	56	59	62	66	69	75
60	0.386	64	66	69	73	77	81	88
70	0.394	74	77	80	83	88	92	100
80	0.401	84	87	90	94	99	104	112
90	0.406	94	97	101	105	110	115	123
100	0.411	104	107	111	115	121	126	135

This table was adapted from a similar table that appeared earlier in Kupper and Hafner (1989). Adapted with the permission of the American Statistical Association

Table A.27 Smallest Sample Size Needed to Estimate a Normal Distribution Standard Deviation Such That $\Pr(s < k\sigma) \geq 1 - \alpha$

				$1 - \alpha$			
k	0.80	0.85	0.90	0.95	0.98	0.99	0.999
1.01	3488	5323	8174	13507	21094	27084	47841
1.02	859	1319	2035	3372	5275	6778	11984
1.03	376	582	901	1497	2346	3016	5337
1.04	209	325	505	841	1320	1698	3008
1.05	132	206	322	538	845	1088	1929
1.06	90	142	223	373	588	757	1343
1.07	66	104	163	274	432	557	989
1.08	50	79	125	210	331	427	759
1.09	39	62	98	166	262	338	601
1.10	31	50	79	134	212	274	488
1.11	25	41	66	111	176	227	404
1.12	21	34	55	93	148	191	340
1.13	18	29	47	80	126	163	291
1.14	15	25	40	69	109	141	251
1.15	13	22	35	60	95	123	219
1.16	11	19	31	53	84	108	193
1.17	10	17	27	47	74	96	172
1.18	9	15	24	42	66	86	154
1.19	8	13	22	38	60	77	138
1.20	7	12	20	34	54	70	125
1.21	2	11	18	31	49	64	114
1.22	2	10	16	28	45	58	104
1.23	2	9	15	26	41	53	95
1.24	2	8	14	24	38	49	88
1.25	2	8	13	22	35	45	81
1.30	2	5	9	16	25	32	57
1 35	2	4	7	12	19	24	43
1.40	2	3	5	9	14	19	33
1.45	2	2	4	7	12	15	27
1.50	2	2	4	6	10	12	22
1.60	2	2	3	5	7	9	16
1.70	2	2	2	4	6	7	12
1.80	2	2	2	3	5	6	10
1.90	2	2	2	3	4	5	8
2.00	2	2	2	2	3	4	7
2.25	2	2	2	2	3	3	5
2.50	2	2	2	2	2	3	4
3.00	2	2	2	2	2	2	3

A table similar to Table A.27 appeared earlier in Dixon and Massey (1969). Adapted with permission of McGraw-Hill, Inc.

Table A.28 Smallest Sample Size for Normal Distribution Two-Sided 100(1 − α)% Tolerance Intervals Such That the Probability of Enclosing at Least the Population Proportion p′ Is 1 − α and the Probability of Enclosing at Least p* Is No More Than δ

p′ = 0.500

1 − α	p* = 0.540				p* = 0.560				p* = 0.580				p* = 0.600			
δ:	0.20	0.10	0.05	0.01	0.20	0.10	0.05	0.01	0.20	0.10	0.05	0.01	0.20	0.10	0.05	0.01
0.80	174	272	370	594	80	124	167	265	47	71	96	151	31	47	62	98
0.90	279	401	517	776	128	182	235	351	75	105	135	200	50	69	88	130
0.95	383	526	654	942	177	239	298	429	103	139	172	245	68	91	113	160
0.99	628	801	962	1313	288	368	442	597	169	213	255	343	112	141	168	224

p′ = 0.750

1 − α	p* = 0.790				p* = 0.810				p* = 0.830				p* = 0.850			
δ:	0.20	0.10	0.05	0.01	0.20	0.10	0.05	0.01	0.20	0.10	0.05	0.01	0.20	0.10	0.05	0.01
0.80	194	304	413	664	85	132	179	285	48	73	98	155	30	46	61	95
0.90	311	447	578	869	137	195	251	376	76	108	138	205	48	68	86	126
0.95	429	586	735	1060	189	257	321	459	106	142	176	251	67	89	110	155
0.99	700	899	1081	1469	310	395	473	639	173	219	261	351	109	137	163	218

p′ = 0.900

1 − α	p* = 0.930				p* = 0.940				p* = 0.950				p* = 0.960			
δ:	0.20	0.10	0.05	0.01	0.20	0.10	0.05	0.01	0.20	0.10	0.05	0.01	0.20	0.10	0.05	0.01
0.80	154	240	326	522	81	125	169	270	48	74	99	157	31	47	62	97
0.90	246	353	456	685	130	185	238	355	77	109	140	207	49	69	88	129
0.95	340	463	581	836	179	243	303	435	107	144	179	254	68	91	112	158
0.99	555	711	855	1159	293	374	448	605	175	221	264	355	111	140	167	222

p′ = 0.950

1 − α	p* = 0.970				p* = 0.975				p* = 0.980				p* = 0.985			
δ:	0.20	0.10	0.05	0.01	0.20	0.10	0.05	0.01	0.20	0.10	0.05	0.01	0.20	0.10	0.05	0.01
0.80	139	216	294	471	81	125	169	269	50	77	104	164	33	49	66	103
0.90	223	319	412	617	130	185	238	354	81	114	146	217	52	73	92	136
0.95	307	419	524	755	179	243	303	434	111	150	187	266	72	96	118	167
0.99	502	641	772	1046	293	373	447	604	182	231	276	371	117	148	176	235

p′ = 0.990

1 − α	p* = 0.996				p* = 0.997				p* = 0.998				p* = 0.999			
δ:	0.20	0.10	0.05	0.01	0.20	0.10	0.05	0.01	0.20	0.10	0.05	0.01	0.20	0.10	0.05	0.01
0.80	117	182	247	396	73	112	152	241	45	69	92	145	26	39	51	80
0.90	188	269	347	519	117	166	213	318	72	101	130	192	41	57	72	106
0.95	259	353	442	634	161	218	272	390	99	133	166	235	57	75	93	130
0.99	424	542	651	881	264	336	402	543	162	206	245	329	92	116	138	183

Table A.28 is similar to a table that appeared earlier in Faulkenberry and Daly (1970). Adapted with permission of the American Statistical Association.

Table A.29 Smallest Sample Size for Normal Distribution One-Sided 100(1 − α)% Tolerance Bounds Such That the Probability of Enclosing at Least the Population Proportion p' Is $1 - α$ and the Probability of Enclosing at Least $p*$ Is No More Than δ

$p' = 0.500$

$1 - α$	$p* = 0.560$				$p* = 0.580$				$p* = 0.600$				$p* = 0.620$			
δ:	0.20	0.10	0.05	0.01	0.20	0.10	0.05	0.01	0.20	0.10	0.05	0.01	0.20	0.10	0.05	0.01
0.80	125	199	272	441	70	111	153	247	45	71	97	157	31	49	67	108
0.90	199	290	377	572	112	162	211	321	72	104	135	204	50	72	93	141
0.95	273	378	477	694	154	212	267	389	98	135	170	248	68	94	118	171
0.99	444	574	695	953	249	323	390	534	160	206	249	340	111	143	172	235

$p' = 0.750$

$1 - α$	$p* = 0.790$				$p* = 0.810$				$p* = 0.830$				$p* = 0.850$			
δ:	0.20	0.10	0.05	0.01	0.20	0.10	0.05	0.01	0.20	0.10	0.05	0.01	0.20	0.10	0.05	0.01
0.80	208	328	448	723	90	141	192	303	49	76	103	166	30	47	63	101
0.90	333	482	625	944	144	207	268	404	79	113	145	217	49	69	89	133
0.95	459	630	793	1148	199	272	341	492	109	148	185	266	68	91	114	162
0.99	748	964	1162	1585	325	417	502	682	178	228	273	369	111	141	168	226

$p' = 0.900$

$1 - α$	$p* = 0.930$				$p* = 0.940$				$p* = 0.950$				$p* = 0.960$			
δ:	0.20	0.10	0.05	0.01	0.20	0.10	0.05	0.01	0.20	0.10	0.05	0.01	0.20	0.10	0.05	0.01
0.80	148	231	315	506	77	120	163	261	46	70	95	151	29	44	59	93
0.90	237	341	441	663	124	178	229	343	73	104	134	199	46	65	83	123
0.95	327	447	560	808	172	233	292	419	101	137	171	244	64	86	107	151
0.99	534	685	824	1120	281	359	431	583	166	212	253	341	105	133	159	212

$p' = 0.950$

$1 - α$	$p* = 0.970$				$p* = 0.975$				$p* = 0.980$				$p* = 0.985$			
δ:	0.20	0.10	0.05	0.01	0.20	0.10	0.05	0.01	0.20	0.10	0.05	0.01	0.20	0.10	0.05	0.01
0.80	131	205	279	448	76	118	160	256	47	73	98	156	30	46	62	97
0.90	211	303	391	587	123	175	225	337	76	108	138	205	49	68	87	129
0.95	291	397	498	716	169	230	287	412	105	142	177	252	67	90	112	158
0.99	476	610	733	994	277	354	424	573	172	219	262	352	111	140	166	222

$p' = 0.990$

$1 - α$	$p* = 0.996$				$p* = 0.997$				$p* = 0.998$				$p* = 0.999$			
δ:	0.20	0.10	0.05	0.01	0.20	0.10	0.05	0.01	0.20	0.10	0.05	0.01	0.20	0.10	0.05	0.01
0.80	111	173	235	376	69	106	144	229	42	65	87	138	24	36	48	76
0.90	178	255	329	493	111	157	202	302	68	96	123	182	39	54	68	100
0.95	246	335	419	603	153	207	258	370	94	127	157	224	54	71	88	124
0.99	402	515	618	837	250	319	382	515	154	195	233	313	88	110	131	174

Table A.29 is similar to a table that appeared earlier in Faulkenberry and Daly (1970). Adapted with permission of the American Statistical Association.

Table A.30 Smallest Sample Size for Distribution-Free 100(1 − α)% Tolerance Intervals and Tolerance Bounds Such That the Probability of Enclosing at Least the Population Proportion p′ Is 1 − α and the Probability of Enclosing at Least p* Is No More Than δ

p′ = 0.500

1 − α	p* = 0.560				p* = 0.580				p* = 0.600				p* = 0.620			
δ:	0.20	0.10	0.05	0.01	0.20	0.10	0.05	0.01	0.20	0.10	0.05	0.01	0.20	0.10	0.05	0.01
0.80	199	314	435	704	112	183	249	392	77	112	158	253	49	81	108	177
0.90	316	463	601	906	176	260	337	506	115	168	214	322	81	115	149	227
0.95	438	602	748	1098	245	340	423	614	158	213	268	392	106	147	188	275
0.99	701	903	1092	1497	391	510	615	843	252	325	391	535	174	224	274	370

p′ = 0.750

1 − α	p* = 0.790				p* = 0.810				p* = 0.830				p* = 0.850			
δ:	0.20	0.10	0.05	0.01	0.20	0.10	0.05	0.01	0.20	0.10	0.05	0.01	0.20	0.10	0.05	0.01
0.80	325	503	689	1107	140	216	291	474	75	118	161	258	49	75	101	157
0.90	508	730	955	1434	219	318	411	613	118	171	223	335	73	109	136	206
0.95	697	958	1205	1745	299	416	514	752	167	229	282	407	103	139	176	251
0.99	1128	1450	1758	2398	490	630	760	1030	271	344	415	560	164	211	253	344

p′ = 0.900

1 − α	p* = 0.930				p* = 0.940				p* = 0.950				p* = 0.960			
δ:	0.20	0.10	0.05	0.01	0.20	0.10	0.05	0.01	0.20	0.10	0.05	0.01	0.20	0.10	0.05	0.01
0.80	245	385	534	849	135	213	278	449	78	124	169	267	54	78	113	169
0.90	402	566	739	1102	210	301	390	588	128	187	233	346	91	116	152	222
0.95	549	748	934	1344	298	402	493	715	179	239	298	425	116	154	191	275
0.99	887	1134	1367	1860	482	612	727	989	287	362	435	589	197	236	287	374

p′ = 0.950

1 − α	p* = 0.970				p* = 0.975				p* = 0.980				p* = 0.985			
δ:	0.20	0.10	0.05	0.01	0.20	0.10	0.05	0.01	0.20	0.10	0.05	0.01	0.20	0.10	0.05	0.01
0.80	272	427	580	923	180	249	339	536	110	157	226	339	85	110	157	226
0.90	446	628	807	1202	282	377	492	718	184	258	306	446	132	158	209	306
0.95	601	832	1036	1481	361	506	624	877	234	311	386	554	153	208	260	361
0.99	997	1279	1511	2057	606	755	901	1233	398	504	580	779	259	344	398	529

p′ = 0.990

1 − α	p* = 0.996				p* = 0.997				p* = 0.998				p* = 0.999			
δ:	0.20	0.10	0.05	0.01	0.20	0.10	0.05	0.01	0.20	0.10	0.05	0.01	0.20	0.10	0.05	0.01
0.80	551	906	1137	1811	427	551	790	1137	299	427	551	790	132	299	299	427
0.90	926	1297	1538	2358	667	798	1051	1538	388	531	667	1051	388	388	531	667
0.95	1182	1693	2064	2902	913	1050	1312	1941	628	773	913	1312	473	473	628	773
0.99	2010	2539	3052	4047	1453	1736	2010	2669	1001	1157	1453	1736	662	838	1001	1157

APPENDIX B

Summary of Notation

This appendix outlines most of the notation that is used in this book. Some symbols, when they are only used within a particular section, are not listed here but are defined where used. For statistical intervals, we show the lower and upper limits corresponding to a two-sided interval. Notation for one-sided bounds is not given explicitly but is denoted by using the appropriate endpoint of the two-sided interval.

$B(x; n, p)$	cumulative binomial distribution probability of observing x or fewer nonconforming units in a sample of size n units when the probability of a nonconforming unit is p
c	critical number of nonconforming items in an "attribute" demonstration test. If the observed number is more than c, the demonstration is unsuccessful.
$c_{(\cdot)}, c'_{(\cdot)}, c_{L(\cdot)}, c_{U(\cdot)}, c'_{L(\cdot)}, c'_{U(\cdot)}$	factors for computing statistical intervals for a normal distribution
d	desired confidence interval half-width
$\text{EPKM}(n, \ell, u, m, k)$	gives the probability that the interval defined by the order statistics $[x_{(\ell)}, x_{(u)}]$ from a random sample of size n will contain at least k observations from a subsequent independent random sample of size m from the same population. For one-sided lower (upper) bounds, set $u = n + 1$ ($\ell = 0$).
$\text{EPTI}(n, \ell, u, p)$	gives the probability that the interval defined by the order statistics $[x_{(\ell)}, x_{(u)}]$ from a random sample of size n will cover at least $100p\%$ of the population. For one-sided lower (upper) bounds, set $u - n + 1$ ($\ell = 0$).
$\text{EPYJ}(n, \ell, u, m, j)$	gives the probability that the interval defined by the order statistics $[x_{(\ell)}, x_{(u)}]$ from

353

	a random sample of size n will contain $y_{(j)}$, the jth largest observation from a subsequent independent random sample of size m from the same population. For one-sided lower (upper) bounds, set $u = n + 1$ ($\ell = 0$).
EPYP(n, ℓ, u, p)	gives the probability that the interval defined by the order statistics $[x_{(\ell)}, x_{(u)}]$ from a random sample of size n will contain the $100p$th percentile of the population. For one-sided lower (upper) bounds, set $u = n + 1$ ($\ell = 0$).
$F_{(\gamma; r_1, r_2)}$	the 100γ percentile of the F distribution with r_1 numerator and r_2 denominator degrees of freedom
$g_{(\gamma; p, n)}$	factors from Table A.10 used to compute $100\gamma\%$ two-sided tolerance intervals to control the center of a normal population
$g'_{(\gamma; p, n)}$	factors from Table A.12 used to compute $100\gamma\%$ one-sided tolerance bounds or confidence limits for percentiles from a normal distribution
$g''_{(\gamma; p, n)}$	factors from Table A.11 used to compute $100\gamma\%$ two-sided tolerance intervals to control both tails of a normal population
$G_{(\cdot)}$	factor for computing confidence intervals for a Poisson occurrence rate
$h_{(\gamma; p, n)}$	endpoints for $100\gamma\%$ confidence limits for normal distribution tail probabilities (from Table 7 of Odeh and Owen 1980)
$H(x; n, D, N)$	cumulative hypergeometric distribution probability of obtaining x or fewer nonconforming units in a sample of size n drawn at random, without replacement, from a population of size N units which initially contains D nonconforming units
$H^*(L, U; D, n, N)$	the "inverse hypergeometric distribution" [see Guenther (1975)]
k	critical value in a "variables" demonstration test
ℓ, u	indices for a particular ordered observation
m	sample size of a future sample
n	sample size of a previous sample
p or p_i	the probability of a single randomly selected unit being nonconforming or the probability of some other particular event. If there is

	more than one such probability in a problem, we use subscripts (like i) to distinguish them from each other
$\hat{p}$	estimate of p
$[\tilde{p}, p]$	confidence interval or bounds for p
$P(x; \lambda)$	cumulative Poisson distribution probability of observing x or fewer occurrences in a unit of time or space where λ is the constant occurrence rate per unit of time or space
$r_{(\gamma; m, n)}$	factors from Table A.13 used to compute $100\gamma\%$ two-sided simultaneous prediction intervals for m observations from a normal distribution
$r'_{(\gamma; m, n)}$	factors from Table A.14 used to compute $100\gamma\%$ one-sided simultaneous prediction bounds for m observations from a normal distribution
$s(s_n)$	sample standard deviation of a previous sample (of size n) and an estimate for σ
s_m	sample standard deviation of a future sample of size m (estimate for σ)
$[\underline{s}_m, \tilde{s}_m]$	prediction interval or bounds to contain the standard deviation computed from m future observations
$t_{(\gamma; r)}$	the 100γ percentile of the Student t distribution with r degrees of freedom
$[\underline{T}_p, \tilde{T}_p]$	tolerance interval or bounds to enclose $100p\%$ of the population
ν	number of extreme observations to be removed from the end (ends) of a sample to construct a distribution-free confidence, tolerance, or prediction interval
x	number of binomial events or Poisson nonconforming units (occurrences) in a sample
$x_1, x_2, \ldots, x_n$	previous sample observations
$x_{(1)}, x_{(2)}, \ldots, x_{(n)}$	order statistics; previous sample observations, ordered from smallest $x_{(1)}$ to largest $x_{(n)}$
$\bar{x}$	sample mean of previous sample, estimate of μ
y	number of binomial (Poisson) nonconforming (occurrences) in a future sample
$y_1, y_2, \ldots, y_m$	future sample observations
$y_{(1)}, y_{(2)}, \ldots, y_{(m)}$	future sample observations, ordered from smallest to largest (order statistics)

$\bar{y}_m$	sample mean of a future sample of m observations
$[\underline{\bar{y}}_m, \tilde{\bar{y}}_m]$	prediction interval or bounds to contain the mean of a future sample of m observations
$[\underline{y}_m, \tilde{y}_m]$	simultaneous prediction interval or bounds to contain all m future observations
$[\underline{y}_{k;m}, \tilde{y}_{k;m}]$	simultaneous prediction interval or bounds to contain at least k of m future observations
$[\underline{y}_{(j)}, \tilde{y}_{(j)}]$	prediction interval or bounds to contain the jth largest observation (the jth order statistic) from a future sample of m observations
$[\underline{y}, \tilde{y}]$	prediction interval or bounds to contain a single future observation
Y_p	$100p$th percentile of a population
$\hat{Y}_p$	estimator for the $100p$th percentile of a population
$[\underline{\hat{Y}}_p, \tilde{\hat{Y}}_p]$	prediction interval or bounds to contain the estimate of Y_p
$[\underline{Y}_p, \tilde{Y}_p]$	confidence interval or bounds for Y_p
$z_{(\gamma)}$	the 100γ percentile of the standard normal distribution
$\chi^2_{(\gamma;r)}$	the 100γ percentile of the χ^2 distribution with r degrees of freedom
λ	Poisson distribution parameter
$\hat{\lambda}$	estimate of λ
$[\underline{\lambda}, \tilde{\lambda}]$	confidence interval or bounds for λ
μ	mean of a population
$[\underline{\mu}, \tilde{\mu}]$	confidence interval or bounds for μ
$\tilde{\sigma}$	standard deviation of a population
$[\underline{\sigma}, \tilde{\sigma}]$	confidence interval or bounds for σ
$\Phi(z)$	the standard normal cdf; i.e., $\Phi(z)$ is the probability that a normal random variable with mean 0 and standard deviation 1 will be less than z.
$100(1 - \alpha)$	confidence level (percent)
superscript *	indicates "guess value" for a parameter, to be used in choosing an appropriate sample size
superscript s	indicates a specified limit for a parameter (e.g., one might need to demonstrate that $Y_p \geq Y_p^s$)

Listing of Computer Subroutines for Distribution-Free Statistical Intervals

```
c*
c*
c*    fortran subroutines, example main program, and output for computing
c*    probabilities associated with nonparametric statistical intervals.
c*
      implicit real*8(a-h,o-z)
c #
c #set variables for calling the functions
c #
c #set the sample size
      n=100
c #set the print level
      kprint=1
c #compute the table of log factorials
      call setlnf
c #set the other input parameters and call the functions
      p=0.90
      ir=2
      is=98
      call eptiww(n,ir,is,p,eptixx,kprint)
      ir=45
      is=55
      p=0.50
      call epypww(n,ir,is,p,epypxx,kprint)
      ir=1
      is=10
      j=2
      m=50
      call epyjww(n,ir,is,m,j,epyjxx,kprint)
      ir=10
      is=90
      k=8
      m=10
      call epkmww(n,ir,is,m,k,epkmxx,kprint)
c #write the results
      write(6,45)eptixx,epypxx,epyjxx,epkmxx
45    format(' epti=',f8.4/
     &       ' epyp=',f8.4/
     &       ' epyj=',f8.4/
     &       ' epkm=',f8.4/
     &)
      stop
      end
c-----------------------------------------------------------------------
      subroutine eptiww(n,ir,is,p,prob,kprint)
c-----------------------------------------------------------------------
c***********************************************************************
c*
c*  this subroutine gives the probability that the interval defined
```

```
c*  by the order statistics [x(ir), x(is)] from a random
c*  sample of size n will cover at least 100p percent
c*  of the population.
c*
c*  input parameters:
c*
c*      n               sample size
c*
c*      ir              index of the order statistic corresponding to the
c*                      lower endpoint of the tolernace interval.
c*
c*      is              index of the order statistic corresponding to the
c*                      upper endpoint of the tolerance interval.
c*
c*              set is=n+1 for a one-sided lower tolerance interval
c*              and ir=0 for a one-sided upper tolerance interval.
c*
c*      p               is the proportion of the population to be enclosed
c*
c*      kprint          is the level of printing
c*
c*                          kprint
c*                      ----------------------
c*                          0               no printing
c*                          1               print result below
c*                          3               debug print
c*
c*  output parameter:
c*
c*      prob            confidence level associated with the tolerance interval
c*
c******************************************************************************
      implicit real*8(a-h,o-z)
      data zero,one/0.0d00,1.0d00/
      if(kprint.ge.2)write(6,422)n,ir,is,p,kprint
422   format(' eptiww**2**n,ir,is,p,kprint=',3i10,f10.3,i10)
      if(n.lt.1)go to 991
      if(ir.lt.0.or.ir.gt.n)go to 992
      if(ir.ge.is.or.is.gt.(n+1))go to 993
      if((p.le.zero).or.(p.ge.one))go to 994
      hp=100.0d00*p
      if(kprint.ge.1)write(6,42)hp,ir,is,n
42    format(' computing the probability that at least',
     &' a proportion ',f7.1,' of the population will be'/
     &           ' enclosed by x(',i4,') and x(',i4,') of',i5,
     &           ' observations.')
      prob=binodf(is-ir-1,n,p)
      if(kprint.ge.1)write(6,43)prob
43    format(' probability=',f8.4/)
      return
991   call err(12991)
      return
992   call err(12992)
      return
```

```
993   call err(12993)
      return
994   call err(12994)
      return
995   call err(12995)
      return
      end
c----------------------------------------------------------------------
      subroutine epypww(n,ir,is,p,prob,kprint)
c----------------------------------------------------------------------
c**********************************************************************
c*
c*   this subroutine gives the probability that the interval defined
c*   by the order statistics [x(ir), x(is)] from a random
c*   sample of size n will contain the pth percentile of the sampled
c*   population.
c*
c*   input parameters:
c*
c*      n           sample size
c*
c*      ir          index of the order statistic corresponding to the
c*                  lower endpoint of the confidence interval.
c*
c*      is          index of the order statistic corresponding to the
c*                  upper endpoint of the confidence interval.
c*
c*                  set is=n+1 for a one-sided lower tolerance interval
c*                  and ir=0 for a one-sided upper tolerance interval.
c*
c*      p           is the percentile of the population that is to be
c*                  contained in the confidence interval
c*
c*      kprint      is the level of printing
c*
c*                      kprint
c*                      ----------------------
c*                         0           no printing
c*                         1           print result below
c*                         3           debug print
c*
c*   output parameter:
c*
c*      prob        confidence level associated with the confidence interval
c*
c*
c**********************************************************************
      implicit real*8(a-h,o-z)
      data zero,one/0.0d00,1.0d00/
      if(kprint.ge.2)write(6,422)n,ir,is,p,kprint
422   format(' epypww**2**n,ir,is,p,kprint=',3i10,f10.3,i10)
      if(n.lt.1)go to 991
      if(ir.lt.0.or.ir.gt.n)go to 992
      if(ir.ge.is.or.is.gt.(n+1))go to 993
```

```
      if((p.le.zero).or.(p.ge.one))go to 994
      hp=100.0d00*p
      if(kprint.ge.1)write(6,42)hp,ir,is,n
42    format(' computing the probability that ',
     &' the ',f7.1,' percentile of the population will be'/
     &            ' enclosed by x(',i4,') and x(',i4,') of',i5,
     &            ' observations.')
      prbis=binodf(is-1,n,p)
      prbir=binodf(ir-1,n,p)
      prob=prbis-prbir
      if(kprint.ge.1)write(6,43)prob
43    format(' probability=',f8.4/)
      return
991   call err(12991)
      return
992   call err(12992)
      return
993   call err(12993)
      return
994   call err(12994)
      return
995   call err(12995)
      return
      end
c----------------------------------------------------------------------
      subroutine epkmww(n,ir,is,m,k,probp,kprint)
c----------------------------------------------------------------------
c**********************************************************************************
c*
c*  this subroutine gives the probability that the interval defined
c*  by the order statistics [x(ir), x(is)] from a random
c*  sample of size n will contain k of m future observations
c*  from the same population.
c*
c*  input parameters:
c*
c*      n           sample size of the first sample
c*
c*      ir          index of the order statistic corresponding to the
c*                  lower endpoint of the confidence interval.
c*
c*      is          index of the order statistic corresponding to the
c*                  upper endpoint of the confidence interval.
c*
c*              set is=n+1 for a one-sided lower tolerance interval
c*              and ir=0 for a one-sided upper tolerance interval.
c*
c*      m           sample size of the second sample
c*
c*      k           the number of observations to be enclosed in the
c*                  prediction interval
c*
c*      kprint      is the level of printing
c*
```

```
c*                      kprint
c*                      ----------------------
c*                 0          no printing
c*                 1          print result below
c*                 3          debug print
c*
c*   output parameter:
c*
c*        prob       confidence level associated with the confidence interval
c*
c******************************************************************************
      implicit real*8(a-h,o-z)
      data zero,one/0.0d00,1.0d00/
      if(kprint.ge.2)write(6,422)n,ir,is,m,k,kprint
422   format(' epkmww**2**n,ir,is,m,k,kprint=',6i10)
      if(n.lt.1)go to 991
      if(ir.lt.0.or.ir.gt.n)go to 992
      if(ir.ge.is.or.is.gt.(n+1))go to 993
      if(m.lt.1)go to 994
      if(k.lt.0.or.k.gt.m)go to 995
      if(kprint.ge.1)write(6,42)k,m,ir,is,n
42    format(' computing the probability that at least',i4,' of',i5,
     &           ' future observations'/
     &           ' will fall between x(',i4,') and x(',i4,') of',i5,
     &           ' past observations.')
      prob=hstar(k,m,is-ir,m,n+m,kprint)
      if(kprint.ge.1)write(6,43)prob
      probp=prob
43    format(' probability=',f8.4/)
      return
991   call err(12991)
      return
992   call err(12992)
      return
993   call err(12993)
      return
994   call err(12994)
      return
995   call err(12995)
      return
      end
c---------------------------------------------------------------
      subroutine epyjww(n,ir,is,m,j,probp,kprint)
c---------------------------------------------------------------
c******************************************************************************
c*
c*  this subroutine gives the probability that the interval defined
c*  by the order statistics [x(ir), x(is)] from a random
c*  sample of size n will contain the jth largest of m future observations
c*  from the same population.
c*
c*  input parameters:
c*
c*      n           sample size of the first sample
```

```
c*
c*      ir              index of the order statistic corresponding to the
c*                      lower endpoint of the confidence interval.
c*
c*      is              index of the order statistic corresponding to the
c*                      upper endpoint of the confidence interval.
c*
c*              set is=n+1 for a one-sided lower tolerance interval
c*              and ir=0 for a one-sided upper tolerance interval.
c*
c*      m               sample size of the second sample (m>0)
c*
c*      j               the particular order statistic to be enclosed in the
c*                      interval (1<=j<=m)
c*
c*      kprint          is the level of printing
c*
c*                          kprint
c*                      ----------------------
c*                          0           no printing
c*                          1           print result below
c*                          3           debug print
c*
c*   output parameter:
c*
c*      prob            confidence level associated with the confidence interval
c*
c*******************************************************************************
      implicit real*8(a-h,o-z)
      if(kprint.ge.2)write(6,422)n,ir,is,m,j,kprint
422   format(' epyjww**2**n,ir,is,m,j,kprint=',6i10)
      if(n.lt.1)go to 991
      if(ir.lt.0.or.ir.gt.n)go to 992
      if(ir.ge.is.or.is.gt.(n+1))go to 993
      if(m.lt.1)go to 994
      if(j.lt.1.or.j.gt.m)go to 995
      if(kprint.ge.1)write(6,42)j,m,ir,is,n
42    format(' computing the probability that y(',i3,') of',i4,
     &            ' future observations'/
     &            ' will fall between x(',i3,') and x(',i3,') of',i4,
     &            ' past observations.')
      prob=hstar(ir,is-1,j,n,n+m,kprint)
      if(kprint.ge.1)write(6,43)prob
      probp=prob
43    format(' probability=',f8.4/)
      return
991   call err(13991)
      return
992   call err(13992)
      return
993   call err(13993)
      return
994   call err(13994)
      return
```

```
995   call err(13995)
      return
      end
c-----------------------------------------------------------------------
      double precision function binodf(ix,n,p)
c-----------------------------------------------------------------------
c  #
x  #subroutine to return cumulative binomial probabilities
c  #
      implicit real*8(a-h,o-z)
      data zero,one/0.0d00,1.0d00/
      binodf=zero
      if(p.le.zero)go to 991
      if(p.ge.one)go to 992
      if(ix.ge.n)go to 992
      if(ix.lt.0)go to 991
      if(n.lt.1)go to 991
      np2=n/2
      if(ix.lt.np2)go to 50
      lsw=1
      ixp=n-ix-1
      pp=one-p
      go to 51
50    pp=p
      ixp=ix
      lsw=0
51    sum=zero
      ixpp=ixp+1
      dlogp=dlog(pp)
      dlogpm=dlog(one-pp)
      do 22 ip=1,ixpp
      inow=ip-1
      sum=sum+dexpc(405,bicof(n,inow)+dfloat(inow)*dlogp
     &+dfloat(n-inow)*dlogpm)
22    continue
      binodf=sum
      if(lsw.eq.1)binodf=one-sum
      return
991   binodf=zero
      return
992   binodf=one
      return
      end
c-----------------------------------------------------------------------
      double precision function hsmall(i,id,isn,ibn,kprint)
c-----------------------------------------------------------------------
c  #compute the inverse hypergeometric distribution
      implicit real*8 (a-h,o-z)
      hsmall=dexpc(405,bicof(id-1,i)+bicof(ibn-id,isn-i)-
     @bicof(ibn,isn))
      if(kprint.ge.6)write(6,47)hsmall
47    format(1h+,87x,f15.6)
      return
      end
```

```
c------------------------------------------------------------------------
      . double precision function hstar(ir,is,id,isn,ibn,kprint)
c------------------------------------------------------------------------
c  #conmpute the cumulative inverse hypergeometric
      implicit real*8(a-h,o-z)
      if(kprint.ge.4)write(6,42)ir,is,id,isn,ibn
42    format(' hstar**4**',5i5)
      hstar=0.0d00
      irp=ir+1
      isp=is+1
      do 22 ip=irp,isp
      i=ip-1
      hstar=hstar+hsmall(i,id+i,isn,ibn,kprint)
22    continue
      if(kprint.ge.4)write(6,47)hstar
47    format(1h+,45x,f15.6)
      return
      end
c------------------------------------------------------------------------
      subroutine setlnf
c------------------------------------------------------------------------
c  #
c  #    generate and save a table of log factorials
c  #
      double precision d,zero
      double precision dlog,dfloat
c  #    set the dimension of d and the value of nmaxp = nmax+1
c  #    where nmax is the largest argument of flnf
      common/lnfval/d(2001)
      data nmaxp/2001/
      data zero/0.0d00/
      d(1)=zero
      d(2)=zero
      do 100 i=3,nmaxp
      d(i)=d(i-1)+dlog(dfloat(i-1))
100   continue
      return
      end
c------------------------------------------------------------------------
      double precision function flnf(j)
c------------------------------------------------------------------------
c  #
c  #    return log factorials through a table lookup
c  #
      double precision d
c  #    set the dimension of d to nmax+1 where nmax
c  #    is the largest argument to be used in flnf
      common/lnfval/d(2001)
      flnf=d(j+1)
      return
      end
c------------------------------------------------------------------------
      function bicof(n,k)
c------------------------------------------------------------------------
```

```
c
c        return log binomial coefficient
c
      real*8 bicof,flnf
      bicof=flnf(n)-flnf(k)-flnf(n-k)
      return
      end
c----------------------------------------------------------------------
      subroutine err(mark)
c----------------------------------------------------------------------
c  #print an error message if the arguement is positive
c  #otherwise a warning message is printed
      common/trap/iercc
      marker=iabs(mark)
      if(mark.lt.0) go to 225
      write(6,42)marker
42    format(1h ,4(1h*),'***error','*** ',i6,1x,50(1h*))
      iercc=1
      return
225   write(6,43)marker
43    format(1h ,4(1h*),' warning','*** ',i6,1x,50(1h*))
      return
      end
c----------------------------------------------------------------------
      function dexpc(mark,x)
c----------------------------------------------------------------------
c  #exponential with overflow/underflow check
      implicit real*8(a-h,o-z)
      data small,big/-70.0d00,70.0d00/
      data smexp,bigexp/3.975d-31,2.515d30/
      if(x.gt.big) go to 101
      if(x.lt.small)go to 102
      dexpc=dexp(x)
      return
101   dexpc=bigexp
      if(mark.le.0)return
      call err(67000+iabs(mark))
      return
102   dexpc=smexp
      return
      end
```

Output from the Program:

computing the probability that at least a proportion 90.0 of the population
will be enclosed by x(2) and x(98) of 100 observations.
probability= 0.9763

computing the probability that the   50.0 percentile of the population
will be enclosed by x(45) and x(55) of 100 observations.
probability= 0.6803

computing the probability that y(2) of 50 future observations
will fall between x(1) and x(10) of 100 past observations.
probability= 0.8227

computing the probability that at least   8 of 10 future observations
will fall between x(10) and x(90) of 100 past observations.
probability= 0.6521

epti=   0.9763
epyp=   0.6803
epyj=   0.8227
epkm=   0.6521

References

Aitchison, J., and Brown, J. A. C., (1957), *The Lognormal Distribution*, London: Cambridge University Press.

Aitchison, J., and Dunsmore, I. R. (1975), *Statistical Prediction Analysis*, Cambridge: Cambridge University Press.

Aitchison, J., and Sculthorpe, D. (1965), Some problems of statistical prediction, *Biometrika* **52**, 469–483.

Antle, C. E., and Rademaker, F. (1972), An upper confidence limit on the maximum of *m* future observations from a Type I extreme value distribution, *Biometrika* **59**, 475–477.

Ascher, H., and Feingold, H. (1984), *Repairable Systems Reliability*, New York: Marcel Dekker, Inc.

Atwood, C. L. (1984), Approximate tolerance intervals, based on maximum likelihood estimates, *Journal of the American Statistical Association* **79**, 459–465.

Bain, L. J. (1978), *Statistical Analysis of Reliability and Life-Testing Models*, New York: Marcel Dekker, Inc.

Bain, L. J., Balakrishnan, N., Eastman, J., Engelhardt, M., and Antle, C. (1991), Reliability Estimation Based on MLE's for Complete and Censored Samples, Chapter 5 in *The Logistic Distribution*, N. Balakrishnan, Editor, New York: Marcel Dekker, Inc. (to appear).

Bain, L. J., and Englehardt, M. (1973), Interval estimation for the two-parameter double exponential distribution, *Technometrics* **15**, 875–887.

Bain, L. J., and Englehardt, M. (1981), Simple approximate distributional results for confidence and tolerance limits for the Weibull distribution based on maximum likelihood estimators, *Technometrics* **23**, 15–20.

Bain, L. J., and Weeks, D. L. (1965), Tolerance limits for the generalized gamma distribution, *Journal of the American Statistical Association* **60**, 1142–1152.

Balakrishnan, N., Editor (1991), *The Logistic Distribution*, New York: Marcel Dekker, Inc. (to appear).

Balakrishnan, N., and Fung, K. (1991), Tolerance Limits and Sampling Plans Based on Censored Samples, Chapter 14 in *The Logistic Distribution*, N. Balakrishnan, Editor, New York: Marcel Dekker, Inc. (to appear).

368

Bates, D. M., and Watts, D. G. (1988), *Nonlinear Regression Analysis and its Applications*, New York: John Wiley & Sons, Inc.

Becker, R. A., Chambers, J. M., and Wilks, A. R., (1988), *The New S Language*, Pacific Grove, CA: Wadsworth and Brooks/Cole.

Beyer, W. H., Editor (1968), *Handbook of Tables for Probability and Statistics*, Cleveland, OH: The Chemical Rubber Company.

Billman, B. R., Antle, C. E., and Bain, L. J. (1972), Statistical inference from censored Weibull samples, *Technometrics* **14**, 831–40.

Blyth, C. R. (1986), Approximate binomial confidence limits, *Journal of the American Statistical Association* **81**, 843–855.

Blyth, C. R., and Still, H. A. (1983), Binomial confidence intervals, *Journal of the American Statistical Association* **78**, 108–116.

Bowden, D. C. (1968), Tolerance intervals in regression (query), *Technometrics* **10**, 207–209.

Bowker, A. H., and Lieberman, G. J. (1972), *Engineering Statistics* (Second Edition), Englewood Cliffs, NJ: Prentice-Hall, Inc.

Box, G. E. P., and Cox, D. R. (1964), An analysis of transformations (with discussion), *Journal of the Royal Statistical Society, Series B* **26**, 211–252.

Box, G. E. P., Hunter, W. G., and Hunter, J. S. (1978), *Statistics for Experimenters*, New York: John Wiley & Sons, Inc.

Box, G. E. P., and Jenkins, G. W. (1976), *Time Series Analysis: Forecasting and Control* (Revised Edition), San Francisco: Holden-Day, Inc.

Box, G. E. P., and Tiao, G. C. (1973), *Bayesian Inference in Statistical Analysis*, Reading, MA: Addison-Wesley, Inc.

Bratcher, T. L., Schucany, W. R., and Hunt, W. R. (1971), Bayesian prediction and population size assumptions, *Technometrics* **13**, 678–681.

Broemeling, L. D. (1985), *Bayesian Analysis of Linear Models*, New York: Marcel Dekker, Inc.

Brush, G. G. (1988), *How to Choose the Proper Sample Size*, Volume 12 of the ASQC Basic References in Quality Control: Statistical Techniques, Milwaukee, WI: The American Society for Quality Control.

Buckland, S. T. (1984), Monte Carlo confidence intervals, *Biometrics* **40**, 811–817.

Buckland, S. T. (1985), Calculation of Monte Carlo confidence intervals, *Applied Statistics* **34**, 296–301.

Buonaccorsi, J. P. (1987), A note on confidence intervals for proportions in finite populations, *The American Statistician* **41**, 215–218.

Canavos, G. C., and Kautauelis, I. A. (1984), The robustness of two-sided tolerance limits for normal distributions, *Journal of Quality Technology* **16**, 144–149.

Carroll, R. J., and Ruppert, D. (1988), *Transformation and Weighting in Regression*, London: Chapman and Hall.

Casella, G. (1985), An introduction to empirical Bayes data analysis, *American Statistician* **39**, 83–87.

Chambers, J. M., Clevelend, W. S., Kleiner, B., and Tukey, P. A. (1983), *Graphical Methods for Data Analysis*, Belmont, CA: Wadsworth, Inc.

Chandra, R., and Hahn, G. J. (1981), Confidence bounds on the probability of meeting a Poisson or binomial distribution specification limit, *Journal of Quality Technology* **13**, 241–248.

Chew, V. (1966), Confidence, prediction, and tolerance regions for the multivariate normal distribution, *Journal of the American Statistical Association* **61**, 605–617.

Chew, V. (1968), Simultaneous prediction intervals, *Technometrics* **10**, 323–330.

Chhikara, R. S., and Folks, J. L. (1989), *The Inverse Gaussian Distribution*, New York: Marcel Dekker, Inc.

Chhikara, R. S., and Guttman, I. (1982), Prediction limits for the inverse Gaussian distribution, *Technometrics* **24**, 319–324.

Clopper, C. J., and Pearson, E. S. (1934), The use of confidence or fiducial limits illustrated in the case of the binomial, *Biometrika* **26**, 404–413.

Cochran, W. G. (1977), *Sampling Techniques* (Third Edition), New York: John Wiley & Sons, Inc.

Cohen, A. C., and Whitten, B. J. (1988), *Parameter Estimation in Reliability and Life Span Models*, New York: Dekker.

Cook, R. D., and Weisberg, S. (1988), Confidence curves in nonlinear regression, Manuscript, Department of Applied Statistics, University of Minnesota, St. Paul, MN.

Cox, D. R., and Hinkley, D. V. (1974), *Theoretical Statistics*, London: Chapman and Hall.

Cox, D. R., and Lewis, P. A. W. (1966), *The Statistical Analysis of a Series of Events*, London: Methuen.

Crow, E. L., and Shimizu, K. (1988), *Lognormal Distributions: Theory and Applications*, New York: Marcel Dekker, Inc.

Crow, L. H. (1982), Confidence interval procedures for the Weibull process with applications to reliability growth, *Technometrics* **24**, 67–72.

Danziger, L., and Davis, S. A. (1964), Tables of distribution-free tolerance limits, *Annals of Mathematical Statistics* **35**, 1361–1365.

David, H. A. (1981), *Order Statistics* (Second Edition), New York: John Wiley & Sons, Inc.

Deming, W. E. (1950), *Some Theory of Sampling*, New York: John Wiley & Sons, Inc.

Deming, W. E. (1953), On the distinction between enumerative and analytic surveys, *Journal of the American Statistical Association* **48**, 244–255.

Deming, W. E. (1975), On probability as a basis for action, *The American Statistician* **29**, 146–152.

Deming, W. E. (1976), On the use of judgment samples, *Reports of Statistical Applications, Japanese Union of Scientists and Engineers* **23**, 25–31.

Dixon, W. J., and Massey, F. J., Jr. (1969), *Introduction to Statistical Analysis* (Third Edition) New York: McGraw-Hill Book Co.

Dodge, H. F., and Romig, H. G. (1944), *Sampling Inspection Tables*, New York: John Wiley & Sons, Inc.

Donaldson, J. R., and Schnabel, R. B. (1987), Computational experience with confidence regions and confidence intervals for nonlinear least squares, *Technometrics* **29**, 67–82.

Dongarra, J. J., and Grosse, E. (1985), Distribution of mathematical software via electronic mail, *Signum Newsletter* **20**, 45–47.

Draper, N. R., and Smith, H. (1981), *Applied Regression Analysis* (Second Edition), New York: John Wiley & Sons, Inc.

Drenick, R. F. (1960), The failure law of complex equipment, *Journal of the Society of Industrial and Applied Mathematics* **8**, 680–690.

Duncan, A. J. (1974), *Quality Control and Industrial Statistics* (Fourth Edition), Homewood, IL: Richard D. Irwin, Inc.

Dunsmore, I. R. (1978), Some approximations for tolerance factors for the two parameter exponential distribution, *Technometrics* **20**, 317–318.

Easterling, R. G. (1976), Goodness of fit and parameter estimation, *Technometrics* **18**, 1–9.

Eberhardt, K. R., Mee, R. W., and Reeve, C. P. (1989), Computing factors for exact two-sided tolerance limits for a normal distribution, *Communications in Statistics, Simulation and Computation* **18**, 397–413.

Efron, B. (1985), Bootstrap confidence intervals for a class of parametric problems, *Biometrika* **72**, 45–58.

Efron, B. (1987), Better bootstrap confidence intervals (with discussion), *Journal of the American Statistical Association* **82**, 171–200.

Efron, B., and Gong, G. (1983), A leisurely look at the bootstrap, the jackknife, and cross validation, *The American Statistician* **37**, 36–48.

Efron, B., and Tibshirani, R. (1986), The bootstrap (with discussion), *Statistical Science* **1**, 54–78.

Englehardt, M., and Bain, L. J. (1978a), Tolerance limits and confidence limits on reliability for the two-parameter exponential distribution, *Technometrics* **20**, 37–39.

Englehardt, M., and Bain, L. J. (1978b), Prediction intervals for the Weibull process, *Technometrics* **20**, 167–169.

Englehardt, M., and Bain, L. J. (1979), Prediction limits and two-sample problems with complete or censored Weibull data, *Technometrics* **21**, 233–237.

Englehardt, M., and Bain, L. J. (1982), On prediction limits for samples from a Weibull or extreme-value distribution, *Technometrics* **24**, 147–150.

Evans, I. G., Jones, D. A., and Owen, R. J. (1976), Relative frequency interpretation of Bayesian predictive distributions, *The American Statistician* **30**, 145.

Farewell, V. T., and Prentice, R. L. (1977), A study of distributional shape in life testing, *Technometrics* **19**, 69–75.

Faulkenberry, G. D., and Weeks, D. L. (1968), Sample size determination for tolerance limits, *Technometrics* **10**, 343–348.

Faulkenberry, G. D., and Daly, J. C. (1970), Sample size for tolerance limits on a normal distribution, *Technometrics* **12**, 813–821.

Fertig, K. W., and Mann, N. R. (1977), One-sided prediction intervals for at least *p* out of *m* future observations from a normal population, *Technometrics* **19**, 167–177.

Fertig, K. W., Meyer, E., and Mann, N. R. (1980), On constructing prediction intervals for samples from a Weibull or extreme value distribution, *Technometrics* **22**, 567–573.

Finkelstein, J. M. (1976), Confidence bounds on the parameters of the Weibull process, *Technometrics* **18**, 115–117.

Fligner, M. A., and Wolfe, D. A. (1976), Some applications of sample analogues to the probability integral transform and a coverage property, *The American Statistician* **30**, 78–85.

Fligner, M. A., and Wolfe, D. A. (1979a), Nonparametric prediction intervals for a future sample median, *Journal of the American Statistical Association* **74**, 453–456.

Fligner, M. A., and Wolfe, D. A. (1979b), Methods for obtaining a distribution-free prediction interval for the median of a future sample, *Journal of Quality Technology* **11**, 192–198.

Folks, J. L., and Chhikara, R. S. (1978), The inverse Gaussian distribution and its statistical application—a review, *Journal of the Royal Statistical Society Series B* **40**, 263–289.

Fujino, Y. (1980), Approximate binomial confidence limits, *Biometrika* **67**, 677–681.

Fuller, W. A. (1976), *Introduction to Statistical Time Series*, New York: John Wiley & Sons, Inc.

Fuller, W. A. (1987), *Measurement Error Models*, New York: John Wiley & Sons, Inc.

Galambos, J. (1978), *The Asymptotic Theory of Extreme Order Statistics*, New York: John Wiley & Sons, Inc.

Gallant, R. A. (1987), *Nonlinear Statistical Models*, New York: John Wiley & Sons, Inc.

Gardner, M. J., and Altman, D. G. (1989), *Statistics with Confidence*, London: British Medical Journal.

Gefland, A. E., and Smith, A. F. M. (1990), Sampling-based approaches to calculating marginal densities, *Journal of the American Statistical Association* **85**, 398–409.

General Electric Company, Defense Systems Department (1962), *Tables of the Individual Terms of Poisson Distribution*, Princeton, NJ: D. Van Nostrand Co.

Gibbons, J. D. (1971), *Nonparametric Statistical Inference*, New York: McGraw-Hill Book Co.

Gibbons, J. D. (1975), *Nonparametric Methods for Quantitative Analysis*, New York: Holt, Rinehart, & Winston.

Gitlow, H., Gitlow, S., Oppenheim, A., and Oppenheim, R. (1989), *Tools and Methods for the Improvement of Quality*, Homewood, IL: Irwin.

Goodman, L. A., and Madansky, A. (1962), Parameter-free and nonparametric tolerance limits: the exponential case, *Technometrics* **4**, 75–95.

Grant, E. L., and Leavenworth, R. S. (1988), *Statistical Quality Control* (Sixth Edition), New York: McGraw-Hill Book Co.

Graybill, F. A. (1976), *Theory and Application of the Linear Model*, North Scituate MA: Duxbury Press.

Greenwood, J. A., and Sandomire, M. M. (1950), Sample size required for estimating the standard deviation as a percent of its true value, *Journal of the American Statistical Association* **45**, 257–260.

Griffiths, P., and Hill, I. D., Editors (1985), *Applied Statistics Algorithms*, Chichester: Ellis Horwood Limited.

Guenther, W. C. (1972), Tolerance intervals for univariate distributions, *Naval Research Logistics Quarterly* **19**, 309–333.

Guenther, W. C. (1975), The inverse hypergeometric—a useful model, *Statistica Neerlandica* **29**, 129–144.

Guenther, W. C., Patil, S. A., and Uppuluri, V. R. R. (1976), One-sided β-content tolerance factors for the two parameter exponential distribution, *Technometrics* **18**, 333–340.

Guilbaud, O. (1983), Nonparametric prediction intervals for sample medians in the general case, *Journal of the American Statistical Association* **78**, 937–941.

Gumbel, E. J. (1958), *Statistics of Extremes*, New York: Columbia University Press.

Guttman, I. (1970), *Statistical Tolerance Regions: Classical and Bayesian*, London: Charles Griffin & Co., Ltd.

Guttman, I., Wilks, S. S., and Hunter, J. S. (1982), *Introductory Engineering Statistics* (Third Edition), New York: John Wiley & Sons, Inc.

Hahn, G. J. (1969), Factors for calculating two-sided prediction intervals for samples from a normal distribution, *Journal of the American Statistical Association* **64**, 878–888.

Hahn, G. J. (1970a), Additional factors for calculating prediction intervals for samples from a normal distribution, *Journal of the American Statistical Association* **65**, 1668–1676.

Hahn, G. J. (1970b), Statistical intervals for a normal population. Part I. Tables, examples and applications, *Journal of Quality Technology* **2**, 115–125.

Hahn, G. J. (1971), How abnormal is normality? *Journal of Quality Technology* **3**, 18–22.

Hahn, G. J. (1972a), Simultaneous prediction intervals to contain the standard deviations or ranges of future samples from a normal distribution, *Journal of the American Statistical Association* **67**, 938–942.

Hahn, G. J. (1972b), Prediction intervals for a regression model, *Technometrics* **14**, 203–213.

Hahn, G. J. (1974), Don't let statistical significance fool you!, *Chemtech* **4**, 16–18.

Hahn, G. J. (1975), A simultaneous prediction limit on the means of future samples from an exponential distribution, *Technometrics* **17**, 341–345.

Hahn, G. J. (1977), A prediction interval on the difference between two future sample means and its application to a claim of product superiority, *Technometrics* **19**, 131–134.

Hahn, G. J. (1979), Sample size determines precision, *Chemtech*, **9**, 294–295.

Hahn, G. J. (1982), Removing measurement error in assessing conformance to specifications, *Journal of Quality Technology* **14**, 117–121.

Hahn, G. J., and Chandra, R. (1981), Tolerance intervals for Poisson and binomial variables, *Journal of Quality Technology* **13**, 100–110.

Hahn, G. J., and Meeker, W. Q. (1984), An engineer's guide to books on statistics and data analysis, *Journal of Quality Technology* **16**, 196–218.

Hahn, G. J., and Nelson, W. (1973), A survey of prediction intervals and their applications, *Journal of Quality Technology* **5**, 178–188.

Hahn, G. J., and Raghunathan, T. E. (1988), Combining information from various sources: A prediction problem and other industrial applications, *Technometrics* **30**, 41–52.

Hahn, G. J., and Shapiro, S. S. (1967), *Statistical Models in Engineering*, New York: John Wiley & Sons, Inc.

Hald, A. (1952), *Statistical Tables and Formulas*, New York: John Wiley & Sons, Inc.

Hall, D. B. (1989), Analysis of surveillance data: A rationale for statistical tests with comments on confidence intervals and statistical models, *Statistics in Medicine* **8**, 481–493.

Hall, I. J. (1975), One-sided tolerance limits for a logistic distribution based on censored samples, *Biometrics* **31**, 873–880.

Hall, I. J., and Prairie, R. R. (1973), One-sided prediction intervals to contain at least *m* out of *k* future observations, *Technometrics* **15**, 897–914.

Hall, I. J., Prairie, R. R., and Motlagh, C. K. (1975), Nonparametric prediction intervals, *Journal of Quality Technology* **7**, 109–114.

Hall, I. J., and Sheldon, D. D. (1979), Improved bivariate normal tolerance regions with some applications, *Journal of Quality Technology* **11**, 13–19.

Hall, P. (1987), On the bootstrap and likelihood-based confidence regions, *Biometrika* **74**, 481–493.

Hall, W. J., and Wellner, J. A. (1980), Confidence bands for the Kaplan-Meier survival curve estimate, *Biometrika* **67**, 133–143.

Hanson, D. L., and Koopmans, L. H. (1964), Tolerance limits for the class of distributions with increasing hazard rate, *Annals of Mathematical Statistics* **35**, 1561–1570.

Harvard University Computations Laboratory (1955), *Tables of the Cumulative Binomial Probability Distribution*, Cambridge, MA: Harvard University Press.

Hogg, R. V., and Craig, A. T. (1978), *Introduction to Mathematical Statistics* (Fourth Edition), New York: Macmillan.

IMSL (1987), *IMSL STAT/LIBRARY User's Manual* (Version 1.0), Houston, TX: IMSL, Inc.

Jaech, J. (1984), Removing the effects of measurement errors in constructing statistical tolerance intervals, *Journal of Quality Technology* **16**, 69–73.

Jilek, M. (1981), A bibliography of statistical tolerance regions, *Statistics* **12**, 441–456.

Jilek, M., and Ackerman, H. (1989), A bibliography of statistical tolerance regions, II, *Statistics* **20**, 165–172.

Johns, M. V., Jr., and Lieberman, G. J. (1966), An exact asymptotically efficient confidence bound for reliability in the case of the Weibull distribution, *Technometrics* **8**, 135–175.

Johnson, N. L., and Kotz, S. (1969), *Distributions in Statistics—Discrete Distributions*, New York: John Wiley & Sons, Inc.

Johnson, N. L., and Kotz, S. (1970a), *Distributions in Statistics—Continuous Univariate Distributions*, *Vol. 1*, New York: John Wiley & Sons, Inc.

Johnson, N. L., and Kotz, S. (1970b), *Distributions in Statistics—Continuous Univariate Distributions, Vol. 2*, New York: John Wiley & Sons, Inc.

Johnson, R. A., and Wichern, D. W. (1988), *Applied Multivariate Analysis* (Second Edition), Englewood Cliffs, NJ: Prentice-Hall.

Jones, R. A., and Scholtz, F. W. (1983), Tolerance limits for the three parameter Weibull distribution, mathematics and modeling, Technical Report MM-TR-11, Boeing Computer Services Company, Seattle, WA.

Kaminsky, K. S. (1977), Comparison of prediction intervals for failure times when life is exponential, *Technometrics* **19**, 83–86.

Kapperman, R. F. (1977), Tolerance limits for the double exponential distribution, *Journal of the American Statistical Association* **72**, 908–909.

Kempthorne, O. (1971), The classical problem of inference–goodness of fit, *Proceedings of the Fifth Berkeley Symposium* **1**, 235–249.

Kempthorne, O., and Folks, L. (1971), *Probability, Statistics, and Data Analysis*, Ames, IA: Iowa State University Press.

Kendall, M. G., and Stuart, A. (1973), *The Advanced Theory of Statistics* (Third Edition), New York: Hafner Publishing Company.

Kennedy, W. J., and Gentle, J. E. (1980), *Statistical Computing*, New York: Marcel Dekker Inc.

Kupper, L. L., and Hafner, K. B. (1989), How appropriate are popular sample size formulas?, *The American Statistician* **43**, 101–105.

Kushary, D., and Mee, R. (1990), Prediction limits for the Weibull distribution utilizing simulation, Manuscript, University of Tennessee, Knoxville, TN.

Larson, H. R. (1966), A nomograph of the cumulative binomial distribution, *Industrial Quality Control* **23**, 270–278.

Lawless, J. F. (1971), A prediction problem concerning samples from the exponential distribution with applications in life testing, *Technometrics* **13**, 725–730.

Lawless, J. F. (1973), On estimation of safe life when the underlying distribution is Weibull, *Technometrics* **15**, 857–865.

Lawless, J. F. (1975), Construction of tolerance bounds for the extreme-value and Weibull distributions, *Technometrics* **17**, 255–261.

Lawless, J. F. (1977), Prediction intervals for the two parameter exponential distribution, *Technometrics* **19**, 469–472.

Lawless, J. F. (1982), *Statistical Models and Methods for Life Time Data*, New York: John Wiley & Sons, Inc.

Lee, L., and Lee, S. K. (1978), Some results on inference for the Weibull process, *Technometrics* **20**, 41–45.

Lemon, G. H. (1977), Factors for one-sided tolerance limits for balanced one-way-ANOVA random-effects model, *Journal of the American Statistical Association* **72**, 676–680.

Lieberman, G. J. (1961), Prediction regions for several predictions from a single regression line, *Technometrics* **3**, 21–27.

Lieberman, G. J., and Miller, R. G., Jr. (1963), Simultaneous tolerance intervals in regression, *Biometrika* **50**, 155–168.

Lieberman, G. J., and Owen, D. B. (1961), *Tables of the Hypergeometric Probability Distribution*, Palo Alto, CA: Stanford University Press.

Lieblein J., and Zelen, M. (1956), Statistical investigation of the fatigue life of deep-groove ball bearings, *Journal of Research, National Bureau of Standards* **57**, 273–316.

Likes, J. (1974), Prediction of the sth ordered observation for the two-parameter exponential distribution, *Technometrics* **16**, 241–244.

Limam, M. M. T., and Thomas, D. R. (1988a), Simultaneous tolerance intervals for the linear regression model, *Journal of the American Statistical Association* **83**, 801–804.

Limam, M. M. T., and Thomas, D. R. (1988b), Simultaneous tolerance intervals in the one-way model with covariates, *Communications in Statistics—Simulation and Computation* **17**, 1007–1019.

Lindley, D. V. (1965), *Introduction to Probability and Statistics from a Bayesian Viewpoint. Part 2, Inference*, Cambridge: Cambridge University Press.

Lindley, D. V. (1970), *Bayesian Statistics, A Review*, Philadelphia: Society for Industrial and Applied Mathematics.

Mace, A. E. (1964), *Sample Size Determination*, New York: Reinhold Publishing, Inc.

Mann, N. R. (1970), Warranty periods based on three ordered sample observations from a Weibull population, *IEEE Transactions on Reliability* **R-19**, 167–171.

Mann, N. R., and Fertig, K. W. (1973), Tables for obtaining Weibull confidence bounds and tolerance bounds based on best linear invariant estimates of parameters of the extreme value distribution, *Technometrics* **15**, 87–101.

Mann, N. R., and Fertig, K. W. (1977), Efficient unbiased quantile estimates for moderate size samples from extreme value and Weibull distributions: Confidence bounds and tolerance and prediction intervals, *Technometrics* **19**, 87–93.

Mann, N. R., and Saunders, S. C. (1969), On evaluation of warranty assurance when life has a Weibull distribution, *Biometrika* **56**, 615–625.

Mann, N. R., Schafer, R. E., and Singpurwalla, N. D. (1974), *Methods for Statistical Analysis of Reliability and Life Data*, New York: John Wiley & Sons, Inc.

Maritz, J. S. (1970), *Empirical Bayes Methods*, London: Methuen and Co., Ltd.

Martz, H. F., and Waller, R. A. (1982), *Bayesian Reliability Analysis*, New York: John Wiley & Sons, Inc.

Mee, R. W. (1984a), β-expectation and β-content tolerance limits for balanced one-way ANOVA random model, *Technometrics* **26**, 251–254.

Mee, R. W. (1984b), Tolerance limits and bounds for proportions based on data subject to measurement error, *Journal of Quality Technology* **16**, 74–80.

Mee, R. W. (1989), Normal distribution tolerance intervals for stratified random samples, *Technometrics* **31**, 99–105.

Mee, R. W., Eberhardt, K. R., and Reeve, C. P. (1989), Calibration and simultaneous tolerance intervals for regression, Manuscript, University of Tennessee, Knoxville, TN.

Mee, R. W., and Owen, D. B. (1983), Improved factors for one-sided tolerance limits for balanced one-way ANOVA random model, *Journal of the American Statistical Association* **78**, 901–905.

Meeker, W. Q. (1987), Limited failure population life tests: Application to integrated circuit reliability, *Technometrics* **29**, 51–65.

Meeker, W. Q., and Duke, S. D. (1981), CENSOR—A user-oriented computer program for life data analysis, *The American Statistician* **35**, 112.

Meeker, W. Q., and Hahn, G. J. (1980), Prediction intervals for the ratios of normal distribution sample variances and exponential distribution means, *Technometrics* **22**, 357–366.

Meeker, W. Q., and Hahn, G. J. (1982), Sample sizes for prediction intervals, *Journal of Quality Technology* **14**, 201–206.

Mendenhall, W. (1968), *Introduction to Linear Models and the Design and Analysis of Experiments*, Belmont, CA: Duxbury Press.

Miller, R. G., Jr. (1981), *Simultaneous Statistical Inference* (Second Edition), New York: Springer-Verlag.

Miller, R. W. (1989), Parametric empirical Bayes tolerance intervals, *Technometrics* **31**, 449–459.

Mood, A. M., Graybill, F. A., and Boes, D. C. (1974), *Introduction to the Theory of Statistics* (Third Edition), New York: McGraw-Hill Book Company.

Muench, J. O. (1984), Cumulative binomial distribution computer, SAND 76–0016, Reliability Department, Sandia National Laboratories, Albuquerque, NM.

Murphy, R. B. (1948), Non-parametric tolerance limits, *Annals of Mathematical Statistics* **19**, 581–589.

NAG (1988), *The NAG Fortran Library Manual—Mark 13*, Oxford, U.K.: The Numerical Algorithms Group.

Nair, V. N. (1981), Confidence bands for survival functions with censored data, *Biometrika* **68**, 99–103.

Nair, V. N. (1984), Confidence bands for survival functions with censored data: A comparative study, *Technometrics* **26**, 265–275.

Natrella, M. G. (1963), *Experimental Statistics*, National Bureau of Standards, Handbook No. 91, U.S. Government Printing Office, Washington, D.C.

Nelson, P. R. (1985), Computation of some common discrete distributions, *Journal of Quality Technology* **17**, 160–166.

Nelson, W. (1968), Two-sample prediction, TIS Report 68-C-404, General Electric Company, Corporate Research and Development, Schenectady, NY.

Nelson, W. (1970), Confidence intervals for the ratio of two Poisson means and Poisson predictor intervals, *IEEE Transactions on Reliability* **R-19**, 42–49.

Nelson, W. (1972a), A short life test for comparing a sample with previous accelerated test results, *Technometrics* **14**, 175–186.

Nelson, W. (1972b), Charts for confidence limits and test for failure rates, *Journal of Quality Technology* **4**, 190–195.

Nelson, W. (1982), *Applied Life Data Analysis*, New York: John Wiley & Sons, Inc.

Nelson, W. (1983), *How to Analyze Reliability Data*, Volume 6 of the ASQC Basic References in Quality Control: Statistical Techniques, Milwaukee: American Society for Quality Control.

Nelson, W. (1984), Fitting of fatigue curves with nonconstant standard deviation to data with runouts, *Journal of Testing and Evaluation* **12**, 69–77.

Nelson, W. (1985), Weibull analysis of reliability data with few or no failures, *Journal of Quality Technology* **17**, 140–146.

Nelson, W. (1990), *Accelerated Testing: Statistical Models, Test Plans, and Data Analyses*, New York: John Wiley & Sons, Inc.

Nelson, W., and Schmee, J. (1981), Prediction limits for the last failure of a (log) normal sample from early failures, *IEEE Transactions on Reliability* **R-30**, 461–463.

Neter, J., Wasserman, W., and Kutner, M. H. (1990), *Applied Linear Statistical Models* (Third Edition), Homewood, IL: Richard D. Irwin, Inc.

Ng, V. (1984), Prediction intervals for the 2-parameter exponential distribution using incomplete data, *IEEE Transactions on Reliability* **R-33**, 188–191.

Nolan, T. W. (1988), Analytic Studies. Paper presented at the 1988 NYU Deming Seminar for Statisticians.

Odeh, R. E. (1990), Two-sided prediction intervals to contain at least k out of m future observations from a normal distribution, *Technometrics* **32**, 203–216.

Odeh, R. E., and Fox, M. (1975), *Sample Size Choice*, New York: Marcel Dekker, Inc.

Odeh, R. E., and Owen, D. B. (1980), *Tables for Normal Tolerance Limits, Sampling Plans, and Screening*, New York: Marcel Dekker, Inc.

Odeh, R. E., and Owen, D. B. (1983), *Attribute Sampling Plans, Tables of Tests and Confidence Limits for Proportions*, New York: Marcel Dekker, Inc.

Odeh, R. E., Owen, D. B., Birnbaum, Z. W., and Fisher, L. (1977), *Pocket Book of Statistical Tables*, New York: Marcel Dekker, Inc.

Oppenlander, J. E., Schmee, J., and Hahn, G. J. (1988), Some simple robust estimators of normal distribution tail percentiles and their properties, *Communications in Statistics—Theory and Methods* **17**, 2279–2301.

Ostle, B., and Mensing, R. W. (1975), *Statistics in Research*, Ames, IA: The Iowa State University Press.

Ostrouchov, G., and Meeker, W. Q. (1988), Accuracy of approximate confidence bounds computed from interval censored Weibull and lognormal data, *Journal of Statistical Computation and Simulation* **29**, 43–76.

Owen, D. B., Li, L., and Chou, Y. (1981), Prediction intervals for screening using a measured correlated variate, *Technometrics* **23**, 165–170.

Padgett, W. J. (1982), An approximate prediction interval for the mean of future observations from the inverse Gaussian distribution, *Journal of Statistical Computation and Simulation* **14**, 191–199.

Padgett, W. J., and Tsoi, S. K. (1986), Prediction intervals for future observations from the inverse Gaussian distribution, *IEEE Transactions on Reliability* **R-35**, 406–408.

Pankratz, A. (1983), *Forecasting with Univariate Box–Jenkins Models*, New York: John Wiley & Sons.

Patel, J. K. (1980a), Tolerance intervals for a class of IFR distributions with a threshold parameter, *IEEE Transactions on Reliability* **R-29**, 154–157.

Patel, J. K. (1980b), Prediction intervals for IFR distributions, *IEEE Transactions on Reliability* **R-29**, 406–409.

Patel, J. K. (1986), Tolerance intervals—a review, *Communications in Statistics— Theory and Methods* **15**, 2719–2762.

Patel, J. K. (1989), Prediction intervals—a review, *Communications in Statistics— Theory and Methods* **18**, 2393–2465.

Patel, J. K., Kapadia, C. H., and Owen, D. B. (1976), *Handbook of Statistical Distributions*, New York: Marcel Dekker, Inc.

Pearson, E. S., and Hartley, H. O. (1954), *Biometrika Tables for Statisticians*, Vol. 1, London: Cambridge University Press.

Posten, H. O. (1982), Computer algorithms for the classical distribution functions, in *Proceedings of the First International Conference on the Teaching of Statistics*, Sheffield, U.K.: University of Sheffield Printing Unit.

Press, S. J. (1989), *Bayesian Statistics: Principles, Models, and Applications*, New York: John Wiley & Sons, Inc.

Provost, L. P. (1989), Interpretation of Results of Analytical Studies. Paper presented at the 1989 NYU Deming Seminar for Statisticians.

Raghunathan, T. E., and Rubin, D. B. (1988), An application of Bayesian statistics using sampling/importance resampling to a deceptively simple problem in quality control, Manuscript, Department of Statistics, Harvard University, Cambridge, MA.

Rand Corporation (1955), *A Million Random Digits with 100,000 Normal Deviates*, New York: The Free Press.

Randles, R. H., and Wolfe, D. A. (1979), *Introduction to the Theory of Nonparametric Statistics*, New York: John Wiley & Sons, Inc.

Rubin, D. B. (1988), Using the SIR algorithm to simulate posterior distributions, in *Bayesian Statistics*, *Vol. 3*, Bernardo, DeGroot, Lindley, and Smith, Editors, London: Oxford University Press, pp. 395–402.

Saunders, S. C. (1968), On the determination of a safe life for distributions classified by failure rate, *Technometrics* **10**, 361–377.

Scheaffer, R. L. (1976), A note on approximate confidence limits for the negative binomial model, *Communications in Statistics A—Theory and Methods* **5**, 149–158.

Scheaffer, R. L., Mendenhall, W., and Ott, L. (1979), *Elementary Survey Sampling* (Second Edition), North Scituate, MA: Duxbury Press.

Scheffé, H. (1959), *The Analysis of Variance*, New York: John Wiley & Sons, Inc.

Scheuer, E. M. (1990), Let's teach more about prediction, *Proceedings of the ASA Section on Statistical Education*, Washington, D.C.: American Statistical Association.

Schmee, J., Gladstein, D., and Nelson, W. (1985), Confidence limits for parameters of a normal distribution from singly censored samples, using maximum likelihood, *Technometrics* **27**, 119–128.

Schneider, H. (1986), *Truncated and Censored Samples for Normal Populations*, New York: Marcel Dekker, Inc.

Seber, G. A. F. (1977), *Linear Regression Analysis*, New York: John Wiley and Sons, Inc.

Seber, G. A. F., and Wild, C. J. (1989), *Nonlinear Regression*, New York: John Wiley & Sons, Inc.

Shiue, W., and Bain, L. J. (1986), Prediction intervals for the gamma distribution, in *Computer Science and Statistics: Proceedings of the 18th Symposium on the Interface*, pp. 405–410.

Snedecor, G. W., and Cochran, W. G. (1967), *Statistical Methods* (Sixth Edition), Ames, IA: Iowa State University Press.

Somerville, P. N. (1958), Tables for obtaining non-parametric tolerance limits, *Annals of Mathematical Statistics* **29**, 599–601.

Stein, C. (1945), A two-sample test for a linear hypothesis whose power is independent of the variance, *Annals of Mathematical Statistics* **16**, 243–258.

Sukhatme, P. V., Sukhatme, B. V., Sukhatme, S., and Asok, C. (1984), *Sampling Theory of Surveys with Applications* (Third Edition), Ames, IA: Iowa State University Press.

Thatcher, A. R. (1964), Relationships between Bayesian and confidence limits for prediction (with discussion), *Journal of the Royal Statistical Society* **B26**, 176–210.

Thisted, R. A. (1988), *Elements of Statistical Computing, Numerical Computations*, London: Chapman and Hall.

Thomas, D. L., and Thomas, D. R. (1986), Confidence bands for percentiles in the linear regression model, *Journal of the American Statistical Association* **81**, 705–708.

Tietjen, G. L., and Johnson, M. E. (1979), Exact statistical tolerance limits for sample variances, *Technometrics* **21**, 107–110.

Tomsky, J. L., Nakano, K., and Iwashika, M. (1979), Confidence limits for the number of defectives in a lot, *Journal of Quality Technology* **11**, 199–204.

Tsokos, C. P., and Shimi, I. N., Editors (1977a), *The Theory and Applications of Reliability with Emphasis on Bayesian and Nonparametric Methods, Volume 1*, New York: Academic Press.

Tsokos, C. P., and Shimi, I. N., Editors (1977b), *The Theory and Applications of Reliability with Emphasis on Bayesian and Nonparametric Methods, Volume 2*, New York: Academic Press.

Turner, D. L., and Bowden, D. C. (1977), Simultaneous confidence bands for percentile lines in the general linear model, *Journal of the American Statistical Association* **72**, 886–889.

Turner, D. L., and Bowden, D. C. (1979), Sharp confidence bands for percentile lines and tolerance bands for simple linear models, *Journal of the American Statistical Association* **74**, 885–888.

Vander Wiel, S. A., and Meeker, W. Q. (1990), Accuracy of approximate confidence bounds using censored Weibull regression data from accelerated life tests, *IEEE Transactions on Reliability* **R-39**, 346–351.

Vardeman, S. (1990), What about the other intervals? Preprint 90-29, Department of Statistics, Iowa State University, Ames, IA.

Velleman, P. F., and Hoaglin, D. C. (1981), *Applications, Basics, and Computing of Exploratory Data Analysis*, Boston, MA: Duxbury Press.

Venzon, D. J., and Moolgavkar, S. H. (1988), A method of computing profile-likelihood-based confidence intervals, *Applied Statistics* **37**, 87–94.

Wallis, W. A. (1951), Tolerance intervals for linear regression, *Proceedings of the Second Berkeley Symposium on Mathematical Statistics and Probability*, J. Neyman, Editor, University of California Press, Los Angeles and Berkeley, pp. 43–51.

Weiss, L. (1961), *Statistical Decision Theory*, New York: McGraw-Hill Book Co.

Weston, S. A., and Meeker, W. Q. (1991), Coverage probabilities of nonparametric simultaneous confidence bands for a survival function, *Journal of Statistical Computation and Simulation* **32**, 83–97.

Williams, B. (1978), *A Sampler on Sampling*, New York: Wiley-Interscience, Inc.

Zellner, A. (1971), *An Introduction to Bayesian Inference in Econometrics*, New York: John Wiley & Sons, Inc.

Author Index

Subject Index

Absolute sample size, 19
Analytic study, 13–16, 22–23
Assignable cause variation, 14
Assumed distribution, 75
Assumptions:
    analytic study, 13–16
    enumerative *vs.* analytic study, 6–9
    enumerative study, 9–13
    normal distribution, 24–25, 41, 64–74
    population size, 19
    practical, 5–6, 20–23
    random sample, 10
    representative data, 5–6
    sample data, 3–5
    sampled population, 9–10
    sample size, relative to population size, 19
    in sampling people, 18–19
    statistical distribution, 24–25
    target population, 9–10
Attribute sampling plan, 186

Bayesian confidence interval, 243–244
Bayesian intervals for regression, 244
Bayesian prediction interval, 244
Bayesian statistical interval, 235, 242–244
Bayesian tolerance interval, 244
β-expectation tolerance interval, 204
Bibliography:
    prediction intervals, 206
    tolerance intervals, 206
Binomial coefficient, 78
Binomial distribution, 229
    application, 100
    confidence interval, 103–108
    cumulative distribution function, 78
    point estimate, 103
    prediction interval, 113–114
    probability function, 78, 101
    random variable, 101

sample size, 134–135
    confidence interval, 134–135, 144–145
    specification limits, 102
    tolerance interval, 111–112
Blocks, removed from extremes of sample, 90
BMDP®, 236–237
Bonferroni approximation, 62, 202, 203, 242
Bonferroni bound, *see* Bonferroni
    approximation
Bootstrap, 241–242
Box–Cox transformation, 72–74, 209
Box plot, 67

Calibration, 231
CENSOR, 237, 240
Censored data, 204–205, 230, 236, 241, 280–281
Central limit theorem, 41, 65
Chi-square distribution, 43, 54, 55–56
Classical inference methods, 235, 243
Cluster, 12
Cluster sample, 11–13
Common cause variation, 14
Comparing populations, 232–234
Comparing processes, 232–234
Computing packages, *see* Statistical
    computing packages
Confidence interval or bound:
    based on bootstrap, 241–242
    based on inverting goodness of fit tests,
      241
    based on inverting likelihood ratio tests,
      240–241
    based on large sample normal distribution
      theory, 238–240
    based on resampling, 241
    Bayesian, 243–244
    binomial probability for future sample,
      109
    differences in location parameters, 233

387

*Now available in a lower priced paperback edition in the Wiley Classics Library.